The Chemistry of
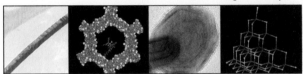
Nanostructured Materials

The Chemistry of

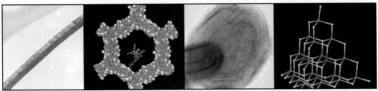

Nanostructured Materials

Editor

Peidong Yang

University of California, Berkeley, USA

NEW JERSEY • LONDON • SINGAPORE • SHANGHAI • HONG KONG • TAIPEI • BANGALORE

Published by

World Scientific Publishing Co. Pte. Ltd.
5 Toh Tuck Link, Singapore 596224
USA office: Suite 202, 1060 Main Street, River Edge, NJ 07661
UK office: 57 Shelton Street, Covent Garden, London WC2H 9HE

British Library Cataloguing-in-Publication Data
A catalogue record for this book is available from the British Library.

THE CHEMISTRY OF NANOSTRUCTURED MATERIALS

Copyright © 2003 by World Scientific Publishing Co. Pte. Ltd.

All rights reserved. This book, or parts thereof, may not be reproduced in any form or by any means, electronic or mechanical, including photocopying, recording or any information storage and retrieval system now known or to be invented, without written permission from the Publisher.

For photocopying of material in this volume, please pay a copying fee through the Copyright Clearance Center, Inc., 222 Rosewood Drive, Danvers, MA 01923, USA. In this case permission to photocopy is not required from the publisher.

ISBN 981-238-405-7
ISBN 981-238-565-7 (pbk)

Printed in Singapore by World Scientific Printers (S) Pte Ltd

FOREWORD

Nanostructured material has been a very exciting research topic in the past two decades. The impact of these researches to both fundamental science and potential industrial application has been tremendous and is still growing. There are many exciting examples of nanostructured materials in the past decades including colloidal nanocrystal, bucky ball C_{60}, carbon nanotube, semiconductor nanowire, and porous material. The field is quickly evolving and is now intricately interfacing with many different scientific disciplines, from chemistry to physics, to materials science, engineer and to biology. The research topics have been extremely diverse. The papers in the literature on related subjects have been overwhelming and is still increasing significantly each year.

The research on nanostructured materials is highly interdisciplinary because of different synthetic methodologies involved, as well as many different physical characterization techniques used. The success of the nanostructured material research is increasingly relying upon the collective efforts from various disciplines. Despite the fact that the practitioners in the field are coming from all different scientific disciplines, the fundamental of this increasing important research theme is unarguably about how to make such nanostructured materials. For this reason, chemists are playing a significant role since the synthesis of nanostructured materials is certainly about how to assemble atoms or molecules into nanostructures of desired coordination environment, sizes, and shapes. A notable trend is that many physicists and engineers are also moving towards such molecular based synthetic routes.

The exploding information in this general area of nanostructured materials also made it very difficult for newcomers to get a quick and precise grasp of the status of the field itself. This is particularly true for graduate students and undergraduates who have interest to do research in the area. The purpose of this book is to serve as a step-stone for people who want to get a glimpse of the field, particularly for the graduate students and undergraduate students in chemistry major. Physics and engineering researchers would also find this book useful since it provides an interesting collection of novel nanostructured materials, both in terms of their preparative methodologies and their structural and physical property characterization.

The book includes thirteen authoritative accounts written by experts in the field. The materials covered here include porous materials, carbon nanotubes, coordination networks, semiconductor nanowires, nanocrystals, Inorganic Fullerene, block copolymer, interfaces, catalysis and nanocomposites. Many of these materials represent the most exciting, and cutting edge research in the recent years.

While we have been able to cover some of these key areas, the coverage of book is certainly far from comprehensive as this wide-ranging subject deserves. Nevertheless, we hope the readers will find this an interesting and useful book.

Feb. 2003

Peidong Yang
Berkeley, California

CONTENTS

Foreword	v
Crystalline Microporous and Open Framework Materials *Xianhui Bu and Pingyun Feng*	1
Mesoporous Materials *Abdelhamid Sayari*	39
Macroporous Materials Containing Three-Dimensionally Periodic Structures *Younan Xia, Yu Lu, Kaori Kamata, Byron Gates and Yadong Yin*	69
CVD Synthesis of Single-Walled Carbon Nanotubes *Bo Zheng and Jie Liu*	101
Nanocrystals *M. P. Pileni*	127
Inorganic Fullerene-Like Structures and Inorganic Nanotubes from 2-D Layered Compounds *R. Tenne*	147
Semiconductor Nanowires: Functional Building Blocks for Nanotechnology *Haoquan Yan and Peidong Yang*	183
Harnessing Synthetic Versatility Toward Intelligent Interfacial Design: Organic Functionalization of Nanostructured Silicon Surfaces *Lon A. Porter and Jillian M. Buriak*	227
Molecular Networks as Novel Materials *Wenbin Lin and Helen L. Ngo*	261
Molecular Cluster Magnets *Jeffrey R. Long*	291
Block Copolymers in Nanotechnology *Nitash P. Balsara and Hyeok Hahn*	317

The Expanding World of Nanoparticle and Nanoporous Catalysts 329
Robert Raja and John Meurig Thomas

Nanocomposites 359
Walter Caseri

CRYSTALLINE MICROPOROUS AND OPEN FRAMEWORK MATERIALS

XIANHUI BU
Chemistry Department, University of California, CA93106, USA

PINGYUN FENG
Chemistry Department, University of California, Riverside, CA92521, USA

A variety of crystalline microporous and open framework materials have been synthesized and characterized over the past 50 years. Currently, microporous materials find applications primarily as shape or size selective adsorbents, ion exchangers, and catalysts. The recent progress in the synthesis of new crystalline microporous materials with novel compositional and topological characteristics promises new and advanced applications. The development of crystalline microporous materials started with the preparation of synthetic aluminosilicate zeolites in late 1940s and in the past two decades has been extended to include a variety of other compositions such as phosphates, chalcogenides, and metal-organic frameworks. In addition to such compositional diversity, synthetic efforts have also been directed towards the control of topological features such as pore size and channel dimensionality. In particular, the expansion of the pore size beyond 10Å has been one of the most important goals in the pursuit of new crystalline microporous materials.

1 Introduction

Microporous materials are porous solids with pore size below 20Å [1,2,3,4]. Porous solids with pore size between 20 and 500Å are called mesoporous materials. Macroporous materials are solids with pore size larger than 500Å. Mesoporous and macroporous materials have undergone rapid development in the past decade and they are covered in other chapters of this book. A frequently used term in the field of microporous materials is "molecular sieves" [5] that refers to a class of porous materials that can distinguish molecules on the basis of size and shape. This chapter focuses on crystalline microporous materials with a three-dimensional framework and will not discuss amorphous microporous materials such as carbon molecular sieves. However, it should be kept in mind that some amorphous microporous materials can also display shape or size selectivity and have important industrial applications such as air separation [6].

The development of crystalline microporous materials started in late 1940s with the synthesis of synthetic zeolites by Barrer, Milton, Breck and their coworkers [7,8]. Some commercially important microporous materials such as zeolites A, X, and Y were made in the first several years of Milton and Breck's work. In the following thirty years, zeolites with various topologies and chemical compositions (e.g., Si/Al ratios) were prepared, culminating with the synthesis of ZSM-5 [9] and aluminum-free pure silica polymorph silicalite [10] in 1970s. A breakthrough leading to an extension of crystalline microporous materials to non-aluminosilicates occurred in 1982 when Flanigen *et al.* reported the synthesis of aluminophosphate molecular sieves [11,12]. This breakthrough was followed by the development of substituted aluminophosphates. Since late 1980s and the early 1990s, crystalline microporous materials have been made in many other compositions including chalcogenides and metal-organic frameworks [13,14].

Crystalline microporous materials usually consist of a rigid three-dimensional framework with hydrated inorganic cations or organic molecules located in the cages or cavities of the inorganic or hybrid inorganic-organic host framework. Organic guest molecules can be protonated amines, quaternary ammonium cations, or neutral solvent molecules. Dehydration (or desolvation) and calcination of organic molecules are two methods frequently used to remove extra-framework species and generate microporosity.

Crystalline microporous materials generally have a narrow pore size distribution. This makes it possible for a microporous material to selectively allow some molecules to enter its pores and reject some other molecules that are either too large or have a shape that does not match with the shape of the pore. A number of applications involving microporous materials utilize such size and shape selectivity.

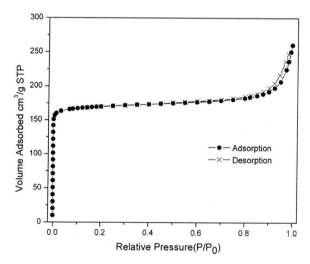

Figure 1. Nitrogen adsorption and desorption isotherms typical of a microporous material. Data were measured at 77K on a Micromeritics ASAP 2010 Micropore Analyzer for Molecular Sieve 13X. The structure of 13X is shown in Fig. 3. The sample was supplied by Micromeritics.

Two important properties of microporous materials are ion exchange and gas sorption. The ion exchange is the exchange of ions held in the cavity of microporous materials with ions in the external solutions. The gas sorption is the ability of a microporous material to reversibly take in molecules into its void volume (Fig. 1). For a material to be called microporous, it is generally necessary to demonstrate the gas sorption property.

The report by Davis *et al.* of a hydrated aluminophosphate VPI-5 with pore size larger than 10Å in 1988 generated great enthusiam toward the synthesis of extra-large pore materials [15]. The expansion of the pore size is an important goal of the current research on microporous materials [16]. Even though microporous materials include those with pore sizes between 10 to 20Å, The vast majority of known crystalline microporous materials have a pore size <10Å. The synthesis of microporous materials with pore size between 10 and 20Å is desirable for applications involving molecules in such size regime and remains a significant synthetic challenge today.

In the following sections, we will first review oxide-based microporous materials followed by a review on related chalcogenides. We will then discuss metal-organic frameworks, in which the framework is a hybrid between inorganic and organic units. The research on metal-organic frameworks is a rapidly developing area. These metal-organic materials are being studied not only for their porosity, but also for other properties such as chirality and non-linear optical activity [17]. The last section gives a discussion on materials with extra-large pore sizes. There exist many excellent reviews and books from which readers can find detailed information on various zeolite and phosphate topics [1,4,13,18,19,20,21,22,23,24,25].

2 Microporous Silicates

From a commercial perspective, the most important microporous materials are zeolites, a special class of microporous silicates. A strict definition of zeolites is difficult [5] because both chemical compositions and geometric features are involved. Zeolites can be loosely considered as crystalline three-dimensional aluminosilicates with open channels or cages. Not all zeolites are microporous because some are unable to retain their framework once extra-framework species (e.g., water or organic molecules) are removed. The stability of zeolites varies greatly depending on framework topologies and chemical compositions such as the Si/Al ratio and the type of charge-balancing cations. In addition to aluminum, many other metals have been found to form microporous silicates such as gallosilicates [26], titanosilicates [27,28], and zincosilicates [16]. Some microporous frameworks can even be made as pure silica polymorphs, SiO_2 [10].

2.1 Chemical compositions and framework structures of zeolites

Natural zeolites are crystalline hydrated aluminosilicates of group IA and group IIA elements such as Na^+, K^+, Mg^{2+}, and Ca^{2+}. Chemically, they are represented by the empirical formula: $M_{2/n}O \cdot Al_2O_3 \cdot ySiO_2 \cdot wH_2O$ where y is 2 or larger, n is the cation valence, and w represents the water contained in the voids of the zeolite. An empirical rule, Loewenstein rule [29], suggests that in zeolites, only Si-O-Si and Si-O-Al linkages be allowed. In other words, the Al-O-Al linkage does not occur in zeolites and the Si/Al molar ratio is ≥ 1.

Synthetic zeolites fall into two families on the basis of extra-framework species. One family is similar to natural zeolites in chemical compositions. These zeolites have a low Si/Al ratio that is usually less than 5. The other family of zeolites are made with organic structure-directing agents and they generally have a Si/Al ratio larger than 5.

In the absence of the framework interruption, the overall framework formula of a zeolite is AO_2 just like SiO_2. When A is Si^{4+}, no framework charge is produced. However, for each Al^{3+}, a negative charge develops on the framework. The negative charge is balanced by either inorganic or organic cations located in channels or cages of the framework. The charge-balancing cations are usually mobile and can undergo ion exchange.

Frameworks of zeolites are based on the three-dimensional, four-connected network of AlO_4 and SiO_4 tetrahedra linked together through the corner-sharing of oxygen anions. In a zeolite framework, oxygen atoms are bi-coordinated between two tetrahedral cations. When describing a zeolite framework, oxygen atoms are often omitted and only the connectivity among tetrahedral atoms is taken into consideration (Fig. 2).

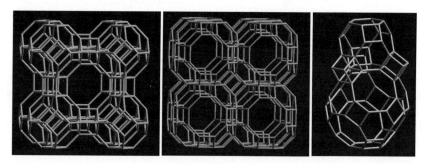

Figure 2. The three-dimensional framework of small-pore zeolite A (LTA) showing connectivity among framework tetrahedral atoms. (Left) viewed as sodalite cages linked together through double 4-rings (D4R); (middle) viewed as α-cages linked together by sharing single 8-rings; (right) three different cage units in zeolite A. The cage on top is called the β (or sodalite) cage and is built from 24 tetrahedral atoms. The cage at bottom is called the α cage and has 48 tetrahedral atoms. Also shown are three D4R's. Reprinted with permission from http://www.iza-structures.org/ and reference [30].

Zeolites and zeolite-like oxides are classified according to their framework types. A framework type is determined based on the connectivity of tetrahedral atoms and is independent of chemical compositions, types of extra-framework species, crystal symmetry, unit cell dimensions, or any other chemical and physical properties. In theory, there are numerous ways to connect tetrahedral atoms into a three-dimensional, four-connected network. However, in practice, only a very limited number of topological types have been found. In the past two decades, new framework topologies have been found mainly in non-zeolites such as open framework phosphates.

Even taking into consideration of both zeolites and non-zeolites, synthetic and natural solids, there are only 133 framework types listed in the "Atlas of Zeolite Framework Types" published by the structure commission of the International Zeolite Association [30]. These framework types are also published on the internet at http://www.iza-structures.org/. Each framework type in the ATLAS is assigned a three capital letter code. For example, FAU designates the framework type of a whole family of materials (e.g., SAPO-37, [Co-Al-P-O]-FAU, zeolites X and Y) with the same topology as the mineral faujasite (Fig. 3) [30]. Those codes help to clear the confusion resulting from many different names given to materials with different chemical compositions, but with the same topology. Sometimes even the same material can have different names assigned by different laboratories.

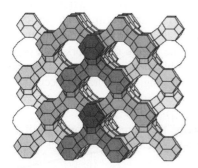 

Figure 3. (left) The three-dimensional framework of the mineral faujasite (FAU). Zeolites X and Y have the same topology as faujasite, but zeolite Y has a higher Si/Al ratio than zeolite X. Reprinted with permission from http://www.iza-structures.org/ and reference [30]. (right) The faujasite supercage with 48 tetrahedral atoms. The cage can be assembled from four 6-rings and six 4-rings. Four 12-ring windows are arranged tetrahedrally.

An important structural parameter is the size of the pore opening through which molecules diffuse into channels and cages of a zeolite. The pore size is related to the ring size defined as the number of tetrahedral atoms forming the pore. In the literature, zeolites with 8-ring, 10-ring, and 12-ring windows are often called small-pore, medium-pore, and large-pore zeolites, respectively. In addition to the ring size, the pore size is affected by other factors such as the ring shape, the size of tetrahedral atoms, the type of non-framework cations. For example, molecular sieves 3A, 4A, and 5A all have the same zeolite A (LTA) structure and the difference in the pore size is caused by different extra-framework cations (K^+, Na^+, and Ca^{2+}, respectively).

The pore volume of a zeolite is related to the framework density defined as the number of tetrahedral atoms per 1000Å^3. For zeolites, the observed values range from 12.7 for faujasite to 20.6 for cesium aluminosilicate (CAS) [30]. In general, the framework density does not reflect the size of the pore openings. For example, CIT-5 has an extra-large pore size with 14-ring windows, but its framework density is 18.3, significantly larger than that of faujasite (12.7) with 12-ring windows [30]. In general, large pore sizes, large cages, and multidimensional channel systems are three important factors that contribute to a low framework density for a four-connected, three-dimensional framework.

The framework density has been increasingly used to describe non-zeolites. The care must be taken when comparing the framework density of two compounds because the framework density can be significantly altered by framework interruptions (e.g., terminal OH$^-$ groups) that can lead to a substantial decrease in the framework density. Even for the same framework topology, a change in the chemical composition will lead to a change in bond distances and consequently in unit cell volumes. This will result in either an increase or decrease in the framework density.

All zeolites are built from TO$_4$ tetrahedra, called primary (or basic) building units. Larger finite units with three to sixteen tetrahedra (called Secondary Building Units or SBU's) are often used to describe the zeolite framework [30]. A SBU is a finite structural unit that can alone or in combination with another one build up the whole framework. The smallest SBU is a 3-ring, but it rarely occurs in zeolite framework types. Instead, 4-rings and 6-rings are most common in zeolite and zeolite-like structures.

There are several other ways to describe the framework topology of a zeolite. For example, structural units larger than SBU's can be used. In this way, zeolites

Figure 4. The wall structure of UCSB-7. UCSB-7 is one of a number of zeolite or zeolite-like structures that can be described using a minimal surface. UCSB-7 can be readily synthesized as germanate or arsenate, but has not been found as silicate or phosphate.

can be described as packing of small cages or clusters, cross-linking of chains, and stacking of layers with various sequences [31]. Some zeolite and zeolite-like frameworks can also be described using minimal surfaces (Fig. 4) [32].

When zeolite structures are described using clusters or cage units, these clusters and cages can be considered as large artificial atoms. Under such circumstances, structures of zeolites can be simplified to some of the simplest structures such as diamond and metals (e.g., fcc, ccp, and bcp). For examples, zeolite A is built from the simple cubic packing of sodalite cages and zeolite X has the diamond-type structure with the center of the sodalite cages occupying the tetrahedral carbon sites in diamond. Because these artificial atoms (clusters or cages) often have lower symmetry than a real spherical atom, the overall crystal symmetry can be lower than the parent compounds.

2.2 High silica or pure silica molecular sieves

In the past three decades, synthetic efforts directly related to aluminosilicate zeolites are generally in the area of high silica (Si/Al > 5) or pure silica molecular sieves [33]. The use of organic bases has had a significant impact on the development of high silica zeolites. The Si/Al ratio in the framework is increased because of the low charge to volume ratios of organic molecules. In general, the crystallization temperature (about 100-200°C) is higher than that required for the synthesis of hydrated zeolites. Alkali-metal ions, in addition to the organic materials, are usually used to help control the pH and promote the crystallization of high silica zeolites.

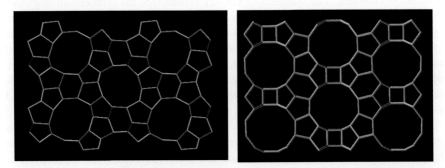

Figure 5. (Left) The framework of ZSM-5 projected down the [010] direction showing the 10-ring straight channels. ZSM-5 is thus far the most important crystalline microporous material discovered by using the organic structure-directing agent. It also has a large number of 5-rings that are common in high silica zeolites. (right) the framework of zeolite beta (polymorph A) projected down the [100] direction. Zeolite beta is an important zeolite because its framework is chiral and because it has a three-dimensional 12-ring channel system. Reprinted with permission from http://www.iza-structures.org/ and reference [30].

One of the most important zeolites created by this approach is ZSM-5 (Fig. 5), originally prepared using tetrapropylammonium cations as the structure-directing agent [9]. ZSM-5 (MFI) has a high catalytic activity and selectivity for various reactions. The pure silica form of ZSM-5 is called silicalite [10]. Another important zeolite is zeolite beta shown in Figure 5.

The use of fluoride media has been found to generate some new phases [34]. Frequently, crystals prepared from the fluoride medium have better quality and larger size compared to those made from the hydroxide medium [35]. In addition to serve as the mineralizing agent, F^- anions can also be occluded in the cavities or attached to the framework cations. This helps to balance the positive charge of organic cations. Upon calcination of high silica or pure silica phases, F^- anions are usually removed together with organic cations.

Among recently created high silica or pure silica molecular sieves are a series of materials denoted as ITQ-n synthesized from the fluoride medium. By employing H_2O/SiO_2 ratios lower than those typically used in the synthesis of zeolites in F^- or OH^- medium, a series of low-density silica phases were prepared [36]. Some of these (i.e., ITQ-3, ITQ-4, and ITQ-7) possess framework topologies not previously known in either natural or synthetic zeolites [37,38,39]. Another structure with a novel topology is germanium-containing ITQ-21 [40]. Similar to faujasite, ITQ-21 is also a large pore and large cage molecular sieve with a three-dimensional channel system. However, the cage in ITQ-21 is accessible through six 12-ring windows compared to four in faujasite.

The double 4-ring unit (D4R) as found in zeolite A often leads to a highly open architecture. However, for the aluminosilicate composition, it is a strained unit and does not occur often. The synthesis of ITQ-21 is related to the synthetic strategy that the incorporation of germanium helps stabilize the D4R. Similarly, during the synthesis of ITQ-7, the incorporation of germanium substantially reduced the crystallization time from 7 days to 12 hours [41]. The use of germanium has also led to the synthesis of the pure polymorph C of zeolite beta (BEC) even in the absence of the fluoride medium that is generally believed to assist in the formation of D4R units [42]. Both ITQ-7 and the polymorph C of zeolite beta contain D4R units and their syntheses were strongly affected by the presence of germanium.

The effect of germanium in the synthesis of D4R-containing high silica molecular sieves reflects a more general observation that there is a correlation between the framework composition and the preferred framework topology. For example, UCSB-7 can be easily synthesized in germanate or arsenate compositions [32], but has never been made in the silicate composition.

In general, large T-O distances and small T-O-T angles tend to favor more strained SBU's such as 3-rings and D4R units. It has already been observed that the germanate composition favors 3-rings and D4R units [43,44]. This observation can be extended to non-oxide open framework materials such as halides (e.g., CZX-2) [45], sulfides, and selenides with four-connected, three-dimensional topologies [46].

In these compositions, the T-X-T (X = Cl, S, and Se) angles are around 109° and three-rings become common. The presence of 3-rings is desirable because it could lead to highly open frameworks [30].

2.3 Low and intermediate silica molecular sieves

Low (Si/Al ≤ 2) and intermediate (2 < Si/Al ≤ 5) silica zeolites [18] are used as ion exchangers and have also found use as adsorbents for applications such as air separation. Syntheses of low and intermediate zeolites are usually performed under hydrothermal conditions using reactive alkali-metal aluminosilicate gels at low temperatures (~100°C and autogenous pressures). The synthesis procedure involves combining alkali hydroxide, reactive forms of alumina and silica, and H_2O to form a gel. Crystallization of the gel to the zeolite phase occurs at a temperature near 100°C. Two most important zeolites prepared by this approach are zeolites A and X [47]. The framework topology of zeolite A has not been found in nature. Zeolite X is compositionally different but topologically the same as mineral faujasite. Both zeolite A and zeolite X are built from packing of sodalite cages. In zeolite A, sodalite cages are joined together through 4-rings (Fig. 2) whereas in zeolite X, sodalite cages are coupled through 6-rings (Fig. 3).

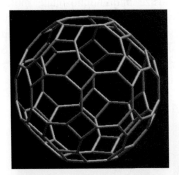 

Figure 6. (left) The tschortnerite cage built from 96 tetrahedral atoms. Reprinted with permission from http://www.iza-structures.org/ and reference [30]. (right) The UCSB-8 cage built from 64 tetrahedral atoms [30].

Few synthetic low and intermediate silica zeolites with new framework types have been reported in the past three decades. However, some new topologies have been found in natural zeolites. The most interesting one is a recently discovered mineral tschortnerite [48] with a Si/Al ratio of 1. This structure consists of several well-known structural units in zeolites including double 6-rings, double 8-rings, α-cages, and β-cages. Of particular interest is the presence of a cage (tschortnerite cage) with 96 tetrahedral atoms (Fig. 6), the largest known cage in four-connected, three-dimensional networks. In terms of the number of tetrahedral atoms, the tschortnerite cage is twice as large as the supercage in faujasite. However, the tschortnerite cage is accessible through 8-rings that are smaller than the 12-ring windows in faujasite.

The difficulty involving the creation of new low and intermediate silica molecular sieves is in part because of the limited choice in structure-directing agents. Traditionally, inorganic cations such as Na^+ are employed and it has not been possible to synthesize zeolites with a Si/Al ratio smaller than 5 with organic cations. However, recent results demonstrate that organic cations can template the formation of M^{2+} substituted alumino- (gallo-)phosphate open frameworks in which the M^{2+}/M^{3+} molar ratio is 3 [49,50]. In terms of the framework charge per tetrahedral unit, this is equivalent to aluminosilicates with a Si/Al ratio 3. Thus, it might be feasible to prepare low and intermediate silica zeolites using amines as structure-directing agents.

3 Microporous and Open Framework Phosphates

Because of the structural similarity between dense SiO_2 and $AlPO_4$ phases, the research in the 1970s on high silica or pure silica molecular sieves quickly led to the realization that it might be possible to synthesize aluminophosphate molecular sieves using the method similar to that employed for the synthesis of silicalite. In 1982, Flanigen *et al.* reported a major discovery of a new class of aluminophosphate molecular sieves ($AlPO_4$-n) [11,12]. Unlike zeolites that are capable of various Si/Al ratios, the framework of these aluminophosphates consists of alternating Al^{3+} and P^{5+} sites and the overall framework is neutral with a general formula of $AlPO_4$.

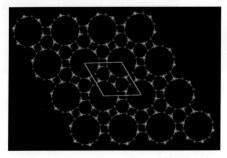 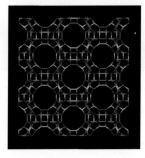

Figure 7. (Left) The three-dimensional framework of $AlPO_4$-5 consists of one-dimensional 12-ring channels. Note the alternating distribution of P and Al sites. Red: P, Yellow: Al. (right) 12-ring channels in metal (Co, Mn, Mg) substituted aluminophosphate UCSB-8.

These aluminophosphates are synthesized hydrothermally using organic amines or quaternary ammonium salts as structure-directing agents. In most cases, organic molecules are occluded into the channels or cages of $AlPO_4$ frameworks. Because the framework is neutral, the positive charge of organic cations is balanced by the simultaneous occlusion of OH⁻ groups. Many of these aluminophosphates have a high thermal stability and remain crystalline after calcination at temperatures between 400-600°C. In addition to framework types already known in zeolites, new topologies have also been found in some structures including $AlPO_4$-5 (AFI) that has a one-dimensional 12-ring channel (Fig. 7) [51].

The next family of new molecular sieves consists of a series of silicon substituted aluminophosphates [52] called silicoaluminophosphates (SAPO-n). To avoid the Si-O-P linkage, Si^{4+} cations tend to replace P^{5+} sites or both Al^{3+} and P^{5+} sites. The substitution of P^{5+} sites by Si^{4+} cations produces negatively charged frameworks with cation exchange properties and acidic properties. The SAPO family includes two new framework types, SAPO-40 (AFR) and SAPO-56 (AFX), not previously known in aluminosilicates, pure silica polymorphs, or aluminophosphates [30].

In addition to silicon, other elements can also be incorporated into aluminophosphates. In 1989, Wilson and Flanigen [53] reported a large family of metal aluminophosphate molecular sieves (MeAPO-n). The metal (Me) species include divalent forms of Mg, Mn, Fe, Co, and Zn (M^{2+}). The MeAPO family represents the first demonstrated synthesis of divalent metal cations in microporous frameworks [53]. In one of these phases, CoAPO-50 (AFY) with a formula of $[(C_3H_7)_2NH_2]_3[Co_3Al_5P_8O_{32}] \cdot 7H_2O$, approximately 37% of Al^{3+} sites are replaced with Co^{2+} cations [30]. For each substitution of Al^{3+} by M^{2+}, a negative charge develops on the framework, which is balanced by protonated amines or quaternary ammonium cations.

For a given framework topology, the framework charge is tunable in aluminosilicates by changing Si/Al ratios. However, it is fixed in binary phosphates such as aluminophosphates or cobalt phosphates [30,54]. The use of ternary compositions as in metal aluminophosphates provides the flexibility in adjusting the framework charge density. Such flexibility contributes to the development of a large variety of new framework types in metal aluminophosphates and has also led to the synthesis of a large number of phosphates with the same framework type as those in zeolites [30,50].

The MeAPSO family further extends the structural diversity and compositional variation found in the SAPO and MeAPO molecular sieves. MeAPSO can be considered as double (Si^{4+} and M^{2+}) substituted aluminophosphates. The MeAPSO family includes one new large pore structure MeAPSO-46 with a formula of $[(C_3H_7)_2NH_2]_8[Mg_6Al_{22}P_{26}Si_2O_{112}] \cdot 14H_2O$ [30]. The quaternary (four different tetrahedral elements at non-trace levels) composition is rare in a microporous framework, but is obviously a promising area for future exploration.

In the two decades following Wilson and Flanigen's original discovery, there has been an explosive growth in the synthesis of open framework phosphates [13,55]. It is apparent that the MeAPO's exhibit much more structural diversity and compositional variation than both SAPO's and MeAPSO's. However, the thermal stability of MeAPO's is generally lower than that of either $AlPO_4$'s or SAPO's. In general, the thermal stability of a metal aluminophosphate decreases with an increase in the concentration of divalent metal cations in the framework.

In addition to the continual exploration of $AlPO_4$ and MeAPO compositions, many other compositions have been investigated including gallophosphates and metal gallophosphates [13]. Of particular interest is the synthesis of a family of extra-large pore phosphates with ring sizes larger than 12 tetrahedral atoms [16]. The use of the fluoride medium [34] and non-aqueous solvents [56] further enriches the structural and compositional diversity of the phosphate-based molecular sieves.

Unlike aluminophosphate molecular sieves developed by Flanigen et al., new generations of phosphates such as phosphates of tin, molybdenum, vanadium [57], iron, titanium, and nickel often consist of metal cations with different coordination numbers ranging from three to six [13]. The variable coordination number helps the generation of many new metal phosphates.

In terms of the framework charge, $AlPO_4$'s, SAPO's, and MeAPO's closely resemble high silica and pure silica molecular sieves. This is not surprising because the synthetic breakthrough in aluminophosphate molecular sieves was based on the earlier synthetic successes in high silica and pure silica phases. However, for certain applications such as N_2 selective adsorbents for air separation, it is desirable to prepare aluminophosphate-based materials that are similar to low or intermediate zeolites. Because each $(AlSi_3O_8)^-$ unit carries the same charge as $(MAlP_2O_8)^-$ (M is a divalent metal cation), the M^{2+}/Al ratio of 1 is equivalent to the Si/Al ratio of 3 in terms of the framework charge per tetrahedral atom. For a Si/Al ratio of 5 as in $(AlSi_5O_{12})^-$, the corresponding M^{2+}/Al ratio is 0.5 as in $(CoAl_2P_3O_{12})^-$. Therefore, to make highly charged aluminophosphates similar to low and intermediate silica, the M^{2+}/Al ratio should be higher than 0.5. Only a very small number of compounds with M^{2+}/Al ratio ≥ 0.5 were known prior to 1997 [30,58,59].

A significant advance occurred in 1997 when a family of highly charged metal aluminophosphates with a $M^{2+}/M^{3+} \geq 1 (M^{2+} = Co^{2+}, Mn^{2+}, Mg^{2+}, Zn^{2+}, M^{3+} = Al^{3+}, Ga^{3+})$ were reported [49,50,60]. After over two decades of extensive research on high silica, pure silica, aluminophosphates, and other open framework materials with low-charged or neutral framework, the synthesis of these highly charged metal aluminophosphates represented a noticeable reversal towards highly charged frameworks often observed in natural zeolites. The recent work on 4-connected, three-dimensional metal sulfides and selenides further increased the framework negative charge to an unprecedented level with a M^{4+}/M^{3+} ratio as low as 0.2 [46].

Three families of open framework phosphates denoted as UCSB-6 (SBS), UCSB-8 (SBE) (Fig. 7), and UCSB-10 (SBT) demonstrate that zeolite-like structures with large pore, large cage, and multidimensional channel systems can be synthesized with a framework charge density much higher than currently known organic-templated silicates [49]. The M^{2+}/M^{3+} ratio in these phases is equal to 1. If these materials could be made as aluminosilicates, the Si/Al ratio would be 3. It is worth noting that until now, no zeolites templated with organic cations only have a Si/Al ratio of 3 or lower. The synthesis of UCSB-6, UCSB-8, UCSB-10, and other highly charged phosphate-based zeolite analogs shows that it might be possible to synthesize low and intermediate silica by templating with organic cations.

While UCSB-6 and UCSB-10 have framework structures similar to EMC-2 (EMT) and faujasite (FAU), respectively, UCSB-8 has an unusual large cage consisting of 64 tetrahedral atoms. Such cage is accessible through four 12-ring windows and two 8-ring windows (Fig. 6). In comparison, the supercage in FAU-type structures is built from 48 T-atoms.

4 Microporous and Open Framework Sulfides

During the development of the above oxide-based microporous materials, two new research directions appeared in late 1980s and early 1990s. One was the synthesis of open framework sulfides initiated by Bedard, Flanigen, and coworkers [61]. Another was the development of metal-organic frameworks in which inorganic metal cations or clusters are connected with organic linkers. Metal-organic frameworks have become an important family of microporous materials and they will be discussed in the next section. Open framework chalcogenides are particularly interesting because of their potential electronic and electrooptic properties, as compared to the usual insulating properties of open framework oxides.

Like in zeolites, the tetrahedral coordination is common in metal sulfides. However, structures of open framework sulfides are substantially different from zeolites. This is mainly because of the coordination geometry of bridging sulfur anions. The typical value for the T-S-T angle in metal sulfides is between 105 and 115 degrees, much smaller than the typical T-O-T angle in zeolites that usually lies between 140 and 150 degrees. In addition, the range of the T-S-T angle is also considerably smaller than that of the T-O-T angle. While the range of the T-S-T angle is approximately between 98 and 120 degrees, the T-O-T angle can extend from about 120 to 180 degrees, depending on the type of tetrahedral atoms.

As the exploratory synthesis in zeolite and zeolite-like materials has progressed from silicates and phosphates to arsenates and germanates [62,63,64], it becomes clear that form a purely geometrical view, the research on open framework sulfides, selenides, and halides continue the trend towards large T-X distances and smaller T-X-T angles (X is an anion such as O, S, and Cl). Such trend has the potential to generate zeolite-like structures with 3-rings and exceptionally large pore sizes.

The tendency for the T-S-T angle to be close to 109 degrees has a fundamental effect on the structure of open framework sulfides. In sulfides with tetrahedral metal cations, all framework elements can adopt tetrahedral coordination. As a result, clusters with structure resembling fragments of zinc blende type lattice can be formed. These clusters are now called supertetrahedral clusters (Fig. 8).

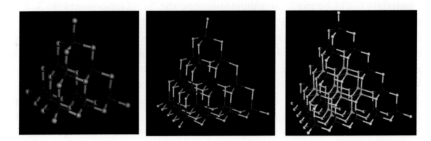

Figure 8. (left) the supertetrahedral T3 cluster, (middle) the T4 cluster. Blue sites are occupied with divalent metal cations. (right) the T5 cluster. Red: In^{3+}; Yellow: S^{2-}; Cyan: the core Cu^+ site. In a given cluster, only four green sites are occupied by Cu^+ ions. The occupation of green sites by Cu^+ ions is not random and follows Pauling's electrostatic valence rule.

Supertetrahedral clusters are regular tetrahedrally shaped fragments of zinc blende type lattice. They are denoted by Yaghi and O'Keeffe as Tn, where n is the number of metal layers [65,66]. One special case is T1 and it simply refers to a tetrahedral cluster such as MS_4, where M is a metal cation. If we add an extra layer, the cluster would be shaped like an adamantane cage with the composition M_4S_{10}, called supertetrahedral T2 cluster because it consists of two metal layers. With the addition of each layer, a new supertetrahedron of a higher order will be obtained. The compositions of supertetrahedral T3, T4, and T5 clusters are $M_{10}X_{20}$ and $M_{20}X_{35}$, and $M_{35}X_{56}$ respectively. When all corners of each cluster are shared through bi-coordinated S^{2-} bridges (as in zeolites), the number of anions per cluster in the overall stoichiometry is reduced by two. While a T2 cluster consists of only bi-coordinated sulfur atoms, a T3 cluster has both bi- and tri-coordinated sulfur atoms. Starting from T4 clusters, tetrahedral coordination begins to occur for sulfur atoms inside the cluster.

At this time, the largest supertetrahedral cluster observed is the T5 cluster (Fig. 8) with the composition of $[Cu_5In_{30}S_{54}]^{13-}$ [67]. This T5 cluster occurs as part of a covalent superlattice in UCR-16 and UCR-17. So far, isolated T5 clusters have not been synthesized. The largest isolated supertetrahedral cluster known to date is T3. Some examples are $[(CH_3)_4N]_4[M_{10}E_4(SPh)_{16}]$, where M = Zn, Cd, E =S, Se, and Ph is a phenyl group [68,69].

With Tn clusters as artificial tetrahedral atoms, it is possible to construct covalent superlattices with framework topologies similar to those found in zeolites. However, the ring size in terms of the number of tetrahedral atoms is increased by *n* times. An increase in the ring size is important because crystalline porous materials with a ring size larger than 12 are rather scarce, but highly desirable for applications involving large molecules.

4.1 Sulfides with tetravalent cations

Some zeolites such as ZSM-5 and sodalite can be made in the neutral SiO_2 form [10,56]. Neutral frameworks have also been found in microporous aluminophosphates [11] and germanates [64,70]. It is therefore reasonable to expect that microporous sulfides with a general framework composition of GeS_2 or SnS_2 may exist. The Ge-S and Sn-S systems were among the earliest compositions explored by Bedard *et al.*, when they reported their work on open framework sulfides in 1989. Thus far, a number of new compounds were found in Ge-S and Sn-S compositions, however, very few have three-dimensional framework structures. Frequently, molecular, one-dimensional, or layered structures are found in these compositions.

In the Ge-S system, the largest observed supertetrahedral cluster is T2 ($Ge_4S_{10}^{4-}$). Larger clusters such as T3 have not been found in the Ge-S system possibly because the charge on germanium is too high to satisfy the coordination environment of tri-coordinated sulfur sites that exist in clusters larger than T2. This is because of Pauling's Electrostatic Valence Rule that suggests the charge on an anion must be balanced locally by neighboring cations.

Isolated T2 clusters ($Ge_4S_{10}^{4-}$) have been found to occur [71,72,73] in the molecular compound [$(CH_3)_4N]_4Ge_4S_{10}$. One-dimensional chains of $Ge_4S_{10}^{4-}$ clusters have also been observed in a compound called DPA-GS-8 [74]. One polymorph of GeS_2, δ-GeS_2, consists of covalently linked $Ge_4S_{10}^{4-}$ clusters with a three-dimensional framework [75]. The framework topology resembles that of the diamond type lattice, however, the extra-framework space is reduced because of the presence of two interpenetrating lattices. As shown in later sections, the interpenetration can be removed by incorporating trivalent metal cations into the cluster to generate negative inorganic frameworks that can be assembled with protonated amines.

In the Sn-S system, layered structures are common [76]. Because of its large size, tin frequently forms non-tetrahedral coordination. In addition, tin may also form oxysulfides, which further complicates the synthetic design of porous tin sulfides. One rare three-dimensional framework [77] based on tin sulfide is $[Sn_5S_9O_2][HN(CH_3)_3]_2$. This material is built from T3 clusters, $[Sn_{10}S_{20}]$. Each T3 cluster has four adamantane-type cavities that can accommodate one oxygen atom per cavity to give a cluster $[Sn_{10}S_{20}O_4]^{8-}$. Because each corner sulfur atom is shared between two clusters. The overall framework formula is $[Sn_{10}S_{18}O_4]^{4-}$. The isolated form of the $[Sn_{10}S_{20}O_4]^{8-}$ cluster is also known in $Cs_8Sn_{10}S_{20}O_4 \cdot 13H_2O$ [78].

4.2 Sulfides with tetravalent and mono- or divalent cations

The early success in the preparation of open framework sulfides depended primarily on the use of mono- or divalent cations (e.g., Cu^+, Mn^{2+}) to join together chalcogenide clusters (e.g., $Ge_4S_{10}^{4-}$). These low-charged mono- or divalent cations help generate negative charges on the framework that are usually charge-balanced by protonated amines or quaternary ammonium cations.

One example was the synthesis of TMA-CoMnGS-2 [61]. Like many other germanium sulfides, the basic structural unit is the T2 cluster. Here, T2 clusters are joined together by three-connected Me(SH)$^+$ (Me = divalent metal cations such as Co^{2+} and Mn^{2+}) units to form a framework structure. Another interesting example was the synthesis of a series of compounds with the general formula of [(CH$_3$)$_4$N]$_2$MGe$_4$S$_{10}$ (M = Mn^{2+}, Fe^{2+}, Cd^{2+}) [73,79,80]. Unlike δ-GeS$_2$ that is an intergrowth of two diamond-type lattice (double-diamond type), [(CH$_3$)$_4$N]$_2$MGe$_4$S$_{10}$ has a non-interpenetrating diamond-type lattice (single-diamond type) in which tetrahedral carbon sites are replaced with alternating T2 and T1 clusters.

In [(CH$_3$)$_4$N]$_2$MGe$_4$S$_{10}$ and TMA-CoMnGS-2, the divalent metal cations join together four and three T2 clusters, respectively. It is also possible for a metal cation to connect to only two T2 clusters. Such is the case in CuGe$_2$S$_5$(C$_2$H$_5$)$_4$N, in which T2 clusters form the single-diamond type lattice with monovalent Cu$^+$ cations bridging between two T2 clusters [81].

The diamond-type lattice is very common for framework structures formed from supertetrahedral clusters. With T2 clusters, amines or ammonium cations are usually big enough to fill the framework cavity. As a result, the interpenetration of two identical lattices does not usually occur. With larger clusters, charge-balancing organic amines are often not enough to fill the extra-framework space and the double-diamond type structure becomes more common.

In addition to the single-diamond type lattice, other types of framework structures are possible. One compound, Dabco-MnGS-SB1 with a formula of MnGe$_4$S$_{10}$· C$_6$H$_{14}$N$_2$· 3H$_2$O, has a framework structure in which T1 and T2 clusters alternate to form the zeolite ABW-type topology with a ring size of 12 tetrahedral atoms [82].

While the use of M^{2+} and M$^+$ cations has led to a number of open framework sulfides, it could have negative effects too. These low-charged metal sites could lower the thermal stability of the framework. The destabilizing effect of divalent cations (e.g. Co^{2+}, Mn^{2+}) in porous aluminophosphates is well known. However, unlike in phosphates, it is difficult to study the destabilizing effect of low-charged cations in open framework sulfides because the incorporation of low-charged cations in sulfides changes both chemical composition and framework type.

4.3 Sulfides with trivalent metal cations

In late 1990s, a new direction appeared when Parise, Yaghi and their coworkers reported several open framework indium sulfides [65,83]. The In-S composition is quite unique because no oxide open frameworks with similar compositions were known before. In fact, the In-O-In and Al-O-Al linkages are not expected to occur in oxides with four-connected, three-dimensional structures. Fortunately, such a restriction does not apply to open framework sulfides.

An interesting structural feature in the In-S system is the occurrence of T3 clusters, $[In_{10}S_{18}]^{6-}$. A T3 cluster has both bi- and tri-coordinated sulfur sites. The lower charge of In^{3+} compared to Ge^{4+} and Sn^{4+} makes it possible to form tri-coordinated sulfur sites. Through the sharing of all corner sulfur atoms, open framework materials with several different framework topologies have been made. These include DMA-InS-SB1 (T3 double-diamond type) [83], ASU-31 (T3-decorated sodalite net), ASU-32 (T3-decorated CrB_4 type) [65], and ASU-34 (T3 single-diamond type) [84].

Very recently, Feng et al. synthesized a series of open framework materials based on T3 gallium sulfide clusters, $[Ga_{10}S_{18}]^{6-}$ [85]. Only the double-diamond type topology has been observed so far in the Ga-S system. In UCR-7GaS, T3 clusters are bridged by a sulfur atom (-S-) whereas in UCR-18GaS, one quarter of the inter-cluster linkage is through the trisulfide group (-S-S-S-).

So far, isolated T3 clusters, $[In_{10}S_{20}]^{10-}$ and $[Ga_{10}S_{20}]^{10-}$, have not been found yet even though isolated T2 clusters, $[In_4S_{10}]^{8-}$ and $[Ga_4S_{10}]^{8-}$, have been known for a while [86]. Regular supertetrahedral clusters larger than T3 have not been found in the binary In-S or Ga-S systems probably because tetrahedral sulfur atoms at the core of these clusters can not accommodate four trivalent metal cations because the positive charge surrounding the tetrahedral sulfur anion would be too high.

4.4 Sulfides with trivalent and mono- or divalent cations

To access clusters larger than T3, mono- or divalent cations need to be incorporated into the Ga-S or In-S compositions. Another motivation to incorporate mono- or divalent cations in the In-S or Ga-S synthesis conditions might be the desire to create new structures in which T3 clusters are joined together by mono- or divalent cations, in a manner similar to the assembly of $[Ge_4S_{10}]^{4+}$ clusters by mono- or divalent cations [73]. So far, mono- and divalent cations have only been observed to occur as part of a supertetrahedral cluster, not as linker units between clusters.

The first T4 cluster, $[Cd_4In_{16}S_{33}]^{10-}$, was synthesized by Yaghi, O'Keffee and coworkers in CdInS-44. In this compound, four Cd^{2+} cations are located around the core tetrahedral sulfur atom (Fig. 8). Because Cd^{2+} and In^{3+} are isoelectronic, it is difficult to distinguish Cd^{2+} and In^{3+} sites through the crystallographic refinement of X-ray diffraction data. Further evidences on the distribution of di- and trivalent cations in a T4 clusters came from UCR-1 and UCR-5 series of compounds that incorporate the first row transition metal cations such as Mn^{2+}, Fe^{2+}, Co^{2+}, and Zn^{2+} [87].

An exciting recent development is the synthesis of two superlattices (UCR-16 and UCR-17) consisting of T5 supertetrahedral clusters, $[Cu_5In_{30}S_{54}]^{13-}$ [67]. There are four tetrahedral core sulfur sites, each of which is surrounded by two In^{3+} and two Cu$^+$ cations. One Cu$^+$ cation is located at the center of the T5 cluster and there is one Cu$^+$ cation on each face of the supertetrahedral cluster (Fig. 8).

Another interesting structural feature is the occurrence of hybrid superlattices. In UCR-19, T3 clusters [Ga$_{10}$S$_{18}$]$^{6-}$ and T4 clusters [Zn$_4$Ga$_{16}$S$_{33}$]$^{10-}$ alternate to form the double-diamond type superlattice [85]. In UCR-15, T3 clusters [Ga$_{10}$S$_{18}$]$^{6-}$ and pseudo-T5 clusters [In$_{34}$S$_{54}$]$^{6-}$ also alternate to form the double-diamond type superlattice [88]. The pseudo-T5 cluster is similar to the regular T5 cluster except that the core metal site is not occupied. The pseudo-T5 cluster has also been found with a different chemical composition in a layered superlattice with the framework composition of [Cd$_6$In$_{28}$S$_{54}$]$^{12-}$ [89].

4.5 Sulfides with tetravalent and trivalent cations

Open framework sulfides based on In-S and Ga-S compositions have open architectures and some have been shown to undergo ion exchange in solutions. However, to generate microporosity, it is necessary to remove a substantial amount of extra-framework species. Open framework sulfides such as indium or gallium sulfides generally do not have sufficient thermal stability to allow the removal of an adequate amount of extra-framework species to generate microporosity.

A general observation in zeolites is that the stability increases with the increasing Si^{4+}/Al^{3+} ratio. It can be expected that the incorporation of tetravalent cations such as Ge^{4+} and Sn^{4+} into In-S or Ga-S compositions could lead to an increase in the thermal stability. Recently, Feng et al. reported a large family of chalcogenide zeolite analogs [46]. These materials were made by simultaneous triple substitutions of O^{2-} with S^{2-} or Se^{2-}, Si^{4+} with Ge^{4+} or Sn^{4+}, and Al^{3+} with Ga^{3+} or In^{3+}. All four possible M^{4+}/M^{3+} combinations (Ga/Ge, Ga/Sn, In/Ge, and In/Sn) could be realized resulting in four zeolite-type topologies.

Based on the topological type, these materials are classified into four families denoted as UCR-20, UCR-21, UCR-22, and UCR-23. Each number refers to a series of materials with the same framework topology, but with different chemical compositions in either framework or extra-framework components. For example, UCR-20 can be made in all four M^{4+}/M^{3+} combinations, giving rise to four sub-families denoted as UCR-20GaGeS, UCR-20GaSnS, UCR-20InGeS, and UCR-20InSnS. An individual compound is specified when both the framework composition and the type of extra-framework species are specified (e.g., UCR-20GaGeS-AEP, AEP = 1-(2-aminoethyl)piperazine).

The extra-large pore size and 3-rings are two interesting features. UCR-22 (Fig. 9) and UCR-23 have 24-ring and 16-ring windows whereas both UCR-20 (Fig. 9) and UCR-21 have 12-ring windows. These inorganic frameworks are strictly 4-connected 3-dimensional networks commonly used for the systematic description of zeolite frameworks. Unlike known zeolite structure types, a key structural feature is the presence of the adamantane-cage shaped building unit, M$_4$S$_{10}$. The M$_4$S$_{10}$ unit

Figure 9. The three-dimensional framework of UCR-20 (left) and UCR-22 (right) families of sulfides.

consists of four 3-rings fused together. For materials reported here, the framework density defined as the number of T-atoms in 1000Å^3 ranges from 4.4 to 6.5.

Although these chalcogenides are strictly zeolite-type tetrahedral frameworks, it is possible to view them as decoration of even simpler tetrahedral frameworks. Here, each M_4S_{10} unit can be treated as a large artificial tetrahedral atom. With this description, UCR-20 has the decorated sodalite-type structure, in which a tetrahedral site in a regular sodalite net is replaced with a M_4S_{10} unit. UCR-21 has the decorated cubic ZnS type structure. UCR-23 has the decorated CrB_4 type network in which tetrahedral boron sites are replaced with M_4S_{10} units.

Upon exchange with Cs^+ ions, the percentage of C, H, and N in UCR-20GaGeS-TAEA was dramatically reduced. The exchanged sample remained highly crystalline as the original sample. The Cs^+ exchanged UCR-20GaGeS-TAEA displayed type I isotherm characteristic of a microporous solid. This sample has a high Langmuir surface area of 807m^2/g and a micropore volume of 0.23cm^3/g despite the presence of much heavier elements (Cs, Ga, Ge, and S) compared to aluminosilicate zeolites.

5   Microporous Metal-Organic Frameworks

Currently, the synthetic design of metal-organic frameworks (also known as coordination polymers) is a very active research area [90,91]. Many new microporous materials synthesized in the past several years belong to this family. Unlike zeolites that have an inorganic host framework, in metal-organic frameworks, the three-dimensional connectivity is established by linking metal cations or clusters with bidentate or multidentate organic ligands. The resulting frameworks are hybrid frameworks between inorganic and organic building units and should be distinguished from microporous materials in which organic amines are encapsulated in the cavities of purely inorganic frameworks.

The development of metal-organic framework materials began in the early 1990s and was apparently an extension of the earlier work on three-dimensional cyanide frameworks [14,92,93]. In $K_2Zn_3[Fe(CN)_6]_2 \cdot xH_2O$ [94], octahedral Fe^{2+} and tetrahedral Zn^{2+} cations are joined together by linear CN^- groups to form a three-dimensional framework with cavities occupied by K^+ cations and water molecules. To generate large cavities, one method is to replace short CN^- ligands with large ligands such as nitriles [93], amines, and carboxylates [95]. A large variety of structural building units are possible with this approach. However, at the early stage of their development, metal-organic frameworks were plagued by problems such as lattice interpenetration and the low stability upon guest removal.

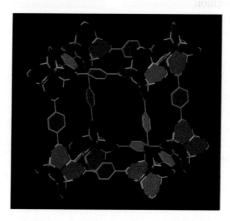

Figure 10. The framework of MOP-5, one of the first microporous metal-organic frameworks [98].

During the past several years, a substantial progress has occurred in the rational synthesis of these materials and a large number of metal-organic frameworks have been made that are capable of supporting microporosity as demonstrated by their gas sorption properties [96,97,98,99]. Such success was in part because of the use of rigid di- and tri-carboxylates and judicious selections of experimental conditions. It is worth noting that despite the wide selection of organic molecules that can serve as bridges between inorganic building units, new metal-organic frameworks are often made by changes in synthesis conditions such as pH, type of solvents, and temperature, instead of using new organic linker molecules. For example, $Zn(BDC)(DMF)(H_2O)$ (denoted as MOF-2, BDC = 1,4-benzenedicarboxylate, DMF = N, N'-dimethylformamide), $Zn_3(BDC)_3 \cdot 6CH_3OH$ (denoted as MOF-3), and $(Zn_4O)(BDC)_3(DMF)_8(C_6H_5Cl)$ (MOF-5) are all made from Zn^{2+} and BDC [100]. Their topological differences are caused by spacing-filling or structure-directing solvent molecules. These compounds clearly show the importance of controlling the synthesis conditions including the selection of solvent. This is somewhat similar to the synthesis in zeolites where the primary building units are the same (i.e., SiO_4 and

AlO$_4$ tetrahedra) in all structures and the difference in secondary building units and three-dimensional topologies is caused by extra-framework structure-directing agents.

Microporous metal-organic materials are complementary to oxide and chalcogenide based microporous materials such as zeolites. There are many fundamental differences between metal-organic materials and zeolites so that rather than competing with each other, they are expected to have different applications. For example, unlike zeolites and chalcogenides that usually have a negative framework, metal-organic frameworks (excluding cyanides) reported so far are usually positive or neutral. Therefore, while zeolites are cation exchangers, metal-organic frameworks can be anion exchangers.

For a given framework topology, the framework of a zeolite or a phosphate can often have a range of different charge density by varying the Si/Al ratio or doping Al^{3+} sites with divalent cations. The difference in the framework charge density in zeolites makes it possible to tune hydrophilicity or hydrophobicity of the framework. Metal-organic structures do not seem to have such flexibility in adjusting the framework charge density, however, the hydrophilicity or hydrophobicity in metal-organic frameworks is tunable by introducing different organic groups as shown in a series of compounds denoted as IRMOF-n [101].

Metal cations in metal-organic frameworks are usually transition metals, while in oxides and chalcogenides, main group elements dominate the framework cationic sites. Therefore, metal-organic frameworks can bind to guest molecules through coordinatively unsaturated transition metal sites [102,103]. Such interaction is not common with main group elements in zeolites or microporous phosphates, even though transition-metal doped zeolites or phosphates might contain active transition metal sites.

One potential with metal-organic frameworks is the possibility to form porous materials with pore size over 10Å by using large inorganic clusters or organic linkers. This potential is evidenced by the recent synthesis of a series of isoreticular MOFs denoted as IRMOF-n (n = 1 through 7, 8, 10, 12, 14, and 16) from different dicarboxylates [101]. These compounds have a calculated aperture size (also called free-diameter) from 3.8 to 19.1Å. IRMOFs also demonstrate the feasibility to have different organic groups in three-dimensional frameworks without a change in the framework topology.

The idea of using chiral structure-directing agents to direct the formation of chiral inorganic frameworks has been around for some time. However, few synthetic successes have been reported. Metal-organic frameworks provide a new opportunity in the design of chiral porous frameworks because chiral organic building units can be directly used for the construction of the framework. One recent example has shown this approach to be highly promising [104].

The recent synthetic success in producing microporous metal-organic frameworks has shifted some focus from the synthetic design to the potential applications. One promising application of metal-organic frameworks is in the area of gas storage. Several metal-organic framework materials have been found to have a high capacity for methane storage [101,105,106]. For example, at 298K and 36atm, the quantity of methane adsorbed by IRMOF-6 is as high as 155cm^3 (STP)/cm^3, considerably higher than other crystalline porous materials such as zeolite 5A (87cm^3/cm^3). Such high adsorption capability is likely related to the more hydrophobic property of organic building units in these metal-organic frameworks, in addition to their high pore volumes and wide pore sizes.

In the following, we discuss in some more details metal-organic frameworks that either have a positive or neutral framework. Excluding cyanide frameworks, metal-organic frameworks with negative charges on the framework are far less common and remain to be explored in the future.

5.1 Cationic metal-organic frameworks

Cationic metal-organic frameworks were among the earliest to be studied. Some early examples of cationic metal-organic frameworks were formed between monovalent metal cations (Cu$^+$ or Ag$^+$) and neutral amines. Metal cations in these compounds can take different coordination geometry such as linear, trigonal or tetrahedral. Interestingly, ligands can also take different geometry. Examples of linear, trigonal, and tetrahedral ligands are 4,4'-bipyridine (4, 4'-bpy), 1,3,5-tricyanobenzene, and 4,4',4'',4'''-tetracyanotetraphenylmethane, respectively.

Examples of compounds with the cationic metal-organic frameworks include Ag(4,4'-bpy)NO$_3$ [107] and Cu(4,4'-bpy)$_2$(PF$_6$) [108]. In Ag(4,4'-bpy)NO$_3$, Ag$^+$ is coordinated to two 4,4'-bpy molecules in a nearly linear configuration and the three-dimensional framework is formed with the help of Ag-Ag (2.977Å) interactions. In Cu(4,4'-bpy)$_2$(PF$_6$), Cu$^+$ ions have tetrahedral coordination and 4,4'-bpy behaves very much like linear CN- groups between two tetrahedral atoms. However, much larger void space forms as a result of longer length of 4,4'-bpy and such void space is reduced by the formation of four interpenetrating diamond-like frameworks in Cu(4,4'-bpy)$_2$(PF$_6$).

Some cationic frameworks have been found to display zeolitic properties such as ion exchange with anions in the solution. However, it has been difficult to remove extra-framework species to produce microporosity. Because of this limitation, there has been an increasing interest in using carboxylates as organic linkers. The current synthetic approach for the synthesis of carboxylate-based metal-organic frameworks usually gives rise to neutral frameworks discussed below.

5.2 Neutral metal-organic frameworks

In oxide and chalcogenide molecular sieves, a low framework charge generally means a high thermal stability. Therefore, neutral metal-organic frameworks should provide the best opportunity for generating microporous metal-organic frameworks. In a metal-organic compound with a neutral framework, the host-guest interaction tends to be weaker than that in a solid with a charged framework. The weak host-guest interaction makes it possible to remove guest solvent molecules at relatively mild conditions. In addition, the neutral framework also tends to be more tolerant of the loss of neutral guest molecules.

Among the first metal-organic frameworks that showed zeolite-like microporosity through reversible gas sorption are MOF-2, $Cu_3(BTC)_2(H_2O)_x$ (denoted as HKUST-1 or Cu-BTC, BTC = 1,3,5-benzenetricarboxylate), and MOF-5 [97,98,109]. A key structural feature of Cu-BTC is the dimeric Cu-Cu (2.628Å) unit. A detailed investigation of sorption properties showed that Cu-BTC may be useable for separation of gas mixtures such as CO_2-CO, CO_2-CH_4, and C_2H_4-C_2H_6 mixtures [110].

The framework structure of MOF-5 is particularly simple with $(Zn_4O)^{6+}$ clusters arranged at eight corners of a cube and linear BDC linkers located on edges of the cube (Fig. 10) [98]. The $(Zn_4O)^{6+}$ cluster has a pseudo-octahedral connectivity because it connects to BDC through six edges of Zn_4 tetrahedra. Even more interesting is the fact that BDC molecules can be replaced by a series of different dicarboxylates without altering the framework topology [101]. This provides an elegant means of adjusting the pore size and framework functionality.

Neutral frameworks can also be prepared from neutral organic ligands. One such example is $CuSiF_6(4,4'$-bpy$)_2 \cdot 8H_2O$ [106]. In this case, Cu^{2+} cations are linked into two-dimensional sheets by 4,4'-bpy ligands and these sheets are then linked into a three-dimensional framework by SiF_6^{2-} anions. This compound is microporous and has a high adsorption capacity for methane.

Metal-organic frameworks can also be created by a combined use of amines and carboxylates. For example, in $[Zn_4(OH)_2(fa)_3(4,4'$-bpy$)_2]$ (fa = fumarate), dicarboxylate and diamine molecules work together to link $Zn_4(OH)_2$ units into an interpenetrating three-dimensional framework [111]. Furthermore, carboxylate-substituted amines can simultaneously use COO- and N to bind to inorganic units to create an extended framework. One example is $Cu(INA)_2 \cdot 2H_2O$ (INA = isonicotinate or pyridine-4-carboxylate) [112].

5.3 Metalloporphyrin-based metal-organic frameworks

A special class of ligands are porphyrins and metalloporphyrins. Metalloporphyrins can form either cationic or neutral frameworks depending on the nature of substituent groups. Two of the earliest examples are Cu(II)(tpp)Cu(I)BF_4(solvent) and Cu(II)(tcp)Cu(I)$BF_4 \cdot$ 17($C_6H_5NO_2$) (tpp = 5,10,15,20-tetra(4-pyridyl)-21H,23H-porphine; tcp = 5,10,15,20-tetrakis(4-cyanophenyl)-21H,23H-porphine) [113]. In both cases, the framework is constructed from equal numbers of tetrahedral (Cu⁻)

and square planar (Cu-tpp or Cu-tcp) centers. Neither of these two compounds is stable upon solvent removal.

A stable metalloporphyrin-based metal-organic framework was recently demonstrated by Suslick et al. [114]. PIZA-1 with a formula of [Co(III)T(p-CO_2)PPCo(II)$_{1.5}$(C_5H_5N)$_3$(H_2O)· C_5H_5N] is formed from carboxylate-substituted tetraphenylporphyrins with cobalt ions. Because of the presence of carboxylate groups, the framework of PIZA-1 is neutral. It is apparent that the ability of transition metals (Cu and Co) to exist in different oxidation states helps the formation of these metalloporphyin-based metal-organic frameworks.

No mixed valency occurs in SMTP-1 [115], a family of layered structures with a general formula of [M(tpp)$_6$]· G (M = Co^{2+}, G =12CH_3COOH· 12H_2O; M = Mn^{2+}, G = 60H_2O; or M = Mn^{2+}, G = 12C_2H_5OH· 24H_2O. SMTP-1 differs from the above metalloporphyrin-based structures. The metal cation in the center of the porphyrin ring is also coordinated to pyridyl groups of other tpp complexes, allowing the creation of an extended layer structure without the use of separate metal cations for crosslinking tpp complexes.

5.4 Metal-organic frameworks from oxide clusters

In metal-organic frameworks, the inorganic unit is often a single transition metal cation (sometimes with some coordinating solvent molecules attached). The diversity of metal-organic frameworks can be greatly increased if inorganic clusters are used as structural building units. The simplest situation is dinuclear units such as Ag_2 in Ag(4,4'-bpy)NO_3 and Cu_2 in Cu-BTC [97,107]. The Zn_2 (2.940Å) unit is found in MOF-2. Clusters containing three or four metal cations are also known. For example, a chiral metal-organic framework called D-POST-1 contains the Zn_3O unit in which the oxygen atom is located at the center of the Zn_3 triangle [104]. Similarly, the Zn_4O unit containing tetrahedrally coordinated oxygen anions was recently found in MOF-5 and IRMOF series of compounds. Much larger units (e.g., Zn_8SiO_4) have also been reported [116,117,118]. In many cases, these inorganic clusters do not occur in the starting materials and they are formed *in situ* during the synthesis of metal-organic frameworks.

5.5 Metal-organic frameworks from chalcogenide clusters

As shown above, the use of organic multidentate ligands to organize inorganic species is an effective method to prepare porous solids with tunable pore sizes. However, inorganic building units are generally limited to individual metal ions (e.g., Zn^{2+}) or their oxide clusters (e.g., Zn_4O^{6+}). To expand applications of porous materials beyond traditional areas such as adsorption and catalysis, metal-organic frameworks based on semiconducting chalcogenide nanoclusters are highly desirable. Recently, Feng et al. reported the organization of the cubic [$Cd_8(SPh)_{12}$]$^{4+}$

clusters by in-situ generated tetradentate dye molecules [119]. The structure consists of three-dimensional inorganic-organic open framework with large uni-dimensional channels. The combination of dye molecules and the inorganic cluster unit in the same material creates a synergistic effect that greatly enhances the emission of the inorganic cluster at 580nm. Such an emission can be excited by an unusually broad spectral range down to the UV, which is believed to result from the absorption of dye molecules and the subsequent energy transfer.

6 Extra-large Pore Crystalline Molecular Sieves

Thus far, an extra-large pore material is conveniently understood as those having a ring size of over 12 tetrahedral atoms [120]. In zeolites, the maximum pore size of a 12-ring pore is about 8Å. The recent progress in metal-organic frameworks has made it possible to obtain porous materials with pore size larger than 8Å by using larger organic linkers rather than by forming pores with more than 12 metal cations.

Among silicates, the extra-large pore has only been found in two high silica zeolites and one beryllosilicate. The first extra-large pore zeolite (UTD-1) was reported in 1996 (Fig. 11) [121,122]. UTD-1 (DON) was synthesized using bis(pentamethylcyclopentadienyl) cobalticinium cations and has a ring size of 14 tetrahedral atoms. It has a one-dimensional channel system with the approximate free diameter of 7.5 x 10Å for the 14-ring pore. Another extra-large pore zeolite (CIT-5) was reported in 1997 [123,124]. Like UTD-1, CIT-5 (CFI) also has a ring size of 14 tetrahedral atoms with a one-dimensional channel system. The effective pore size (6.4Å measured using the Horvath-Kawazoe method) of CIT-5 is similar to that of one-dimensional 12-ring channel in SSZ-24 (AFI) [125]. Very recently, a hydrated potassium beryllosilicate called OSB-1 (OSO) was found to have an extra-large pore size of 14 tetrahedral atoms [30].

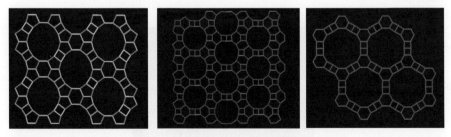

Figure 11. (left) The three-dimensional framework of UTD-1 (DON) with elliptical 14-ring windows; Reprinted with permission from http://www.iza-structures.org/ and reference [30]. (middle) the three-dimensional framework of AlPO$_4$-8 (AET) showing 14-ring windows (right) the three-dimensional framework of VPI-5 (VFI) with 18-ring windows.

Most extra-large pore materials such as cacoxenite, VPI-5, cloverite, and JDF-20 are found in phosphates [15,126,127,128]. While the ring size of only 14 tetrahedral atoms is known in silicates, extra-large pore phosphates come with various ring sizes including 14, 16, 18, 20, and 24. Structures of these phosphates sometimes deviate from those of typical zeolites in several aspects including framework interruptions by terminal OH⁻, F⁻, or H_2O groups and non-tetrahedral coordination. These deviations tend to lower the thermal stability of extra-large pore phosphates. On the other hand, it is often because of these deviations that extra-large pores are formed.

The first synthetic extra-large pore phosphate is VPI-5 with one-dimensional channel defined by 18 oxygen atoms (Fig. 11) [15]. Unlike most aluminophosphate molecular sieves, VPI-5 is a hydrated aluminophosphate and does not contain any organic structure-directing agent. Under suitable heating conditions, VPI-5 can be recrystallized into another extra-large pore phosphate called $AlPO_4$-8 (AET) with a 14-ring pore size (Fig. 11) [129].

Among the most recent development in the area of microporous phosphates is the synthesis of two extra-large pore nickel phosphates denoted as VSB-1 and VSB-5 [130,131]. Similar to VPI-5, both VSB-1 and VSB-5 are hydrates and organic amines used in the syntheses were not occluded into the final structures. VSB-1 and VSB-5 have one-dimensional 24-ring channels and both of them have good thermal stability. The nitrogen adsorption shows the type I isotherms typical of a microporous material.

The synthesis of VPI-5, VSB-1, and VSB-5 demonstrates that neither large nor small organic structure-directing agents are essential for the preparation of extra-large pore sizes. The formation of different pore sizes likely depends on types of small structural units that eventually come together to create the framework and the pore. The structural and synthetic factors that affect the formation of these small structural units may have a substantial effect on the creation of extra-large pore materials.

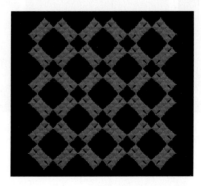

Figure 12. The three-dimensional framework of UCR-23 family of sulfides showing 16-ring channels.

One strategy for the preparation of the extra-large pore size is to generate a large number of small rings, particularly 3-rings. Because the average ring size in a three-dimensional four-connected net is approximately 6, the presence of small rings will be accompanied by large or extra-large rings so that the average ring size will be about 6 [132]. This strategy can be illustrated with the recent discovery of a large family of extra-large pore sulfides.

Because of the large T-S distances and small T-S-T angles, 3-rings often occur in open framework metal sulfides. Correspondingly, large pore and extra-large pore sizes are typically structural features in sulfides. For example, UCR-20, UCR-21, and UCR-22, and UCR-23 consist of adamantane-shaped clusters (T_4S_{10}) with 3-rings. While both UCR-20 and UCR-21 are large-pore sulfides, UCR-23 has three-dimensional intersecting 16-, 12-, and 12-ring channels (Fig. 12) and UCR-22 consists of interpenetrating three-dimensional framework with 24-ring window size.

Other strategies for increasing the pore size include the use of large structural building units such as clusters and the use of long linker molecules between two structural building units. For example, the use of chalcogenide supertetrahedral clusters as large artificial tetrahedral atoms has resulted in a number of three-dimensional frameworks with extra-large pore sizes. Equally successful is the use of dicarboxylates as molecular linkers to join together metal cations or their clusters to generate a series of metal-organic frameworks with pore sizes > 10Å. By using different supertetrahedral clusters and carboxylates, the pore size of the resulting open framework materials can be tuned.

Acknowledgement

We thank Dr. Qisheng Huo for critical evaluation of this manuscript and Nanfeng Zheng for assistance with figures.

References

1. Rouquerol J., Avnir D., Fairbridge C. W., Everett D. H., Haynes J. H., Pernicone N., Ramsay J. D. F., Sing K. S. W. and Unger, K. K. Recommendations for the characterization of porous solids, *Pure Appl. Chem.* **66** (1994) pp. 1739-58.
2. McCusker L. B., Liebau F. and Engelhardt G., Nomenclature of structural and compositional characteristics of ordered microporous and mesoporous materials with inorganic hosts (IUPAC recommendations 2001), *Pure Applied Chem.* **73** (2001) pp. 381-394.
3. Barton T. J., Bull L. M., Klemperer W. G., Loy D. A. McEnaney B., Misono M., Monson P. A., Pez G., Scherer B., Vartuli J. C. and Yaghi O. M., Tailored Porous Materials, *Chem. Mater.* **11** (1999) pp. 2633-2656.
4. Weitkamp J., Zeolites and catalysis, Solid State Ionics 131 (2000) pp. 175-188.

5. Smith J. V., Definition of a Zeolite, *Zeolites* **4** (1984) pp. 309-310.
6. Foley, H. C., Carbogenic molecular sieves: synthesis, properties and applications, *Micro. Mater.* **4** (1995) 407-433.
7. Barrer R. M., Synthesis of a zeolitic mineral with chabazitelike sorptive properties, *J. Chem. Soc.* (1948) pp. 127-132.
8. Milton R. M., Molecular sieve science and technology. A historical perspective, *ACS Symposium Series* (1989), **398** (Zeolite Synth.), pp. 1-10.
9. Argauer R. J. and Landolt G. R. Crystalline Zeolite ZSM-5 and Method of Preparing the Same, US Patent 3,702,886, 1972.
10. Flanigen E. M., Bennett J. M., Grose R. W., Cohen J. P., Patton R. L., Kirchner R. M. and Smith, J. V. Silicalite, a new hydrophobic crystalline silica molecular sieve, *Nature* **271** (1978) pp. 512-516.
11. Wilson S. T., Lok B. M., Messina C. A., Cannan T. R. and Flanigen E. M., Aluminophosphate Molecular Sieves: A New Class of Microporous Crystalline Inorganic Solids, *J. Am. Chem. Soc.* **104** (1982) pp. 1146-1147.
12. Wilson S. T., Lok B. M., Messina C. A., Cannan T. R. and Flanigen, E. M., Aluminophosphate molecular sieves: a new class of microporous crystalline inorganic solids, *ACS Symposium Series* (1983), **218** (Intrazeolite Chem.), pp. 79-106.
13. Cheetham A. K., Ferey G. and Loiseau T., Open-Framework Inorganic Materials, *Angew. Chem. Int. Ed.* **38** (1999) pp. 3268-3292.
14. Bowes C. L. and Ozin G. A., Self-Assembling Frameworks: Beyond Microporous Oxides, *Adv. Mater.* **6** (1996) pp. 13-28.
15. Davis M. E., Saldarriaga C., Montes C., Garces J. and Crwoder C., A molecular sieve with eighteen-membered rings, *Nature* **331** (1988) pp. 698-699.
16. Davis M. E. Ordered porous materials for emerging applications, *Nature* **147** (2002) pp. 813-821.
17. Evans O. R., Lin W., Crystal Engineering of NLO Materials Based on Metal-Organic Coordination Networks, *Acc. Chem. Res.* **35** (2002) pp. 511-522.
18. Davis M. E. and Lobo R. F., Zeolite and Molecular Sieve Synthesis, *Chem. Mater.* **4** (1992) pp. 756-768.
19. Davis M. E., New Vistas in Zeolite and Molecular Sieve Catalysis, *Acc. Chem. Res.* **26** (1993) pp. 111-115.
20. Breck D. W., *Zeolite Molecular Sieves*, Wiley, New York, 1974.
21. Barrer R. M., *Zeolites and Clay Minerals as Sorbents and Molecular Sieves*, Academic Press, 1978.
22. Dyer A., *An Introduction to Zeolite Molecular Sieves*, John Wley & Sons, 1988.
23. van Bekkum H., Flanigen E. M. and Jansen J. C., *Studies in Surface Science and Catalysis*, Vol. **58** (*Introduction to Zeolite Science and Practice*) Elsevier, New York, 1991.

24. Jansen J. C., Stocker M., Karge H. G. and Weitkamp, *Studies in Surface Science and Catalysis*, Vol. **85** (*Advanced Zeolite Science and Applications*), Elsevier, 1994.
25. Atwood J. L., Davies J. E. D., MacNicol D. D. and Vogtle F., *Comprehensive Supramolecular Chemistry*, Vol. **7** (*Solid-State Supramolecular Chemistry: Two- and Three-Dimensional Inorganic Networks*, eds. Alterti G. and Bein T.), Elsevier, 1996.
26. Fricke R., Kosslick H., Lischke G. and Richter M., Incorporation of Gallium into Zeolites: Syntheses, Properties and Catalytic Application. *Chem. Rev.* **100** (2000) pp. 2303-2405.
27. Saxton R. J., Crystalline microporous titanium silicates, *Topics in Catalysis* **9** (1999) pp. 43-57.
28. Rocha J. and Anderson M. W. Microporous Titanosilicates and other Novel Mixed Octahedral-Tetrahedral Framework Oxides, *Eur. J. Inorg. Chem.* (2000) pp. 801-818.
29. Loewenstein W., The distribution of aluminum in the tetrahedra of silicates and aluminates, *Am. Mineralogist* **39**, (1954), pp. 92-96.
30. Baerlocher Ch., Meier W. M., Olson D. H. *Atlas of Zeolite Framework Types*, 2001, Elsevier.
31. Smith J. V., Topochemistry of zeolites and related materials. 1. Topology and geometry, *Chem. Rev.* **88** (1988), pp. 149-82.
32. Gier T. E., Bu X., Feng P. and Stucky G. D., Synthesis and organization of zeolite-like materials with three-dimensional helical pores, *Nature* **395** (1998), pp. 154-157.
33. Zones S. I., Davis M. E., Zeolite materials: Recent discoveries and future prospects. *Curr. Opin. Solid State Mater. Sci.* **1** (1996) pp. 107-117.
34. Kessler H., Patarin J., Schott-Darie C., in The opportunities of the fluoride route in the synthesis of microporous materials, *Studies in Surface Science and Catalysis* (1994), 85(*Advanced Zeolite Science and Applications*), pp. 75-113.
35. Kuperman A., Nadimi S., Oliver S., Ozin G. A., Garces J. M., Olken M. M., Non-aqueous synthesis of giant crystals of zeolites and molecular sieves, *Nature* **365** (1993) pp. 239-242.
36. Camblor M. A., Villaescusa L. A., Diaz-Cabanas M. J., Synthesis of all-silica and high-silica molecular sieves on fluoride media, *Topics in Catalysis* **9** (1999) pp. 59-76.
37. Camblor M. A., Corma A., Lightfoot P., Villaescusa L. A., Wright P. A., Synthesis and structure of ITQ-3, the first pure silica polymorph with a two-dimensional system of straight eight-ring channels, *Angew. Chem. Int. Ed.* **36** (1997) pp. 2659-2661.

38. Barrett A., Camblor M. A., Corma A., Jones H. and Villaescusa, Structure of ITQ-4, a New Pure Silica Polymorph Containing Large Pores and a Large Void Volume, *Chem. Mater.* **9** (1997) pp. 1713-1715.
39. Villaescusa L., Barrett P. and Camblor M. A., ITQ-7: A new Pure Silica Polymorph with a Three-Dimensional System of Large Pore Channels, *Angew. Chem. Int. Ed.* **38** (1999) pp. 1997-2000.
40. Corma A., Diaz-Cabanas M. J., Martinez-Triguero J., Rey F. and Rius J., A large-cavity zeolite with wide pore windows and potential as an oil refining catalyst, *Nature* **418** (2002) pp. 514-517.
41. Blasco T., Corma A., Diaz-Cabanas M. J., Rey F., Vidal-Moya J. A. and Zicovich-Wilson C. M., Preferential Location of Ge in the Double Four-Membered Ring Units of ITQ-7 Zeolite, *J. Phys. Chem. B* **106** (2002) pp. 2634-2642.
42. Corma A., Navarro M. T., Rey F. and Valencia S., Synthesis of pure polymorph C of Beta Zeolite in a fluoride-free system, *Chem. Commun.* (2001) pp. 1486-1487.
43. Bu X., Feng, P. and Stucky G. D. Novel Germanate Zeolite Structures with 3-Rings. *J. Am. Chem. Soc.* **120** (1998) pp. 11204-11205.
44. O'Keeffe M., Yaghi O. M., Germanate zeolites: contrasting the behavior of germanate and silicate structures built from cubic T8O20 units (T = Ge or Si). *Chem. Eur. J.* **5** (1999) pp. 2796-2801.
45. Martin J. D. and Greenwood K. B., Halozeotypes: A New Generation of Zeolite-Type Materials, *Angew. Chem.* **36** (1997) pp. 2072-2075.
46. Zheng N., Bu X., Wang B., Feng P., Microporous and Photoluminescent Chalcogenide Zeolite Analogs, *Science* **298** (2002) pp. 2366-2369.
47. Breck D. W., Eversole W. G., Milton R. M., New synthetic crystalline zeolites, *J. Am. Chem. Soc.* **78** (1956) pp. 2338-9.
48. Effenberger H., Giester G., Krause W. and Bernhardt H.-J. Tschortnerite, a copper-bearing zeolite from the Bellberg volcano, Eifel, Germanay, *Am. Mineralogist* **83** (1998) pp. 607-617.
49. Bu X., Feng P., Stucky G. D. Large-cage zeolite structures with multidimensional 12-ring channels, *Science* **278** (1997) pp. 2080-2085.
50. Feng P., Bu X. and Stucky G. D. Hydrothermal syntheses and structural characterization of zeolite analog compounds based on cobalt phosphate, *Nature* **388** (1997) pp. 735-741.
51. Bennett J. M., Cohen J. P., Flanigen E. M., Pluth J. J., Smith J. V., Crystal structure of tetrapropylammonium hydroxide-aluminum phosphate Number 5, *ACS Symposium Series* **218** (1983) (*Intrazeolite Chem.*), pp. 109-118.
52. Lok B. M., Messina C. A., Patton R. L., Gajek R. T., Cannan T. R. and Flanigen E. M., Silicoaluminophosphate Molecular Sieves: Another New Class of Microporous Crystalline Inorganic Solids, *J. Am. Chem. Soc.* **106** (1984) pp. 6092-6093.

53. Wilson S. T. and Flanigen E. M., Synthesis and Characterization of Metal Aluminophosphate Molecular Sieves, in *Zeolite Synthesis, ACS Symposium Series* **398** (1989), pp. 329-345, Eds. Occelli M. L. and Robson H. E., American Chemical Society, Washington DC.
54. Chen J., Jones R., Natarajan S., Hursthouse M. B. and Thomas J. M., A Novel Open-Framework Cobalt Phosphate Containing a tetrahedral Coordinated Cobalt(II) Center: $CoPO_4 \cdot 0.5C_2H_{10}N_2$, *Angew. Chem. Int. Ed. Engl.* **33** (1994) pp. 639-640.
55. Xiao F., Qiu S., Pang W., Xu R. New Development in Microporous Materials, *Adv. Mater.* **11** (1999) 1091-1099.
56. Bibby D. M. and Dale M. P., Synthesis of silica-sodalite from nonaqueous systems, *Nature* **317** (1985) pp. 157-8.
57. Soghomonian V., Chen Q., Haushalter R. C., Zubieta J. and O'Connor C. J., An Inorganic Double Helix: Hydrothermal Synthesis, Structure, and Magnetism of Chiral $[(CH_3)_2NH_2]K_4[V_{10}O_{10}(H_2O)_2(OH)_4(PO_4)_7] \cdot 4H_2O$, *Science* **259** (1993) pp. 1596-1599.
58. Chippindale A. M., Walton, R. I. Synthesis and characterization of the first three-dimensional framework cobalt-gallium phosphate $[C_5H_5NH]^+[CoGa_2P_3O_{12}]^-$, *J. Chem. Soc., Chem. Commun.* (1994) pp. 2453-2454.
59. Cowley A. R., Chippindale A. M., Synthesis and characterization of $[C_4NH_{10}]^+[CoGaP_2O_8]^-$, a CoGaPO analog of the zeolite gismondine, *Chem. Commu.* (1996) pp. 673-674.
60. Chippindale A. M., Cowley A. R., CoGaPO-5: Synthesis and crystal structure of $(C_6N_2H_{14})_2[Co_4Ga_5P_9O_{36}]$, a microporous cobalt-gallium phosphate with a novel framework topology, *Zeolites* **18** (1997) pp. 176-181.
61. Bedard R. L., Wilson S. T., Vail L. D., Bennett J. M. and Flanigen, E. M. The next generation: synthesis, characterization, and structure of metal sulfide-based microporous solids in *Zeolites: Facts, Figures, Future. Proceedings of the 8th International Zeolite Conference*, (1989) pp. 375-387. eds. Jacobs P. A. and van Santen, R. A. Elsevier, Amsterdam.
62. Feng P., Zhang T., Bu X., Arsenate Zeolite Analogues with 11 Topological Types, *J. Am. Chem. Soc.* **123** (2001) pp. 8608-8609.
63. Bu X., Feng P., Gier T. E., Zhao D. and Stucky G. D., Hydrothermal Synthesis and Structural Characterization of Zeolite-like Structures Based on Gallium and Aluminum Germanates, *J. Am. Chem. Soc.* **120** (1998) pp. 13389-13397.
64. Li H. and Yaghi O. M., Transformation of Germanium Dioxide to Microporous Germanate 4-Connected Nets, *J. Am. Chem. Soc.* **120** (1998) pp. 10569-10570.
65. Li H., Laine A., O'Keeffe M., Yaghi O. M., Supertetrahedral Sulfide Crystals with Giant Cavities and Channels, *Science* **283** (1999) pp. 1145-1147.

66. Cahill C. L., Parise J. B., On the formation of framework indium sulfides, *J. Chem. Soc. Dalton Trans.* (2000) pp. 1475-1482.
67. Bu X., Zheng N., Li Y., Feng P., Pushing Up the Size Limit of Chalcogenide Supertetrahedral Clusters: Two- and Three-Dimensional Photoluminescent Open Frameworks from $(Cu_5In_{30}S_{54})^{13-}$ Clusters, *J. Am. Chem. Soc.* **124** (2002) pp. 12646-12647.
68. Dance I. G., Choy A., Scudder M. L., Syntheses, Properties, and Molecular and Crystal Structures of $(Me_4N)_4[E_4M_{10}(SPh)_{16}]$ (E =S, Se; M = Zn, Cd): Molecular Supertetrahedral Fragments of the Cubic Metal Chalcogenide Lattice, *J. Am. Chem. Soc.* **106** (1984) pp. 6285-6295.
69. Dance I. And Fisher K., Metal Chalcogenide Cluster Chemistry, in *Prog. Inorg. Chem.* (1994) pp. 637-803, Ed. Karlin K. D., John Wiley & Sons, Inc.
70. Conradson T., Dadachov M. S. and Zou X. D., "Synthesis and structure of $(Me_3N)_6[Ge_{32}O_{64}] \cdot (H_2O)_{4.5}$, a thermally stable novel zeolype with 3D interconnected 12-ring channels, *Micro. Meso. Mater.* **41** (2000) pp. 183-191.
71. Krebs B., Thio- and Seleno-Compounds of Main Group Elements- Novel Inorganic Oligomers and Polymers, *Angew. Chem. Int. Ed. Engl.* **22** (1983) pp. 113-134.
72. Pivan J. Y., Achak O. A., Louer M. and Louer D., The Novel Thiogermanate $[(CH_3)_4N]_4Ge_4S_{10}$ with a High Cubic Cell Volume. Ab Initio Structure Determination from Conventional X-ray Powder Diffraction, *Chem. Mater.* **6** (1994) pp. 827-830.
73. Yaghi O. M., Sun Z., Richardson D. A. and Loy T. L. Directed Transformation of Molecules to Solids: Synthesis of a Microporous Sulfide from Molecular Germanium Sulfide Cages, *J. Am. Chem. Soc.* **116** (1994) pp. 807-808.
74. Nellis D. M., Ko Y., Tan K., Koch S. and Parise J. B., A One-dimensional Germanium Sulfide Polymer Akin to the Ionosilicates: Synthesis and Structural Characterization of DPA-GS-8, $Ge_4S_9(C_3H_7)_2NH_2(C_3H_7)NH_2(C_2H_5)$, *J. Chem. Soc., Chem. Commun.* (1995) pp. 541-542.
75. MacLachlan M. J., Petrov S., Bedard R. L., Manners I. And Ozin G. A., Synthesis and Crystal Structure of δ-GeS2, the First Germanium Sulfide with an Expanded Framework Structure, *Angew. Chem. Int. Ed.* **37** (1998) pp. 2076-2079.
76. Jiang T. and Ozin G. O., New directions in tin sulfide chemistry, *J. Mater. Chem.* **8** (1998) pp. 1099-1108.
77. Parise J. B. and Ko Y., Material Consisting of two Interwoven 4-Connected Networks: Hydrothermal Synthesis and Structure of $[Sn_5S_9O_2][NH(CH_3)_3]_2$, *Chem. Mater.* **6** (1994) pp. 718-720.
78. Schiwy W., Krebs B. $Sn_{10}O_4S_{20}^{8-}$: A New Type of Polyanion, *Angew. Chem. Int. Ed.* **14** (1975) p. 436.

79. Achak O., Pivan J. Y., Maunaye M., Louer M. and Louer D. The *ab initio* structure determination of [CH$_3$)$_4$N]$_2$Ge$_4$MnS$_{10}$ from X-ray powder diffraction data, *J. Alloys Compounds* **219** (1995) pp. 111-115.
80. Achak, O., Pivan J. Y., Maunaye M., Louer M., Louer D., Structure Refinement by the Rietveld Methods of the Thiogermanates [(CH$_3$)$_4$N]$_2$MGe$_4$S$_{10}$ (M=Fe, Cd), *J. Solid State Chem.* **121** (1996) pp. 473-478.
81. Tan K., Darovsky A. and Parise J. B., Synthesis of a Novel Open-Framework Sulfide, CuGe$_2$S$_5$· (C$_2$H$_5$)$_4$N, and Its Structure Solution Using Synchrotron Imaging Plate Data, *J. Am. Chem. Soc.* **117** (1995) pp. 7039-7040.
82. Cahill C. L. and Parise J. B. Synthesis and Structure of MnGe$_4$S$_{10}$· (C$_6$H$_{14}$N$_2$)· 3H$_2$O: A Novel Sulfide Framework Analogous to Zeolite Li-A(BW), *Chem. Mater.* **9** (1997) pp. 807-811.
83. Cahill C. L., Ko Y. and Parise J. B., A Novel 3-Dimensional Open Framework Sulfide Based upon the [In$_{10}$S$_{20}$]$^{10-}$ Supertetrahedron: DMA-InS-SB1, *Chem. Mater.* **10** (1998) pp. 19-21.
84. Li H., Eddaoudi M., Laine A., O'Keeffe M. and Yaghi O. M., Noninterpenetrating Indium Sulfide Supertetrahedral Cristobalite Framework, *J. Am. Chem. Soc.* **121** (1999) pp. 6096-6097.
85. Zheng N., Bu X., Feng P., Nonaqueous Synthesis and Selective Crystallization of Gallium Sulfide Clusters into Three-Dimensional Photoluminescent Superlattices, *J. Am. Chem. Soc.* **125** (2003) pp. 1138-1139.
86. Krebs B., Voelker D., Stiller K., Novel Adamantane-like Thio- and Selenoanions from Aqueous Solution: Ga$_4$S$_{10}$$^{8-}$, In$_4S_{10}$$^{8-}$, In$_4Se_{10}$$^{8-}$, *Inorg. Chim. Acta* **65** (1982) pp. L101-L102.
87. Wang C., Li Y., Bu X., Zheng N., Zivkovic O., Yang C., Feng P. Three-Dimensional Superlattices Built from (M$_4$In$_{16}$S$_{33}$)$^{10-}$ (M = Mn, Co, Zn, Cd) Supertetrahedral Clusters, *J. Am. Chem. Soc.* **123** (2002) pp. 11506-11507.
88. Wang C., Bu X., Zheng N., Feng P., Nanocluster with One Missing Core Atom: A Three-Dimensional Hybrid Superlattice Built from Dual-Sized Supertetrahedral Clusters, *J. Am. Chem. Soc.* **124** (2002) pp. 10268-10269.
89. Su W., Huang X., Li J. and Fu H., Crystal of Semiconducting Quantum Dots Built on Covalently Bonded T5 [In$_{28}$Cd$_6$S$_{54}$]$^{-12}$: The Largest Supertetrahedral Clusters in Solid State, *J. Am. Chem. Soc.* **124** (2002) pp. 12944-12945.
90. Ferey, G., Microporous Solids: From Organically Templated Inorganic Skeletons to Hybrid Frameworks...Ecumenism in Chemistry, *Chem. Mater.* **13** (2001) pp. 3084-3098.
91. Batten S. R., Coordination polymers, *Cur. Opin. Solid State Mater. Sci.* **5** (2001) pp. 107-114.
92. Hoskins B. F. and Robson R., Design and Construction of a New Class of Scaffolding-like Materials Comprising Infinite Polymeric Frameworks of 3D-Linked Molecular Rods. A Reappraisal of the Zn(CN)$_2$ and Cd(CN)$_2$ structures and the Synthesis and Structures of the Diamond-Related Frameworks

[N(CH$_3$)$_4$][Cu^{I}Zn^{II}(CN)4 and Cu^I[4,4',4'',4'''-tetracyanotetraphenylmethane] BF$_4$· xC$_6$H$_5$NO$_2$, *J. Am. Chem. Soc.* **112** (1990) pp. 1546-1554.

93. Gardner G. B., Venkataraman D., Moore J. S., Lee S., Spontaneous assembly of a hinged coordination network, Nature 374 (1995) pp. 792-795.
94. Gravereau P. P. and Hardy E. G. A., Les Hexacyanoferrates Zeolithiques: Structure Cristalline de K$_2$Zn$_3$[Fe(CN)$_6$]$_2$· xH$_2$O, *Acta Cryst.* **B35** (1979) pp. 2843-2848.
95. Yaghi O. M. Li G. and Li H., Selective binding and removal of guests in a microporous metal-organic framework, *Nature* **378** (1995) pp. 703-706.
96. Li H., Eddaoudi M., Groy T. L. and Yaghi O. M., Establishing Microporosity in Open Metal-Organic Frameworks: Gas Sorption Isotherm for Zn(BDC) (BDC = 1,4-Benzenedicarboxylate), *J. Am. Chem. Soc.* **120** (1998) pp. 8571-8572.
97. Chui, S. S. –Y, Lo S. M. –F, Charmant J. P. H., Orpen A. G. and Williams I. D. A Chemically Functionalizable Nanoporous Material [Cu$_3$(TMA)$_2$(H$_2$O)$_3$]$_n$, *Science* **283** (1999) pp. 1148-1150.
98. Li H., Eddaoudi M., O'Keeffe M., Yaghi O. M., Design and synthesis of an exceptionally stable and highly porous metal-organic framework, *Nature* **402** (1999) pp. 276-279.
99. Chen B., Eddaoudi M., Hyde S. T., O'Keeffe M. and Yaghi O. M., Interwoven Metal-Organic Framework on a Periodic Minimal Surface with Extra-Large Pores, *Science* **291** (2001) pp. 1021-1023.
100. Eddaoudi M., Li H., Yaghi O. M., Highly Porous and Stable Metal-Organic Frameworks: Structure Design and Sorption Properties, *J. Am. Chem. Soc.* **122** (2000) pp. 1391-1397.
101. Eddaoudi M., Kim J., Rosi N., Vodak D., Wachter J., O'Keeffe M. and Yaghi O. M., Systematic Design of Pore Size and Functionality in Isoreticular MOFs and Their Application In Methane Storage, *Science* **295** (2002) pp. 469-472.
102. Chen B., Eddaoudi M., Reineke, T. M., Kampf, J. W., O'Keeffe M. and Yaghi O. M., Cu$_2$(ATC)· 6H$_2$O: Design of Open Metal Sites in Porous Metal-Organic Crystals (ATC: 1,3,5,7-Admantane Tetracarboxylate), *J. Am. Chem. Soc.* **122** (2000) pp. 11559-11560.
103. Reineke T. M., Eddaoudi M., Fehr M., Kelley D. and Yaghi O. M., From Condensed Lanthanide Coordination Solids to Microporous Frameworks Having Accessible Metal Sites, *J. Am. Chem. Soc.* **121** (1999) pp. 1651-1657.
104. Seo J. S., Whang D., Lee H., Jun S. I., Oh J., Jeon Y. J. and Kim K. A homochiral metal-organic porous material for enantioselective separation and catalysis, *Nature* **404** (2000) pp. 982-986.
105. Seki K., Design of an adsorbent with an ideal pore structure for methane adsorption using metal complexes., *Chem. Commun.* (2001) pp. 1496-1497.

106. Noro S., Kitagawa S., Kondo M. and Seki K. A New Methane Adsorbent, Porous Coordination Polymer [{CuSiF$_6$(4,4'-bipyridine)$_2$}$_n$], *Angew. Chem. Int. Ed.* **39** (2000) pp. 2082-2084.
107. Yaghi O. M. and Li H., T-Shaped Molecular Building Units in the Porous Structure of Ag(4,4'-bpy)· NO$_3$, *J. Am. Chem. Soc.* **118** (1996) pp. 295-296.
108. Yaghi O. M., Li H., Davis C., Richardson D. and Groy T., Synthetic Strategies, Structure Patterns, and Emerging Properties in the Chemistry of Modular Porous Solids, *Acc. Chem. Res.* **31**, pp. 474-484.
109. Eddaoudi M., Moler D. B., Li, H., Chen B., Reineke T. M., O'Keeffe, M. and Yaghi O. M., Modular Chemistry: Secondary Building Units as a Basis for the Design of Highly Porous and Robust Metal-Organic Carboxylate Frameworks. *Acc. Chem. Res.* **34** (2001) pp. 319-330.
110. Wang Q. M., Shen D., Bulow M., Lau M. L., Deng S. Fitch F. R., Lemcoff N. O. and Semanscin J., Metallo-organic molecular sieve for gas separation and purification, *Microporous Mesoporous Mater.* **55** (2002) pp. 217-230.
111. Tao J., Tong M., Shi J., Chen X., Ng S. W., Blue photoluminescent zinc coordination polymers with supertetranuclear cores, *Chem. Commun.* (2000) pp. 2043-2044.
112. Lin C. Z-J., Chui S. S-Y, Lo S. M-F., Shek F. L-Y, Wu M., Suwinska K., Lipkowski J. and Williams I. D., Physical stability vs. chemical stability in microporous metal coordination polymers: a comparison of [Cu(OH)(INA)]$_n$ and [Cu(INA)$_2$]$_n$: INA = 1,4-(NC$_5$H$_4$CO$_2$), *Chem. Commun.* (2002) pp. 1642-1643.
113. Abrahams B. F., Hoskins B. F., Michail D. M. and Robson R., Assembly of phorphyrin building blocks into network structures with large channels, *Nature* **369** (1994) pp. 727-729.
114. Kosal M., Chou J., Wilson S. R., and Suslick K. S., A functional zeolite analogue assembled from metalloporphyrins, *Nature Mater.* **1** (2002) pp. 118-121.
115. Lin K-J, SMTP-1: The First Functionalized Metalloporphyrin Molecular Sieves with Large Channels, *Angew. Chem. Int. Ed.* **38** (1999) pp. 2730-2732.
116. Yang S. Y., Long L. S., Jiang Y. B., Huang R. B., Zheng L. S., An Exceptionally Stable Metal-Organic Framework Constructed from the Zn$_8$(SiO$_4$) Core, *Chem. Mater.* **14** (2002) pp. 3229-3231.
117. Kim J., Chen B., Reineke T. M., Li H., Eddaoudi M., Moler D. B., O'Keeffe M. and Yaghi O. M., Assembly of Metal-Organic Frameworks from Large Organic and Inorganic Secondary Building Units: New Examples and Simplifying Principles for Complex Structures, *J. Am. Chem. Soc.* **123** (2001) pp. 8239-8247.
118. Hagrman D., Hagrman P., and Zubieta J., Solid-State Coordination Chemistry: The Self-Assembly of Microporous Organic-Inorganic Hybrid Frameworks

Constructed from Tetrapyridylporphyrin and Bimetallic Oxide Chains or Oxide Clusters, *Angew. Chem. Int. Ed.* **38** (1999) pp. 3165-3168.
119. Zheng N., Bu X., Feng P., Self-Assembly of Novel Dye Molecules and $[Cd_8(SPh)_{12}]^{4+}$ Cubic Clusters into Three-Dimensional Photoluminescent Superlattices, *J. Am. Chem. Soc.* **124** (2002) pp. 9688-9689.
120. Davis M. E., The Quest for Extra-Large Pore, Crystalline Molecular Sieves, *Chem. Eur. J.* **3** (1997) pp. 1745-1750.
121. Freyhardt C. C., Tsapatsis M., Lobo R. F., Balkus K. J. Jr. and Davis M. E., A high-silica zeolite with a 14-tetrahedral-atom pore opening, *Nature* **381** (1996) pp. 295-298.
122. Lobo R. F., Tsapatsis M., Freyhardt C. C., Khodabandeh S., Wagner P., Chen C., Balkus K. J. Jr., Zones S. I. and Davis M. E., Characterization of the Extra-Large Pore Zeolite UTD-1, *J. Am. Chem. Soc.* **119** (1997) pp. 8474-8484.
123. Wagner P., Yoshikawa M., Lovallo M., Tsuji K., Taspatsis M. and Davis M., CIT-5: a high-silica zeolite with 14-ring pores, *Chem. Commun.* (1997) pp. 2179-2180.
124. Yoshikawa M., Wagner P., Lovallo M., Tsuji K., Takewaki T., Chen C., Beck L. W., Jones C., Tsapatsis M., Zones S. I. and Davis M. E., Synthesis, Characterization, and Structure Solution of CIT-5. A new, High-Silica, Extra-Large-Pore Molecular Sieve, *J. Phys. Chem. B* **102** (1998) pp. 7139-7147.
125. Barrett P. A., Diaz-Cabanas M. J., Camblor M. A. and Jones R. H., Synthesis in fluoride and hydroxide media and structure of the extra-large pore pure silica zeolite CIT-5, *J. Chem. Soc., Faraday Trans.* **94** (1998) pp. 2475-2481.
126. Moore P. B. and Shen J., An X-ray structural study of cacoxenite, a mineral phosphate, *Nature* **306** (1983) pp. 356-358.
127. Estermann M., McCusker L. B., Baerlocher C., Merrouche A., Kessler H., A Synthetic Gallophosphate molecular sieve with a 20-tetrahedral-atom pore opening, *Nature* **352** (1991) pp. 320-323.
128. Jones R. H., Thomas J. M., Chen J. Xu R., Huo Q., Li S., Ma Z.c Chippindale A. M., Structure of an Unusual Aluminum Phosphate $([Al_5P_6O_{24}H]^{2-} 2[N(C_2H_5)_3H]^+ \cdot 2H_2O)$ JDF-20 with large elliptical Apertures, *J. Solid State Chem.* **102** (1993) pp. 204-208.
129. Richardson J. W. Jr. and Vogt E. T. C., Structural determination and Rietveld refinement of aluminophosphate molecular sieve $AlPO_4$-8, *Zeolites* **12** (1992) pp. 13-19.
130. Guillou N., Gao Q., Forster P. M., Chang J., Nogues M., Park S., Ferey G. and Cheetham A. K., Nickle(II) Phosphate VSB-5: A Magnetic Nanoporous Hydrogenation Catalyst with 24-ring Tunnels, *Angew. Chem. Int. Ed.* **40** (2001) pp. 2831-2834.

131. Guillou N., Gao Q., Nogues M., Morris R. E., Hervieu M., Ferey G. and Cheetham A. K., Zeolitic and magnetic properties of a 24-membered ring porous nickle(II) phosphate, VSB-1, *C. R. Acad. Sci. Paris, t. 2, Serie II c* (1999) pp. 387-392.
132. Brunner G. O., "Quantitative zeolite topology" can help to recognize erroneous structures and to plan syntheses, *Zeolites* **13** (1993) pp. 88-91.

MESOPOROUS MATERIALS

ABDELHAMID SAYARI

Department of Chemistry and Centre for Catalysis Research and Innovation
University of Ottawa, Ottawa, Ontario, K1N 6N5, Canada
abdel.sayari@science.uottawa.ca

The area of periodic mesoporous materials has grown tremendously during the last ten years. Remarkable progress has been made in the use of supramolecular templating techniques to synthesize a large variety of periodic inorganic, organic and hybrid mesostructures with tailored framework structures and compositions, pore sizes and architectures, morphologies and surface properties. A wide range of synthesis conditions and a variety of templating surfactants, oligomers and triblock copolymers have been explored and rationalized. Among the most recent developments in this area is the extension of the amphiphile templating techniques to the synthesis of ordered mesoporous organosilicas and the assembly of zeolite seeds. Another important discovery is the use of ordered nanoporous silica and colloidal crystals to create new periodic mesoporous and macroporous materials, including carbons, polymers, metals, and alloys. Combination of different synthesis approaches such as amphiphile, colloidal crystal or microemulsion templating, micromolding and soft lithography led to materials with hierarchically ordered structures. Significant effort was also devoted to the development of potential applications in adsorption, catalysis, separation, environmental cleanup, drug delivery, sensing and optoelectronics.

1 Introduction

Though periodic mesoporous silica had been discovered in 1990 [1], or even much earlier [2], this type of materials did not attract much attention until 1992, upon the publication of two groundbreaking papers [3,4] by a group of Mobil's scientists describing the so-called M41S family of mesoporous silicas. Since then, research in this topic has grown so dramatically that it has developed into a separate field. Well over 3000 papers dealing with such materials have been published and a number of international meetings were also devoted to mesoporous materials. The M41S silicas consisted of three mesophases, namely MCM-41 (hexagonal, *p6mm*), MCM-48 (cubic, $Ia\bar{3}d$) and MCM-50 (lamellar). They were synthesized via a supramolecular templating mechanism using long chain alkyltrimethylammonium surfactants under basic conditions. Currently, periodic mesoporous silicas may be readily prepared under an extremely wide range of conditions. Figure 1 shows schematically, the structural parameters that have been explored in the design and synthesis of periodic mesoporous materials. In addition to cationic alkyltrimethylammonium surfactants, a large variety of amphiphile molecules including cationic, anionic, neutral, zwitterionic, bolaamphiphile, gemini and divalent surfactants as well as many commercially available oligomers and triblock copolymers and appropriate mixtures thereof were found to be suitable for the preparation of periodic mesoporous silicas. The pH and temperature conditions ranged from extremely acidic to highly basic, and from subambient to ca. 170 °C, respectively. This effort led to the discovery of numerous new silica mesophases including SBA-1 and SBA-6 ($Pm\bar{3}n$), SBA-2 and SBA-12 (P6$_3$/mmc), SBA-8 (cmm), SBA-11 ($Pm\bar{3}m$), SBA-16 and FDU-1 ($Im\bar{3}m$), disordered HMS, MSU-n, MSU-V, MSU-G and others. In addition, the supramolecular templating approach was extended to the synthesis of transition and other metal substituted silicas, and to numerous non-silica materials including organosilicates, metals and alloys, transition metal oxides and chalcogenides, metallophosphates and zeolite nanocrystals. Furthermore, silica mesophases were in turn used as templates for the synthesis of a variety of other materials such as nanoporous carbons and polymers, as well as metallic and semiconductor nanowires. This extensive effort in the area of synthesis was paralleled by the development of innovative applications not only in conventional fields such as adsorption, separation and catalysis, but also in the area of advanced materials based on their often unique electronic, magnetic and optical properties, or as hosts for quantum dots and sensing species.

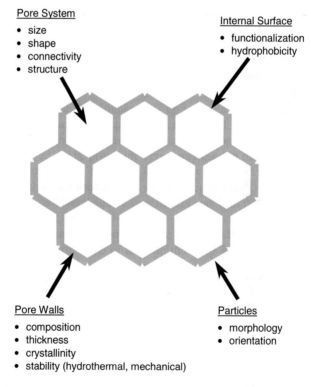

Figure 1. Structural parameters for the design of nanostructured materials.

To keep up with the huge flow of information, experts in the field published a number of authoritative reviews dealing either with the overall area of mesoporous materials [5-9] or with specific segments of this field such as catalysis [10] and adsorption by mesoporous materials [11-13]. A number of reviews dealing with different types of periodic mesoporous materials such as aluminophosphates [14], non-silica materials [15-17], inclusion materials [18,19], surface modified mesoporous silica [20,21] and organic-inorganic mesoporous nanocomposites [22,23] are also readily available. The shear volume of information accumulated up to this date makes any attempt at writing a comprehensive review on periodic mesoporous materials, within the space constraints of this book, rather pointless. Therefore, the current contribution will be deliberately focused on the most significant findings in this research area. However, potential applications of these materials will not be dealt with in any significant detail.

2. Synthesis Mechanisms of Periodic Mesoporous Materials

2.1 Synthesis of Mesoporous Silica under Mobil's Group Conditions

The preparation of M41S silica mesophases was carried out under basic conditions using cationic alkyltrimethylammonium surfactants [3,4]. The synthesis mechanism was originally coined as *liquid crystal templating* mechanism. This could mislead the reader in thinking that silica mesophases are simple replicas of pre-existing liquid crystalline phases. While direct transcription of existing liquid crystals may occur under high concentration of surfactant [24], in practice this is the exception rather than the rule. Indeed, it was possible to synthesize any of the three M41S silicas only by changing the relative amount of silica precursor, everything else being the same. Moreover, mesoporous materials may be prepared using surfactant concentrations well below those required for the formation of liquid crystals, and even below the critical concentration for the formation of rod-like micelles. Some short chain surfactants such as $C_{12}H_{25}N(CH_3)_3OH$ which do not form rod-like micelles in water, were also found to template mesoporous silica. Finally, MCM-41 and MCM-48 silicas were synthesized at temperatures above 70 °C where rod-like micelles are not stable. All these experimental findings are consistent with the fact that the occurrence of a liquid crystalline phase is not a prerequisite for the formation of the corresponding silica mesophase. On the contrary, there is compelling evidence that the inorganic species play a key role in triggering the supramolecular self-assembly process and to some extent, in dictating the nature of the final mesophase [25]. The three-step *cooperative organization* mechanism proposed by Stucky and coworkers [25,26] has proven to be consistent with the behavior of mesoporous materials synthesis mixtures. Figure 2 illustrates such a mechanism for the synthesis of mesoporous silica under basic conditions in the presence of a cationic surfactant. The first step of this process is triggered by *electrostatic interactions*. It corresponds to the displacement of the surfactant counterions by polydentate and polycharged inorganic anions leading to organic-inorganic ion-pairs, which self organize into a liquid crystal-like mesophase. This is followed by cross-linking of the inorganic species, and formation of a rigid replica of the underlying liquid crystalline phase.

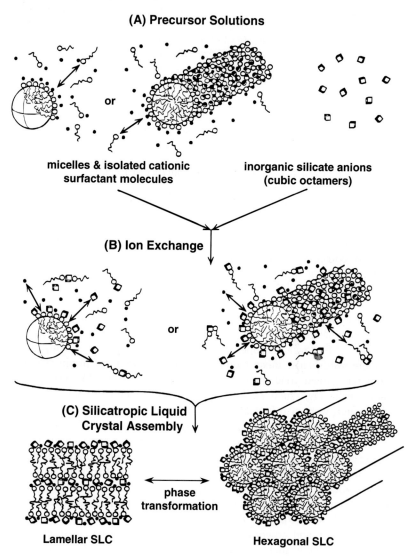

Figure 2. Cooperative templating mechanism [25].

2.2 Generalization of Electrostatically Driven Synthesis Mechanisms

The cooperative organization mechanism described above is not limited to ion-pairs between cationic surfactants (S^+) and anionic inorganic species (I^-), but can be generalized to include three other routes [27,28]. Pathway S^-I^+ involves direct electrostatic interaction between anionic surfactants and cationic inorganic species as in the case of mesostructured antimony or tungsten oxide under acidic conditions. In the two other routes, the interactions occur between similarly charged organic and inorganic species with the mediation of small counterions of opposite charge. These pathways are referred to as $S^+X^-I^+$ ($X^- = Cl^-$, Br^-) and $S^-M^+I^-$ ($M^+ = Na^+$, K^+). The synthesis of silica mesophase at pH < 2, and that of mesostructured zinc oxide at pH > 12.5 are examples of these two routes, respectively.

2.3 Other Supramolecular Templating Mechanisms

Pinnavaia *et al.* [29,30] introduced a *neutral* (N^0I^0) and a *non-ionic* (S^0I^0) templating mechanism using alkylamines and polyethyleneoxide surfactants, respectively. The obtained silicas, referred to as HMS and MSU-n, respectively exhibited non-ordered

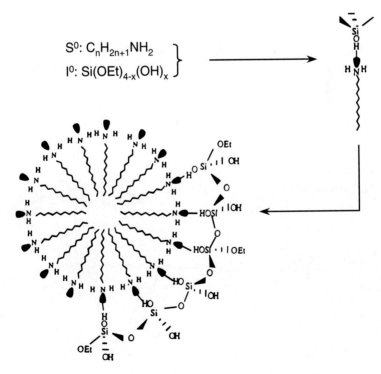

Figure 3. Neutral templating mechanism [29].

systems of cylindrical nanopores with narrow size distributions. Instead of the electrostatic interactions, in the presence of neutral surfactants, hydrogen bonding becomes the predominant driving force in pairing organic and inorganic species. As seen in Figure 3, the neutral species formed by partial hydrolysis of tetraethyl orthosilicate interact with the surfactant amine head group via hydrogen bonding. The obtained organic-inorganic complex may be considered as an amphiphile with a very bulky head group, i.e., small packing factor, thus favoring the formation of high curvature micelles such as rod-like micelles, which exhibit a natural tendency to self-assemble into a hexagonal lyotropic mesophase. This is followed by condensation of silanol groups and formation of rigid silica walls. A related mechanism designated as $(S^0H^+)(X^-I^0)$ was proposed to rationalize the formation of mesoporous silica under strongly acidic conditions in the presence of nonionic alkyl-ethylene oxide surfactants or poly(alkylene oxide) triblock copolymers [31]. In this case, the protonated amphiphile molecules and the cationic silica species are supposed to assemble via a combination of electrostatic, hydrogen bonding and van der Waals interactions.

In addition to the electrostatically driven cooperative assembly pathways, and the non-ionic routes, Antonelli and Ying [32] introduced the so-called *ligand-assisted* templating pathway to explain the formation of a number of periodic

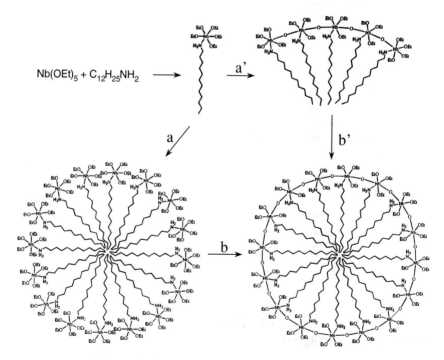

Figure 4. Ligand-assisted templating mechanism.

mesoporous transition metal oxides. This mechanism involves the formation of a covalent bond between organic and inorganic species. As illustrated in Figure 4, the first step in the formation of mesoporous Nb_2O_5, which takes place in the absence of water is the formation of a Nb-N covalent bond between the precursor $Nb(OEt)_5$ and the long-chain amine surfactant. This is followed by intermolecular condensation of these species in the presence of water leading to the formation of cylindrical micelles, and ultimately to a mesophase (route a', b' in Figure 4). Alternatively, the formation of the micelles may precede the condensation step (route a, b in Figure 4). The key feature of this mechanism, i.e., the formation of a covalent M-N bond was inferred from NMR studies on 100% ^{15}N enriched dodecylamine niobium ethoxide precursor in [D_8]toluene and the corresponding as-synthesized Nb_2O_5 mesophase [32].

3 Characterization of Periodic Mesoporous Silica-based Materials

Numerous techniques have been used for the characterization of periodic mesoporous silicates. Despite their long-range order, these materials were found to be amorphous in nature, as conclusively demonstrated by Raman spectroscopy, solid state NMR and other techniques [4,5]. As a result, their X-ray diffraction patterns were dominated by a limited number of peaks at low angles (Figure 5), often providing insufficient information to draw a definite conclusion with regard to the structure of the materials under investigation. In this respect, transmission electron microscopy (Figure 6) and more importantly electron crystallography [33] played a key role in elucidating the structure of various silica mesophases. However, because they are readily available to most researchers, adsorption techniques were, along with XRD, the most frequently used methods for characterizing periodic mesoporous materials [34]. This simple technique provides essential information regarding structural and textural properties of mesoporous materials such as the surface area, the pore volume, the pore size distribution, the occurrence of micropores, the surface hydrophobicity, etc. Moreover, in combination with XRD data, nitrogen adsorption measurements provide an estimate of the pore wall thickness. Figure 7 shows a representative example of nitrogen adsorption desorption isotherms for a MCM-41 silica. Such isotherms are usually of type IV in the IUPAC classification and exhibit characteristic steps on both branches, corresponding to the condensation and evaporation of the adsorbate from mesopores with narrow size distributions. They are usually reversible for MCM-41 silicas with mesopores less than ca. 4 nm in diameter, and exhibit hysteresis loops of different shapes for 2D hexagonal (*p6mm*) silica with pores larger than 4 nm, mesocellular silica foams [35], as well as periodic mesoporous materials with cage-like pores such as the 3D hexagonal (*P6_3/mmc*) SBA-2 [36], the cubic ($Im\overline{3}m$) SBA-16 [31] and FDU-1 [37] mesophases and others [38].

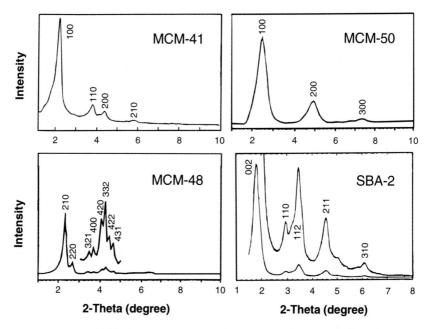

Figure 5. X-ray diffractograms for typical silica mesophases [4,39].

Figure 6. Typical TEM image and electron diffraction pattern for MCM-41 silica.

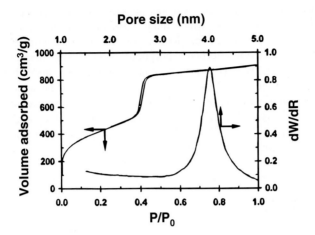

Figure 7. Typical N_2 adsortion-desorption istherm and pore size distribution.

4 Periodic Mesoporous Silicas

4.1 Structure

Remarkable progress has been made regarding the synthesis of new silica mesophases. The structures and synthesis recipes for such mesophases are listed in Table 1. In addition to the three M41S phases mentioned earlier, Yanagisawa et al. [1] synthesized a periodic mesoporous silica referred to as FSM-16 with similar structural characteristics as MCM-41. This mesophase was prepared by treating layered kanemite polysilicate in an aqueous solution of cetyltrimethylammonium bromide at 70 °C. The synthesis mechanism involved the formation of cylindrical surfactant micelles followed by silica sheet folding leading to a highly ordered honeycomb structure.

Stucky and co-workers made an impressive contribution to this field as they synthesized numerous new ordered mesoporous silica mesophases referred to as SBA-n using a variety of amphiphile molecules. Contrary to M41S and FSM-16, most SBA-n mesophases were prepared under strongly acidic conditions. The early members of SBA-n (n = 1, 2, 3) mesophases were obtained in the presence of positively charged surfactants. SBA-1 exhibited a cubic structure with supercage cavities indicative of the involvement of globular micelles [36]. This mesophase was obtained via the $S^+X^-I^+$ pathway under acidic conditions (HCl) in the presence of surfactants with bulky headgroups such as akyltriethylammonium and cetyltriethylpiperidinium, which promote the formation of mesophases with high surface curvature. The nature of the acid used was also found to play an important role, HCl being the most favorable for the formation of SBA-1. The use of HBr resulted in the formation of a 2D hexagonal phase, most likely because the bromide ion is a strongly binding anion, which reduces the effective headgroup area of the surfactant below that is required for globular micelles [28]. Other acids such as HF, H_2SO_4, H_3PO_4 and HNO_3 afforded amorphous silica. Using divalent surfactants $C_nH_{2n+1}N^+(CH_3)_2(CH_2)_sN^+(CH_3)_3$ with very large headgroups, under basic or acidic conditions afforded the SBA-2 mesophase which exhibits 3D hexagonal ($P6_3/mmc$) symmetry, consistent with the occurrence of a hexagonal close packing of globular silica-surfactant arrays [36]. The SBA-3 mesophase exhibits a 2D hexagonal symmetry ($p6mm$) similar to MCM-41. It was also prepared under acidic conditions using either Gemini or alkylammonium surfactants [39].

Table 1. Structure and synthesis conditions for various silica mesophases.

Mesophase	Amphiphile template	pH	Structure	Ref
MCM-41	$C_nH_{2n+1}(CH_3)_3N^+$	basic	2D hexagonal ($p6mm$)	[4]
MCM-48	$C_nH_{2n+1}(CH_3)_3N^+$	basic	cubic ($Ia\bar{3}d$)	[4]
	Gemini C_{n-s-n}[a]			[36]
MCM-50	$C_nH_{2n+1}(CH_3)_3N^+$	basic	lamellar	[4]
FSM-16	$C_{16}H_{31}(CH_3)_3N^+$	basic	2D hexagonal ($p6mm$)	[1]
SBA-1	$C_{18}H_{37}N(C_2H_5)_3^+$	acidic	cubic ($Pm\bar{3}n$)	[36]
SBA-2	Divalent C_{n-s-1}[b]	acidic / basic	3D hexagonal ($P6_3/mmc$)	[36]
SBA-3	$C_nH_{2n+1}N(CH_3)_3^+$	acidic	2D hexagonal ($p6mm$)	[39]
SBA-6	Divalent $18B_{4-3-1}$[c]	basic	cubic ($Pm\bar{3}n$)	[40]
SBA-8	Bolaform[d]	basic	2D rectangular (cmm)	[41]
SBA-11	Brij 56; $C_{16}EO_{10}$	acidic	cubic ($Pm\bar{3}m$)	[31]
SBA-12	Brij 76; $C_{18}EO_{10}$	acidic	3D hex. ($P6_3/mmc$)	[31]
SBA-14	Brij 30; $C_{12}EO_4$	acidic	cubic	[31]
SBA-15	P123; $EO_{20}PO_{70}EO_{20}$	acidic	2D hexagonal ($p6mm$)	[42]
SBA-16	F127; $EO_{106}PO_{70}EO_{106}$	acidic	cubic ($Im\bar{3}m$)	[31]
FDU-1	B50-6600; $EO_{39}BO_{47}EO_{39}$	acidic	cubic ($Im\bar{3}m$)	[37]
MSU-1	Tergitol; $C_{11-15}(EO)_{12}$	neutral	disordered	[30]
MSU-2	TX-114; $C_8Ph(EO)_8$	neutral	disordered	[30]
	TX-100; $C_8Ph(EO)_{10}$			
MSU-3	P64L; $(EO_{13}PO_{30}EO_{13})$	neutral	disordered	[30]
MSU-4	Tween-20, 40, 60, 80	neutral	disordered	[43]
MSU-V	$H_2N(CH_2)_nNH_2$	neutral	lamellar	[44]
MSU-G	$C_nH_{2n+1}NH(CH_2)_2NH_2$	neutral	lamellar	[45]
HMS	$C_nH_{2n+1}NH_2$	neutral	disordered	[29]

(a) Gemini surfactants C_{n-s-n}: $C_nH_{2n+1}N^+(CH_3)_2(CH_2)_sN^+(CH_3)_2C_nH_{2n+1}$.
(b) Divalent surfactants C_{n-s-1}: $C_nH_{2n+1}N^+(CH_3)_2(CH_2)_sN^+(CH_3)_3$.
(c) Divalent surfactant $18B_{4-3-1}$: $C_{18}H_{37}O-C_6H_4-O(CH_2)_4N^+(CH_3)_2(CH_2)_3N^+(CH_3)_3$.
(d) Bolaform surfactants :$(CH_3)_3N^+(CH_2)_nO-C_6H_4-C_6H_4-O(CH_2)_nN^+(CH_3)_3$.

As seen in Table 1, several other SBA-n (n = 11, 12, 14, 15, and 16) mesophases were prepared using commercially available nonionic alkyl poly(ethylene oxide) (PEO) oligomeric surfactants and poly-(alkylene oxide) block copolymers in acid media [31,42]. Of particular interest is the 2D hexagonal (*p6mm*) SBA-15 silica mesophase, which attracted considerable interest. This is due to the fact that it has superior physical attributes and its preparation is simple and reproducible. This mesophase exhibits adjustable pore sizes within a wide range of 5 to 30 nm, high surface area and thick walls (3-6 nm) with enhanced robustness [31,42]. It has also been synthesized reproducibly in many laboratories based on the original recipe using neutral Pluronic P123 ($EO_{20}PO_{70}EO_{20}$) triblock copolymer as structure-directing agent under acidic conditions. Another important silica mesophase is SBA-16, which has a cubic structure ($Im\overline{3}m$) with 5.4 nm cagelike cavities. It was synthesized under acidic conditions, in the presence of a poly(alkylene oxide) triblock copolymer $EO_{106}PO_{70}EO_{106}$ (F127) with large EO moieties, which as shown below, favor the formation of globular aggregates. Using a more hydrophobic poly(butylene oxide) containing triblock copolymer $EO_{39}BO_{47}EO_{39}$ (B50-6600), Zhao et al. [37] synthesized a hydrothermally stable silica mesophase referred to as FDU-1 with similar cubic symmetry ($Im\overline{3}m$) as SBA-16, but with much larger cages (ca. 12 nm).

The structure of a mesophase obtained under a set of experimental conditions is determined by dynamic interplay among organic-inorganic ion-pairs; the charge density, coordination and steric requirements of both the organic and inorganic species at the interface being the main controlling factors. This is ultimately dependent on synthesis parameters such as the composition of the mixture, the pH and the temperature. To rationalize the mesophase structure, the packing factor g = $V/a_0 l$, where V is the total volume of the surfactant chain plus any co-solvent organic molecules between the chains, a_0 is the effective head group area at the micelle surface, and l is the kinetic length of the surfactant tail, was used extensively [28,39]. This dimensionless factor was first introduced to explain the aggregation behavior of amphiphile molecules in aqueous solutions [46]. As the g factor decreases, transitions to higher curvature mesophases occur at the following critical values: g = 1 (lamellar mesophase), 1/2 < g < 2/3 (cubic, $Ia\overline{3}d$), g = 1/2 (hexagonal, *p6mm*) and g = 1/3 (cubic, $Pm\overline{3}n$). Despite its simplicity, the concept of the packing factor proved to be very useful not only to explain experimental findings, but more importantly to design mesoporous materials via direct manipulation of the g factor. For example, addition of co-solvents or polar molecules increases the volume V of the hydrophobic tail [39], which decreases the packing factor and affects the nature of the mesostructure obtained. Another simple and effective way to fine tune the packing factor is through the use of surfactant blends. This approach allowed Kim et al. [47] to prepare high quality silica mesophases having 2D hexagonal (*p6mm*), 3D hexagonal (*P6₃/mmc*) and cubic ($Im\overline{3}m$) structures through systematic variation of the effective head group and tail sizes via appropriate mixtures of amphiphile block copolymers.

Another strategy to control the structure of the mesophase is to use conditions under which the amphiphile molecules form lyotropic liquid crystalline phases behave as a template for the inorganic shell [24]. This methodology was used for the preparation of mesoporous organosilcate [48] and mesoporous silica with different morphologies including monoliths [49], thin films [50] and fibers [51]. In a further development of this approach it was found that by varying the volume ratio between Pluronic P123 triblock copolymer and the inorganic components of the precursor solution, cubic, 2D hexagonal and lamellar silica (and titania) mesophases formed in a predictable manner [52]. Interestingly, the regions over which the three phases were obtained corresponded to those of the water-P123 binary phase diagram when considered in terms of volume fraction of copolymer incorporated. It was concluded that provided the metal oxide does not rapidly cross-link, it is possible to predict the structure of the final mesophase based on the phase diagram of the amphiphile in aqueous solution.

4.2 Pore System

The pore system of periodic mesoporous silicas may consist of a two-dimensional array of uniform channels (e.g., MCM-41), a three-dimensional array of connected channels (e.g. MCM-48) or cage-like pores (e.g. SBA-2, FDU-1). Since the determination of the pore system characteristics such as the surface area, the pore volume, the pore size distribution and the pore connectivity is of paramount importance, gas adsorption has quickly become a key characterization technique for mesoporous materials. However, many of the existing methods for the determination of structural parameters from gas adsorption data such the Horvath-Kawazoe and the Barrett-Joyner-Halenda (BJH) approaches [4] proved to be highly inaccurate [34]. Several methods, all of which are based on the Kelvin equation were developed to overcome these limitations. These methods we recently reviewed and compared [13]. The so-called Kruk-Jaroniec-Sayari (KJS) [34] was the first practical approach for accurate determination of structural parameters. This empirical method was based on the use of a series of high quality MCM-41 silicas with different pore sizes as model adsorbents combined with the determination of their pore sizes by an independent method, the geometrical model. Considering that the 2D hexagonal mesophase is an array of uniform cylindrical parallel tunnels, the following equation was established $w_d = cd_{100}[\rho V_p/(1 + \rho V_p)]^{1/2}$, where w_d is the pore diameter, V_p is the primary mesopore volume, d_{100} is the (100) interplanar spacing, $\rho = 2.2$ g/cm^3 is the density of the silica wall, and $c = 1.213$ is a constant. Because of its simplicity and accuracy, the KJS method is gaining wide acceptance by workers in the field of mesoporous materials [13].

As shown in Table 2, the pore sizes of mesoporous materials can be varied from the low end of mesopore sizes well into the macropore regime. There are two main strategies for pore size engineering. The first is through direct synthesis for example by using surfactants with different chain lengths, adding expanders or by increasing

Table 2. Methods of pore size engineering for mesoporous silicas.

Pore size (nm)	Method	Reference
2-5	Use of surfactants of different chain lengths	
	Charged (alkylammonium)	[3,4]
	Neutral (alkylamine)	[29]
4-10	Use of micelle expanders	
	Aromatic hydrocarbons	[3,4]
	Alkanes	[53]
	Trialkylamines	[54]
	Alkyldimethylamines	[54]
4-7	Hyrothermal post-synthesis treatment	
	In mother liquor	[55,56]
	In water	[39]
2.5-6.6	High temperature synthesis	[57-59]
3.5-25	Water-amine post-synthesis treatment	[54,60]
5-30	Use of triblock copolymer and expander	[42]
> 50	Emulsion templating	[61]
> 200	Colloidal crystal templating	[62,63]

the temperature during the hydrothermal treatment. The second approach is via post-synthesis treatment in the mother liquid, in water or in aqueous suspensions of long chain amines. These techniques were applied mostly to MCM-41 and to a lesser extent with SBA-15 silica. Mesoporous silicas with cubic structure such as MCM-48 [64,65] and SBA-1 [66] showed only limited pore size variations, usually between 2.5 and 4.5 nm. There are however two cubic silica mesophases with pore sizes significantly larger than 4.5 nm. The first is SBA-6 ($Pm\overline{3}n$), which is endowed with a bimodal pore structure consisting of 7.3 and 8.5 nm cages [40]. The second cubic silica mesophase with pores larger than 4.5 nm was prepared in the presence of $EO_{132}PO_{50}EO_{132}$ (F108) block copolymer and a highly charged salt (K_2SO_4) under acidic conditions. It exhibited a cubic symmetry similar to SBA-16 ($Im\overline{3}m$) and a pore size of 7.4 nm [67]. Similar to MCM-48 and SBA-1, silicates with wormhole motifs prepared using the so-called neutral pathway such as HMS [29,68] and MSU-x [69] also showed only limited pore size variations.

The silica mesophases MCM-41 and SBA-15, are by far the most studied periodic mesoporous materials. Though both of them exhibit the same 2D hexagonal symmetry, they do have several important differences. The pores of SBA-15 are usually larger than those of MCM-41 and the pore walls thicker [42]. In addition, while MCM-41 is purely mesoporous in nature [56], SBA-15 has connecting micropores. This was convincingly demonstrated by comparative analysis of gas adsorption data using the so-called α_s-plot method [64] and by synthesizing stable platinum [70] and carbon [71] replicas of SBA-15 indicating that the mesoporous channels are interconnected. The origin of the micropores is believed to be related to penetration of the hydrophilic EO-blocks of the triblock copolymer template within the silica walls [72]. The occurrence of micropores in mesoporous silica prepared in the presence of poly-(alkylene oxide) block copolymers seems to be quite general [73,74]. However, at least as far as SBA-15 is concerned, the micropore volume may be reduced significantly via synthesis above 100 °C [75], in the presence of added salt [76], or via post-synthesis treatment at quite high temperature [70].

An important problem in the area of periodic porous materials is the 1 to 2 nm pore size gap that exists between the largest pore zeolites and the smallest pore silicas prepared by supramolecular templating techniques. Significant progress to fill this gap has recently been achieved from both ends. Two high-silica zeolites with one-dimensional pores circumscribed by 14 tetrahedral atoms with slightly larger pores than faujasite have been reported. These are UTD-1 [77] and CIT-5 [78] with 0.75x1.0 nm and ca. 0.73 nm channels, respectively. Another important discovery in this field is the ITQ-21 germanium silicate with a three-dimensional pore network containing 1.18 nm cavities accessible through six 0.74 nm circular windows [79]. At the other end, silicates [80], organosilicates [81] and other oxides [82] with pores well below 2 nm have been synthesized via self-assembly techniques.

As far as the pore system of periodic mesoporous silicas is concerned it is worth mentioning that several silica mesophases with bimodal pore size distributions have been reported in the literature [63,83,84]. Moreover, combinations of different synthesis approaches such as amphiphile, colloidal crystal or microemulsion templating, micromolding and soft lithography led to materials with hierarchically ordered structures [85].

4.3 Pore Walls

As mentioned earlier, the walls of mesoporous silicas are not only amorphous, but often quite thin [3,4], which poses major problems with regards to their hydrothermal [86] and mechanical [87] stability as well as catalytic properties [5,8,10,88,89]. The initial excitement of using the collective know-how that led to zeolite-based commercial catalysts to generate their mesoporous counterparts has been quickly dampened. For example, it was hoped to design a mesoporous aluminosilicate that would have all the attributes of strong acid zeolites such as USY, in addition to much enhanced accessibility to catalytic sites. Such catalysts would outperform USY for the hydrocracking of heavy oil. However, in addition to their inherent fragility, mesoporous aluminosilicates proved to be much less acidic than USY, a key requirement for breaking hydrocarbon molecules [5,8,10]. Likewise, even though the pore walls of silica mesophases were known to be amorphous, researchers did not resist the temptation of preparing mesoporous titanosilicates that would catalyze the selective oxidation of large organic molecules in the presence of dilute hydrogen peroxide with similar intrinsic activity as TS-1 [90] for small molecules. This was not the case as these materials behaved rather like amorphous titanosilicates with low or no activity for many test reactions [88,89,91,92]. However, it appeared in a recent investigation that the preparation of mesoporous silica-based materials with molecularly ordered frameworks could be possible [93].

The lack of hydrothermal stability is believed to be due to silicate hydrolysis. Several research groups worked on improving the hydrothermal stability of silica mesophases, particularly with MCM-41 structure. Varying degrees of success were achieved via pore wall thickening by adjusting the synthesis conditions or by post-synthesis treatment procedures and/or via enhanced condensation of the silica walls [56,59,94,95]. The hydrothermal stability could also be increased dramatically by trimethylsilylation [96] or by efficient coating of the mesophase with a layer of aluminosilicate via alumination under supercritical conditions [97].

Very recently, another culprit, namely sodium has been identified as being responsible for the limited hydrothermal stability of silica mesostructures [98]. Indeed, it was found that MCM-41 silica prepared in the absence of sodium using fumed silica was much more hydrothermally stable than its counterpart made from sodium silicate. Moreover, doping a sodium-free sample with minute amounts of sodium was shown to have a deleterious effect on its hydrothermal stability. Sodium used during the synthesis was also found to be strongly retained as residual amounts could not be eliminated even by extensive washing or ammonium ion exchange. The entrapped sodium ions are believed to catalyze the collapse of the silica mesostructure upon treatment in water vapor.

Another approach to simultaneously increase the acid strength and the hydrothermal stability of mesoporous aluminosilicates is to crystallize their pore walls, at least partially, into zeolite nanocrystals. This idea was pursued by a number of researchers [99-101] who used a two-step synthesis strategy. The method consisted of preparing a regular MCM-41 aluminosilicate, then use it as precursor for the partial crystallization of the pore walls into zeolitic crystals such as HZSM-5 in the presence of appropriate molecular templates such as tetratrapropylammonium bromide or hydroxide. However, the structure of such materials remains controversial as the size of the zeolite particles sometimes exceeds the wall thickness by more than one order of magnitude [101]. Actually, proceeding the other way around proved to be a much more powerful method for making mesoporous materials with acid strength and hydrothermal stability comparable to those of zeolites [102-106]. The method consists of preparing zeolite nanocrystals using literature recipes, and then assembling them using the supramolecular templating procedures described above. This versatile approach was used for the preparation of a series of mesoporous aluminosilicates assembled from ZSM-5, Y and beta zeolite seeds. Moreover, it was shown that these zeolite seeds could be assembled in the presence of cetyltrimethylammonium bromide under basic conditions into hexagonal mesostructures [103,105,106] and also in the presence of Pluronic P123 triblock copolymer with [104] or without [107] added 1,3,5-trimethylbenzene under acidic conditions into SBA-15 like mesoporous materials or mesostructured cellular foams, respectively [104]. The obtained materials exhibited strong acidity, very high hydrothermal stability and promising catalytic properties.

4.4 Morphology

Periodic mesoporous silica may be prepared in a variety of useful morphologies such as thin films, spheres, fibers and large monoliths. In addition, despite their amorphous nature, cubic silica mesophases with the following space groups $Ia\bar{3}d$ (MCM-48), $Pm\bar{3}n$ (SBA-1) and $Im\bar{3}m$ were prepared as particles with single crystal-like habits [64,108-110]. Cubic ($Pm\bar{3}n$) mesoporous organosilicates were also synthesized as truncated rhombic dodecahedral particles [111,112]. In addition, a large variety of particles with exotic shapes and surface patterns were observed. Particles with toroid, gyroid, diskoide, spiral, spheroid, pinwheel, doughnut, shell, rope morphologies and others have been reported not only for silica mesophases [113-116], but also for organosilicates [112,117] and other oxides [118-120].

4.5 Framework Modification via Cation Isomorphous Substitution

Borrowing from the chemistry of zeolites, a large number of metallic cation substituted silica mesophases were prepared, mostly for the purpose of developing new acid and redox catalysts. The most often used mesophases were MCM-41, MCM-48, SBA-1 and SBA-15. Mesoporous metallosilicates containing trivalent cations such as Al, B, Ga and Fe were prepared as potential acid catalysts. However, most of the catalytic testing used mesoporous aluminosilicates. As mentioned earlier, because of the amorphous nature of such materials, their acidity was much weaker than typical acid zeolites such as USY, HZSM-5 or H-mordenite. However, because of their much more open structure, mesoporous aluminosilicates may find important catalytic applications involving transformation of large molecule substrates requiring mildly acidic sites [5,8,10].

Isomorphous substitution of silicon for tetra and penta-valent cations was practiced on many silica mesophases using a large number of cations including Ti, Sn, V, Mo, Cr and Zr. Again, based on prior knowledge in zeolite chemistry, these materials were expected to be good redox catalysts for large organic substrates in the presence of environmentally benign dilute hydrogen peroxide. However, as mentioned earlier using mesoporous titanosilicate for example, the intrinsic activity was much lower than that of crystalline microporous TS-1. Problems related to the low activity or lack thereof could be circumvented in the same manner as for aluminosilicates by assembling preformed TS-1 seeds under acidic conditions in the presence of P123 triblock copolymer into a hexagonal mesostructure [121]. These materials were found to be excellent selective oxidation catalysts for small as well as large molecules in the presence of dilute hydrogen peroxide. Moreover, as in the case of many transition metal substituted zeolites, mesoporous metallosilicates pose serious leaching problems. This has been well documented in the case of vanadosilicate [122].

4.6 Periodic Mesoporous Organosilicates

Among the most recent innovations in this field is undoubtedly the use of bridged silsesquioxane molecules as precursors for the synthesis of periodic mesoporous organosilicates (PMOs), which were reported for the first time in 1999 [22,23,112,123-125]. The organic moieties in PMOs are built directly in the walls of the channels instead of protruding from them as in surface functionalized mesoporous silica. In addition to the unique features of periodic mesoporous materials such as their high surface area, tunable pore sizes and narrow pore size distributions, PMOs open a wide range of new and exciting opportunities for designing materials with controlled surface properties at the molecular level [22,23]. To date, PMOs with organic spacers R originating from methane, ethane, ethylene, acetylene, butene, benzene, toluene, xylene, dimethoxybenzene, thiophene, bithiophene and ferrocene [23] have been achieved. Successful syntheses of PMOs have been reported to take place under basic, acidic or neutral conditions. Several PMO mesophases have been obtained, including a cubic structure ($Pm\bar{3}n$) akin to SBA-1 [113,123], another cubic structure with possibly $Im\bar{3}m$ space group similar to FDU-1 [38], 2D hexagonal structure ($p6mm$) similar to MCM-41 [113,123-125] or SBA-15 [48,126], and a 3D hexagonal mesophase [112,123] whose space group is yet to be determined.

4.7 Surface Functionalization

The subject of surface modified periodic mesoporous silicates, including their synthesis, characterization and potential applications has recently been thoroughly reviewed [23]. Surface functionalization of periodic mesoporous silicas via covalent bonding of organic molecules may occur in two steps via post-synthesis modifications [4] or in a single step via co-condensation procedures [127,128]. Primary post-synthesis surface modification or direct grafting consists in reacting a suitable organosilane reagent with the silica surface silanols using an appropriate solvent under reflux conditions. Grafting with silane coupling agents such as chlorosilanes, alkoxysilanes and silylamines under anhydrous conditions (or silylation) may also be achieved in the presence of disilazane reagents [$HN(SiR_3)_2$] under mild conditions. Usually, primary surface modifications are carried out using calcined, and often rehydrated mesoporous silica; however, extraction of the surfactant and grafting of organic functionalities without prior calcination may also take place in a single step. Depending on the nature of the desired surface functionality, the material may undergo secondary or higher order modifications, which consist of further reactions of the previously grafted species. Extensive compilation of literature data related to post-synthesis surface functionalization of periodic mesoporous silicas may be found elsewhere [23]. These include materials with grafted alkyl groups, basic or acidic functionalities and chiral species.

One-pot synthesis of organically modified silica via co-condensation of siloxane and organosiloxane precursors by the sol-gel technique has been investigated extensively [129]. In these materials, an organic moiety is covalently linked via a non-hydrolyzable Si-C bond to a siloxane species which hydrolyzes to form a silica network with surface organic functionalities. Burkett et al. [127] and Macquarrie [128] were the first to combine this approach with the supramolecular templating technique to generate in a single step, ordered mesoporous silica-based nanocomposites with covalently linked organic functionalities protruding from the inorganic walls into the pores. Mesoporous silicas whose surface has been functionalized via co-condensation were MCM-41, MCM-48, HMS, MSU and SBA-15. The surface species used included aliphatic and aromatic groups, mercapto and aminopropyl as well as propylsulfonic acid [23].

Whether prepared by post-synthesis treatment or by co-condensation, surface functionalized ordered mesoporous silicas may find many potential applications in selective oxidation, acid, base and chiral catalysis as well as in chiral separation [23]. However, their most promising applications are likely to be in the area of environmental remediation. Extensive work was devoted to the adsorption of heavy metal cations such as mercury in water using mercaptopropyl bearing mesoporous silicas [130,131]. Some materials exhibited remarkably high adsorption capacities of up to 600 mg Hg/g [131]. Adsorption of harmful metallic anions [132] as well as organic pollutants [133] was also achieved on mesoporous silica with proper surface modifications.

5 Periodic Non-Silica Mesoporous Materials

Unlike silica, transition metal compounds often exhibit variable oxidation states, which may be associated with unique magnetic, electronic and optical properties, thus providing great opportunities for innovative applications. This has been a strong driving force behind the effort by many groups to extend the synthesis strategies for silica mesophases to other materials, particularly transition metal oxides and chalcogenides. Progress in this field was covered in several comprehensive reviews [5,15-17].

In their pioneering work on mesostructured transition metal oxides, Stucky et al. [28,134] identified three prerequisite conditions for the formation of an organic-inorganic mesophase to be successful, namely (i) the inorganic precursor should have the ability to form polyanions or polycations allowing multidentate binding to the surfactant, (ii) the polyanions or polycations should be able to condense into rigid walls, and (iii) a charge density matching between the surfactant and the inorganic species should occur. However, all the early attempts to synthesize stable mesoporous transition metal oxides led to either lamellar or unstable hexagonal mesophases [134]. Since then, most of these difficulties have been largely overcome. A wide variety of non-siliceous mesoporous oxides including aluminum, zirconium, hafnium, molybdenum, manganese, tantalum, niobium, tin and titanium oxides as well as numerous mixed oxides such as alumino, vanadophosphates and heteropolyoxides have been successfully prepared [15-17], mostly via the surfactant/inorganic cooperative self assembly pathways described above. However, the true liquid crystal templating route has also been used for example for the assembly of CdS nanoparticles [135] and the synthesis of mesoporous noble metals [136]. In a further development, Stucky et al. [137] discovered a general procedure for the synthesis of ordered large pore mesoporous metal oxides in non aqueous solutions. They used amphiphilic poly(alkylene oxide) block copolymers as structure directing agents and metal salts as precursors rather than alkoxides. Most of the materials exhibited highly ordered pore systems with pore sizes up to 14 nm and thick semicrystalline channel walls.

The preparation of mesostructured chalcogenides is of particular interest because of their potential applications in quantum electronics, photonics and non-linear optics. The most successful synthesis reported to date involves the use of adamantane-like $[Ge_4Q_{10}]^{4-}$ (Q = S, Se, Te) anions as building units for mesostrcutured germanium chalcogenides [138,139]. In these materials, the tetrahedral $[Ge_4Q_{10}]^{4-}$ species are linked together via metallic cations to form a mesostructure whose pores are filled with the surfactant. Using Ni^{2+}, Zn^{2+}, Co^{2+} and Cu^{2+} as linkage cations and formamide as a solvent, Ozin et al. [138] assembled $[Ge_4S_{10}]^{4-}$ into a hexagonal mesophase. Kanadzidis et al. [139,140] carried out extensive work on the synthesis of mesostructured germanium chacogenides using a variety of divalent, trivalent and tetravalent linkage cations. The use of Pt^{2+} in combination with $[Ge_4Q_{10}]^{4-}$ (Q = S, Se) or $[Sn_4Se_{10}]^{4-}$ was found to give rise to very highly ordered materials with adjustable pore sizes depending on the length of the surfactant carbon chain [139]. They also found cetylpyridinium bromide to be particularly suitable for such syntheses. In a further development, Kanadzidis et al. [141] prepared tin selenide mesophase with wormhole, hexagonal and cubic structures via the supramolecular assembly of tetrahedral Zintl anions $[SnSe_4]^{4-}$ in the presence of cetylpyridinium bromide.

Silica mesophases were in turn used as molds to produce nanoporous materials via a replication process referred to as nanocasting [16]. The new materials are prepared within the pore system of the silica, which is then removed by HF or NaOH. This technique has been used for the preparation of periodic mesoporous carbons [71,142], metals [70] and polymers [143]. Making carbon replicas of a mesoporous silica such as SBA-15 was actually shown to be reversible [144].

6 Concluding Remarks

Since the discovery of the M41S family of silica mesophases, the supramolecular templating technique became a powerful method for the synthesis of mesostructured materials with narrow pore size distributions. Figure 1 lists the main experimental parameters that have been explored in the design and synthesis of such materials. On average, every two years since 1992, a new and exciting development occurs in this field. As far as the synthesis of mesostrutured materials is concerned, the most important milestones are (i) generalization of synthesis strategies, (ii) framework and surface modified silicas, (iii) non-silica mesostrutured materials, (iv) mesoporous organosilicates, and (v) assembly of zeolite seeds into mesoporous structures.

Parallel to this remarkable progress in synthesis, a wide variety of potential applications were also investigated. Surface and framework modified mesoporous silicates were tested in a wide range of catalytic processes. The recent discovery that zeolite seeds can be assembled in the presence of amphiphilic molecules into hydrothermally stable mesoporous structures will undoubtedly provide a new impetus to research in acid, redox and photo-catalytic applications using such materials. Moreover, mesoporous silicates were used as molds or nanoreactors for the synthesis of other advanced materials such as metallic or semiconducting nanowires and nanoparticles, as well as nanoporous carbons, metals and polymers. Many mesostructured non-silica oxides and transition metal chalcogenides showed promising applications based on their unique electronic, magnetic or optical properties. The area of synthesis having reached a high level of maturity, it is anticipated that in the future, more focus will be put on developing innovative applications using mesostructured materials.

References

1. Yanagisawa T., Shimizu, T., Kuroda, K. and Kato C., The preparation of Alkyltrimethylammonium-Kanemite complexes and their conversion to mesoporous materials, *Bull. Chem. Soc. Jpn.* **63** (1990) pp. 988-992.
2. Di Renzo F., Cambo H. and Dutartre R., 28-Year old synthesis of micelle-templated mesopoorus silica, *Microporous Mesoporous Mater.* **10** (1997) pp. 283-286.
3. Kresge C.T., Leonowicz M.E., Roth W.J., Vartuli J.C. and Beck J.S., Ordered mesoporous molecular sieves synthesized by a liquid-crystal template mechanism, *Nature* **359** (1992) pp. 710-712.
4. Beck J.S., Vartuli J.C., Roth W.J., Leonowicz M.E., Kresge C.T., Schmitt K.D., Chu, C.T-W., Olson D.H., Sheppard E.W., McCullen S.B., Higgins J.B. and Schlenker J.L., A new family of mesoporous molecular sieves prepared with liquid crystal templates, *J. Am. Chem. Soc.* **114** (1992) pp. 10834-10843.

5. Sayari A., Periodic nanoporous materials: synthesis, characterization and potential applications, in *Recent Advances and New Horizons in Zeolite Science and Technology*; ed. by H. Chong, S.I. Woo and S.E. Park, (Elsevier, Amsterdam, 1996) pp. 1-46.
6. Raman N.K., Anderson M.T. and Brinker C.J., Template-based approaches to the preparation of amorphous, nanoporous silicas, *Chem. Mater.* **8** (1996) pp. 1682-1701.
7. Zhao X. S., Lu G. Q. and Millar G. J., Advances in mesoporous molecular sieve MCM-41, *Ind. Eng. Chem. Res.* **35** (1996) pp. 2075-2090.
8. Corma A., From microporous to mesoporous molecular sieve materials and their use in catalysis, *Chem. Rev.* **97** (1997) pp. 2373-2420.
9. Ying J.Y., Mehnert C.P. and Wong M.S., Synthesis and applications of supramolecular-templated mesoporous materials, *Angew. Chem. Int. Ed.* **38** (1999) pp. 56-77.
10. Sayari A., Catalysis by crystalline mesoporous molecular sieves, *Chem. Mater.* **8** (1996) pp. 1840-1852.
11. Kruk M. and Jaroniec M., Gas adsorption characterization of ordered organic-inorganic nanocomposite materials, *Chem. Mater.* **13** (2001) pp. 3169-3183.
12. Jaroniec M., Kruk M. and Sayari A., The use of ordered mesoporous materials for improving the mesopore size analysis: Current state and future, in *Nanoporous Materials III*, ed. by A. Sayari and M. Jaroniec (Elsevier, Amsterdam, 2002) pp. 437-444.
13. Selvam P., Bhatia S.K. and Sonwane C.G., Recent advances in processing and charaterization of periodic mesoporous MCM-41 silicate molecular sieves, *Ind. Eng. Chem. Res.* **40** (2001) pp. 3237-3261.
14. Tiemann M. and Fröba M., Mesostructured aluminophosphates synthesized with supramolecular structure directors, *Chem. Mater.* **13** (2001) pp. 3211-3217.
15. Sayari A. and Liu P., Non-silica periodic mesostructured materials: recent progress, *Microporous Mater.* **12** (1997) pp. 149-177.
16. Schüth F., Non-siliceous mesostructured and mesoporous materials, *Chem. Mater.* **13** (2001) pp. 3184-3195.
17. He X. and Antonelli D., Recent advances in synthesis and applications of transition metal containing mesoporous molecular sieves, *Angew. Chem. Int. Ed.* **41** (2002) pp. 214-229.
18. Moller K. and Bein T., Inclusion chemistry in periodic mesoporous hosts, *Chem. Mater.* **10** (1998) pp. 2950-2963.
19. Scott B.J., Wirnsberger G. and Stucky G.D., Mesoporous and mesostructured materials for optical applications, *Chem. Mater.* **13** (2001) pp. 3140-3150.
20. Maschmeyer T., Derivatised mesoporous solids, *Curr. Opin. Solid State Mater. Sci.* **3** (1998) pp. 71-78.
21. Brunel D., Functionalised micelle-templated silicas and their use as catalysts for fine chemicals, *Microporous Mesoporous Mater.* **27** (1999) pp. 329-343.

22. Stein A., Melde B.J. and Schroden R.C., Hybrid inorganic-organic mesoporous silicates - nanoscopic reactors coming of age, *Adv. Mater.* **12** (2000) pp. 1403-1419.
23. Sayari A. and Hamoudi S., Periodic mesoporous silica-based organic-inorganic nanocomposite materials, *Chem. Mater.* **13** (2001) pp. 3151-3168.
24. Attard, G.A., Glyde, J.C. and Göltner C.G. Liquid crystalline phases as templates for the synthesis of mesoporous silica, *Nature* **378** (1995) pp. 366-368.
25. Firouzi A., Monnier A., Bull L.M., Besier T., Sieger P., Huo Q., Walker S.A., Zasadzinski J.A., Glinka C., Nicol J., Margolese D., Stucky G.D. and Chmelka B.F., Cooperative organization of inorganic-surfactant and biomimetic assemblies, *Science* **267** (1995) pp. 1138-1143.
26. Firouzi A., Atef F., Oertli A.G., Stucky G.D. and Chmelka B.F., Alkaline lyotropic silicate-surfactant liquid crystals, *J. Am. Chem. Soc.* **119** (1997) pp. 3596-3610.
27. Huo Q., Margolese D.I., Ciesla U., Feng P., Gier T.E., Sieger P., Leon R., Petroff P.M., Schüth F. and Stucky G.D., Generalized synthesis of periodic surfactant/ inorganic composite materials, *Nature* **368** (1994) pp. 317-321.
28. Huo Q., Margolese D.I., Ciesla U., Demuth D.G., Feng P., Gier T.E., Sieger P., Firouzi A., Chmelka B.F., Schüth F. and Stucky G.D., Organization of organic molecules with molecular species into nanocomposite biphase arrays, *Chem. Mater.* **6** (1994) pp. 1176-1191.
29. Tanev P.T. and Pinnavaia T.J., A neutral templating route to mesoporous molecular sieves, *Science* **267** (1995) pp. 865-867.
30. Bagshaw S.A., Prouzet E. and Pinnavaia T.J., Templating of mesoporous molecular sieves by nonionic polyethylene oxide surfactant, *Science* **269** (1995) pp. 1242-1244.
31. Zhao D., Huo Q., Feng J., Chmelka B.F. and Stucky G.D., Nonionic triblock and star diblock copolymer and oligomeric surfactant syntheses of highly ordered, hydrothermally stable, mesoporous silica structure, *J. Am. Chem. Soc.* **120** (1998) pp. 6024-6036.
32. Antonelli D.M. and Ying J.Y., Synthesis of a stable hexagonally packed mesoporous niobium oxide molecular sieve through a noble ligand-assisted templating mechanism, *Angew. Chem. Int. Ed. Engl.* **35** (1996) pp. 426-430.
33. Sakamoto Y., Díaz I., Terasaki O., Zhao D., Pérez-Pariente J., Kim J.M. and Stucky G.D., Three-dimensional cubic mesoporous structures of SBA-12 and related materials by electron crystallography, *J. Phys. Chem. B* **106** (2002) pp. 3118-3123.
34. Kruk M., Jaroniec M. and Sayari A., Application of large pore MCM-41 molecular sieves to improve pore size analysis using nitrogen adsorption measurements, *Langmuir* **13** (1997) pp. 6267-6273.

35. Schmidt-Winkel P., Lukens W.W., Jr., Zhao D., Yang P., Chmelka B.F. and Stucky G.D., Mesocellular siliceous foams with uniformly sized cells and windows, *J. Am. Chem. Soc.* **121** (1999) pp. 254-255.
36. Huo Q., Leon R., Petroff P.M. and Stucky G.D., Mesostructure design with gemini surfactants: supercage formation in a three-dimensional hexagonal array, *Science* **268** (1995) pp. 1324-1327.
37. Yu C., Yu Y. and Zhao D., Highly ordered large caged cubic mesoporous silica structures templated by triblock PEO-PBO-PEO copolymer, *Chem. Commun.* (2000) pp. 575-576.
38. Matos J.R., Kruk M., Mercuri L.P., Jaroniec M., Asefa T., Coombs N., Ozin G.A., Kamiyama, T. and Terasaki, O., Periodic mesoporous organosilica with large cagelike pores, *Chem. Mater.* **14** (2002) pp. 1903-1905.
39. Huo Q., Margolese D.I. and Stucky G.D., Surfactant control of phases in the synthesis of mesoporous silica-based materials, *Chem. Mater.* **8** (1996) pp. 1147-1160.
40. Sakamoto Y., Kaneda M., Terasaki O., Zhao D., Kim J.M., Stucky G.D., Shin H.J. and Ryoo R., Direct imaging of the pores and cages of three-dimensional mesoporous materials, *Nature* **408** (2000) pp. 449-453
41. Zhao D., Huo Q., Feng J., Kim J., Han Y. and Stucky G.D., Novel mesostructure silicates with two-dimensional mesostructure direction using rigid bolaform surfactants, *Chem. Mater.* **11** (1999) pp. 2668-2672.
42. Zhao D., Huo Q., Feng J., Chmelka B.F. and Stucky G.D., Triblock copolymer syntheses of mesoporous silica with periodic 50 to 300 Å pores, *Science* **279** (1998) pp. 548-552.
43. Prouzet E., Cot F., Nabias G., Larbot A., Kooyman P. and Pinnavaia T.J., Assembly of mesoporous silica molecular sieves based on nonionic ethoxylated sorbitan esters as structure directors, *Chem. Mater.* **11** (1999) pp. 1498-1503.
44. Tanev P.T., Liang Y. and Pinnavaia T.J., Assembly of mesoporous lamellar silicas with hierarchical particle architectures, *J. Am. Chem. Soc.* **119** (1997) pp. 8616-8624.
45. Kim S.S., Zhang W. and Pinnavaia T.J., Ultrastable mesostructured silica vesicles, *Science* **282** (1998) pp. 1302-1305.
46. Israelachvili J.N., Mitchell D.J. and Ninham, B.W., Theory of self-assembly of hydrocarbon amphiphiles into micelles and bilayers, *J. Chem. Soc., Faraday Trans. II*, **72** (1976) 1525-1568.
47. Kim J.M., Sakamoto Y., Hwang Y.K., Kwon Y.U., Terasaki O., Park S.E. and Stucky G.D., Structural design of mesoporous silica by micelle-packing control using blends of amphiphilic block copolymers, *J. Phys. Chem. B.* **106** (2002) pp. 2552-2558.
48. Zhu H., Jones D.J., Zajak J., Roziere J. and Dutartre R., Periodic large mesoporous organosilicas from lyotropic liquid crystal polymer templates, *Chem. Comm.* (2001) pp. 2568-2569.

49. Feng P., Bu X., Stucky G.D. and Pines D.J., Monolith mesoporous silica templated by microemulsion liquid crystals, *J. Am. Chem. Soc.* **122** (2000) pp. 994-995.
50. Klotz M., Ayral A., Guizard C. and Cot L., Synthesis conditions for hexagonal mesoporous silica layers, *J. Mater. Chem.* **10** (2000) pp. 663 - 669
51. Yang P., Zhao D., Chmelka B.F. and Stucky G.D., Triblock copolymer direct syntheses of large pore mesoporous silica fibers, *Chem. Mater.* **10** (1998) pp. 2033-2036.
52. Alberius P.C.A., Frindell K.L., Hayward R.C., Kramer E.J., Stucky G.D. and Chmelka B.F., General predictive synthesis of cubic, hexagonal, and lamellar silica and titania mesostructured thin films, *Chem. Mater.* **14** (2002) pp. 3284-3294.
53. Ulagappan N. and Rao C.N.R., Evidence for supramolecular organization of alkane and surfactant molecules in the process of forming mesoporous silica, *Chem. Commun.* (1996) pp. 2759-2760.
54. Sayari A., Kruk M., Jaroniec M. and Moudrakovski I.L., New Approaches to Pore Size Engineering of Mesoporous Silicates, *Adv. Mater.* **10** (1998) pp. 1376-1379.
55. Khushalani D., Kuperman A, Ozin G.A., Tanaka K., Garces J., Olken M.M. and Coombs N., Metamorphic materials: restructuring siliceous mesoporous materials, *Adv. Mater.* **7** (1995) 842-846.
56. Sayari A., Liu P., Kruk M. and Jaroniec M., Characterization of large-pore MCM-41 molecular sieves obtained via hydrothermal restructuring, *Chem. Mater.* **9** (1997) pp. 2499-2506.
57. Corma A., Kan Q., Navarro M. T., Perez-Pariente J. and Rey F., Synthesis of MCM-41 with different pore diameters without addition of auxiliary organics, *Chem. Mater.* **9** (1997) pp. 2123-2126.
58. Sayari A., Yang Y., Kruk M. and Jaroniec M., Expanding the pore size of MCM-41 silicas: Use of Amines as expanders in direct synthesis and postsynthesis procedures, *J. Phys. Chem. B.* **103** (1999) pp. 3651-3658.
59. Kruk M., Jaroniec M. and Sayari A., A unified interpretation of high-temperature pore size expansion processes in MCM-41 mesoporous silicas, *J. Phys. Chem. B* **103** (1999) pp. 4590-4598.
60. Sayari A., Unprecedented expansion of pore size and volume of periodic mesoporous silica, *Angew. Chem. Int. Ed.* **39** (2000) pp. 2920-2922.
61. Imhof A. and Pine D.J., Ordered macroporous materials by emulsion templating, *Nature* **389** (1997) pp. 948-951.
62. Velev O.D., Jede T.A., Lobo R.F. and Lenhoff A.M., Porous silica via colloidal crystallization, *Nature* **389** (1997) pp. 447-448.
63. Holland B.T., Blanford C.F., Do T. and Stein A., Synthesis of highly ordered, three-dimensional, macroporous structures of amorphous or crystalline inorganic oxides, phosphates, and hybrid composites, *Chem. Mater.* **11** (1999) pp. 795-805.

64. Sayari A. Novel synthesis of high quality MCM-48 silica, *J. Am. Chem. Soc.* **122** (2000) pp. 6504-6505.
65. Kruk M., Jaroniec M., Ryoo R. and Joo S. H., Characterization of MCM-48 silicas with tailored pore sizes via a highly efficient procedure, *Chem. Mater.* **12** (2000) pp. 1414-1421.
66. Kim M.J. and Ryoo R., Synthesis and pore size control of cubic mesoporous silica SBA-1, *Chem. Mater.* 11 (1999) pp. 487-491.
67. Yu C., Tian B., Fan J., Stucky G.D. and Zhao D., Nonionic block copolymer synthesis of large-pore cubic mesoporous single crystals by use of inorganic salts, *J. Am Chem. Soc.* **124** (2002) pp. 4556-4557.
68. Pauly T.R. and Pinnavaia T.J., Pore size modification of mesoporous HMS molecular sieve silicas with wormhole framework structures, *Chem. Mater.* **13** (2001) pp. 987-993.
69. Prouzet E and Pinnavaia T.J., Assembly of mesoporous molecular sieves containing wormhole motifs by a nonionic surfactant pathway: control of pore size by synthesis temperature, *Angew. Chem. Int. Ed.* **36** (1997) pp. 516-518.
70. Ryoo R., Ko C.H., Kruk M., Antochshuk V. and Jaroniec M., Block-copolymer-templated ordered mesoporous silica: array of uniform mesopores or mesopore-micropore network? *J. Phys. Chem. B.*, 104 (2000) pp. 11465-11471.
71. Jun S., Joo S.H., Ryoo R., Kruk M., Jaroniec M., Liu Z., Ohsuna T. and Terasaki O., Synthesis of new, nanoporous carbon with hexagonally ordered mesostructure, *J. Am. Chem. Soc.* **122** (2000) pp. 10712-10713.
72. Impéror-Clerc M., Davidson P. and Davidson A. Existence of a microporous corona around the mesopores of silica-based SBA-15 materials templated by triblock copolymers, *J. Am. Chem. Soc.* **122** (2000) pp. 11925-11933.
73. Joo S.H., Ryoo R., Kruk M. and Jaroniec M., Evidence for general nature of pore interconnectivity in 2-dimensional hexagonal mesoporous silicas prepared using block copolymer templates, *J. Phys. Chem. B* **106** (2002) pp. 4640-4646.
74. Van Der Voort P., Benjelloun M. and Vansant E.F., Rationalization of the synthesis of SBA-16: controlling the micro- and mesoporosity *J. Phys. Chem. B.* **106** (2002) pp. 9027-9032.
75. Galarneau A., Cambon H., Martin T., De Menorval L.C., Brunel D., Di Renzo F. and Fajula F., SBA-15 versus MCM-41 : are they the same materials?, in *Nanoporous Materials III*, ed. by A. Sayari and M. Jaroniec (Elsevier, Amsterdam, 2002) pp. 395-402.
76. Newalkar B.L. and Komarneni S., Control over microporosity of ordered microporous-mesoporous silica SBA-15 framework under microwave-hydrothermal conditions: effect of salt addition, *Chem. Mater.* **13** (2001) pp. 4573-4579.
77. Freyhardt C.C., Tsapatsis M., Lobo R.F., Balkus K.J., Jr. and Davis M.E., A high-silica zeolite with a 14-tetrahedral-atom pore opening, *Nature* **381** (1996) pp. 295-298.

78. Yoshikawa M., Wagner P., Lovallo M., Tsuji K., Takewaki T., Chen C.-Y., Beck L.W., Jones C., Tsapatsis M., Zones S.I. and Davis M.E., Synthesis, characterization, and structure solution of CIT-5, a new, high-silica, extra-large-pore molecular sieve, *J. Phys. Chem. B.* **102** (1998) pp. 7139-7147.
79. Corma A., Diaz-Cabanas M.J., Martinez-Triguero J., Rey F. and Rius J., A large-cavity zeolite with wide pore windows and potential as an oil refining catalyst, *Nature* **418** (2002) pp. 514-517.
80. Kruk M., Asefa T., Jaroniec M. and Ozin G.A., Metamorphosis of ordered mesopores to micropores: periodic silica with unprecedented loading of pendant reactive organic groups transforms to periodic microporous silica with tailorable pore size, *J. Am. Chem. Soc.* **124** (2002) pp. 6383-6392.
81. Hamoudi S., Yang Y., Moudrakovski I.L., Lang S. and Sayari, A., Synthesis of porous organosilicates in the presence of alkytrimethylammonium chlorides: effect of the alkyl chain length, *J. Phys. Chem. B.* **105** (2001) pp. 9118-9123.
82. Sun T. and Ying J.Y., Synthesis of microporous transition-metal-oxide molecular sieves by supramolecular templating mechanism, *Nature* **389** (1997) pp. 704-706.
83. Sun J., Shan Z., Maschmeyer T., Moulijn J.A. and Coppens M.O., Synthesis of tailored bimodal mesoporous materials with independent control of the dual pore size distribution, *Chem. Commun.* (2001) pp. 2670-2671.
84. Sayari A., Kruk M. and Jaroniec M., Synthesis and characterization of microporous-mesoporous MCM-41 silicates prepared in the presence of octyltrimethylammonium bromide, *Catal. Lett.* **49** (1997) pp. 147-154.
85. Yang P., Deng T., Zhao D., Feng P., Pine D., Chmelka B.F., Whitesides G.M. and Stucky G.A., Hierarchically ordered oxides, *Nature* **282** (1998) pp. 2244-2246.
86. Cassiers K., Linssen T., Mathieu M., Benjelloun M., Schrijnemakers K., Van Der Voort P., Cool P. and Vansant E.F., A detailed study of thermal, hydrothermal, and mechanical stabilities of a wide range of surfactant assembled mesoporous silicas, *Chem. Mater.* **14** (2002) pp. 2317-2324.
87. Gusev V.Yu., Feng X., Bu Z., Haller G.L. and O'Brien J.A., Mechanical stability of pure silica mesoporous MCM-41 by nitrogen adsorption and small-angle X-ray Diffraction measurements, *J. Phys. Chem.* **100** (1996) pp. 1985-1988.
88. Fraile J.M., García J.I., Mayoral J.A., Vispe E., Brown D.R. and Naderi M., Is MCM-41 really advantageous over amorphous silica? The case of grafted titanium epoxidation catalysts, *Chem. Commun.* (2001) pp. 1510 – 1511.
89. Reddy J.S., Dicko A. and Sayari A., Ti-modified mesoporous molecular sieves, Ti-MCM-41 and Ti-HMS, in *Synthesis of Porous Materials*, ed. by M.L. Occelli and H. Kessler, (Marcel Dekker, Inc., New York, 1997) pp. 405-415.
90. Notari B., Microporous crystalline titanium silicates, *Adv. Catal.* 41 (1996) pp. 253-334.
91. Sayari A., Karra V.R., Reddy J.S. and Moudrakovski I.L., *Mater. Res. Soc. Symp. Proc.* **371** (1995) pp. 81-86.

92. Sayari A., Reddy K.M. and Moudrakovski I., Synthesis of V and Ti modified MCM-41 mesoporous molecular sieves, in *Zeolite Science 1994: Recent Progress and Discussions*, ed. by H.G. Karge and J. Weitkamp, (Elsevier, Amsterdam, 1995) pp. 19-21.
93. Christiansen S.C., Zhao D., Janicke M.T., Landry C.C., Stucky G.D. and Chmelka B.F., *J. Am. Chem. Soc.* **123** (2001) pp. 4519-4529.
94. Mokaya R. Improving the stability of mesoporous MCM-41 silica via thicker more highly condensed pore walls, *J. Phys. Chem. B* **103** (1999) pp. 10204-10208.
95. Mokaya R., Hydrothermally stable restructured mesoporous silica, *Chem. Commun.* (2001) pp. 933-934.
96. Koyano K.A., Tatsumi T., Tanaka Y. and Nakata S., Stabilization of mesoporous molecular sieves by trimethylsilylation, *J. Phys. Chem. B.* **101** (1997) pp. 9436-9440.
97. O'Neil A.S., Mokaya R. and Poliakoff M., Supercritical fluid-mediated alumination of mesoporous silica and its beneficial effect on hydrothermal stability, *J. Am. Chem. Soc.* **124** (2002) pp. 10636-10637.
98. Pauly T.R., Petkov V., Liu Y., Billinge S.J.L. and Pinnavaia T.J., Role of framework sodium versus local framework structure in determining the hydrothermal stability of MCM-41 mesostructures, *J. Am. Chem. Soc.* **124** (2002) pp. 97-103.
99. Kloetstra K.R., van Bekkum H. and Jansen J.C., Mesoporous material containing framework tectosilicate by pore-wall recrystallization, *Chem. Commun.* (1997) pp. 2281-2282.
100. Karlsson A., Stöcker, M. and Schmidt R., Composites of micro- and mesoporous materials: simultaneous syntheses of MFI/MCM-41 like phases by a mixed template approach, *Microporous Mesoporous Mater.* **27** (1999) pp. 181-192.
101. Huang L., Guo W., Deng P., Xue Z. and Li Q., Investigation of synthesizing MCM-41/ZSM-5 composites, *J. Phys. Chem. B.* **104** (2000) pp. 2817-2823.
102. Liu Y., Zhang W. and Pinnavaia T.J., Steam-stable aluminosilicate mesostructures assembled from zeolite type Y seeds, *J. Am. Chem. Soc.* **122** (2000) pp. 8791-8792.
103. Liu Y., Zhang W. and Pinnavaia T.J., Steam-stable MSU-S aluminosilicate mesostructures assembled from zeolite ZSM-5 and zeolite beta seeds, *Angew. Chem. Int. Ed.* **40** (2001) pp. 1255-1258.
104. Liu Y. and Pinnavaia T.J., Assembly of hydrothermally stable aluminosilicate foams and large-pore hexagonal mesostructures from zeolite seeds under strongly acidic conditions, *Chem. Mater.* **14** (2002) pp. 3-5.
105. Zhang Z., Han Y., Xiao F.-S., Qiu S., Zhu L., Wang R., Yu Y., Zhang Z., Zou B., Wang Y., Sun H., Zhao D. and Wei Y., Mesoporous aluminosilicates with ordered hexagonal structure, strong acidity, and extraordinary hydrothermal stability at high temperatures, *J. Am. Chem. Soc.* **123** (2001) pp. 5014-5021.

106. Zhang Z., Han Y., Zhu L., Wang R., Yu Y., Qiu S., Zhao D. and Xiao F.-S., Strongly acidic and high-temperature hydrothermally stable mesoporous aluminosilicates with ordered hexagonal structure, *Angew. Chem. Int. Ed.* **40** (2001) pp. 1258-1262.
107. Han Y., Wu S., Sun Y., Li D., Xiao F.-S., Liu J. and Zhang X., Hydrothermally stable ordered hexagonal mesoporous aluminosilicates assembled from a triblock copolymer and preformed aluminosilicate precursors in strongly acidic media, *Chem. Mater.* **14** (2002) pp. 1144-1148.
108. Kim J.M., Kim S.K. and Ryoo R., Synthesis of MCM-48 single crystals, *Chem. Commun.* (1998) pp. 259-260.
109. Che S., Sakamoto Y., Terasaki O. and Tatsumi T., Control of crystal morphology of SBA-1 mesoporous silica, *Chem. Mater.* **13** (2001) pp. 2237-2239.
110. Yu C., Tian B., Fan J., Stucky G.D. and Zhao D., Nonionic block copolymer synthesis of large-pore cubic mesoporous single crystals by use of inorganic salts, *J. Am. Chem. Soc.* **124** (2002) pp. 4556-4557.
111. Guan S., Inagaki S., Ohsuna T. and Terasaki O., Cubic hybrid organic-inorganic mesoporous crystal with a decaoctahedral shape, *J. Am. Chem. Soc.* **122** (2000) pp. 5660-5661.
112. Sayari A., Hamoudi S., Yang Y., Moudrakovski I.L. and Ripmeester J.R., New insights into the synthesis, morphology, and growth of periodic mesoporous organosilicas, *Chem. Mater.* **12** (2000) pp. 3857-3863.
113. Yang H., Coombs N. and Ozin G.A., Morphogenesis of shapes and surface patterns in mesoporous silica, *Nature* **386** (1997) pp. 692-695.
114. Lin H.-P. and Mou C.-Y., "Tubules-within-a-tubule" hierarchical order of mesoporous molecular sieves in MCM-41, *Science* **273** (1996) pp. 765-768.
115. Lin H.-P. and Mou C.-Y., Structural and morphological control of cationic surfactant-templated mesoporous silica, *Acc. Chem. Res.* (2002) ASAP Article.
116. Zhao D., Sun J., Li Q. and Stucky G.D., Morphological control of highly ordered mesoporous silica SBA-15, *Chem. Mater.* **12** (2000) pp. 275-279.
117. Park S.S., Lee C.H., Cheon J.H. and Park D.H., Formation mechanism of PMO with rope- and gyroid-based morphologies via close packing of secondary building units, *J. Mater. Chem.* (2001) pp. 3397-3403.
118. Ozin G.A., Morphogenesis of biomineral and morphosynthesis of biomimetic forms, *Acc. Chem. Res.* **30** (1997) pp. 17-27.
119. Chenite A., Le Page Y., Karra V.R. and Sayari A., Concentric bilayer growth evidenced by TEM: A novel phase in inorganic-surfactant systems, *J. Chem. Soc., Chem. Commun.* (1996) pp. 413-415.
120. Yada M., Hiyoshi H., Ohe K., Machida M. and Kijima T., Synthesis of aluminum-based surfactant mesophases morphologically controlled through a layer to hexagonal transition, *Inorg. Chem.* **36** (1997) pp. 5565-5569.

121. Xiao F.-S., Han Y., Yu Y., Meng X., Yang M. and Wu S., Hydrothermally stable ordered mesoporous titanosilicates with highly active catalytic sites, *J. Am. Chem. Soc.* **124** (2002) pp. 888-889.
122. Reddy J.S., Liu P. and Sayari A., Vanadium containing crystalline mesoporous molecular sieves: leaching of vanadium in liquid phase reactions, *Appl. Catal.* **148** (1996) pp. 7-21.
123. Inagaki S., Guan S., Fukushima Y., Ohsuna T. and Terasaki O, Novel mesoporous materials with a uniform distribution of organic groups and inorganic oxide in their frameworks, *J. Am. Chem. Soc.* **121** (1999) pp. 9611-9614.
124. Melde B.J., Holland B.T., Blanford C.F. and Stein A., Mesoporous sieves with unified hybrid inorganic/organic frameworks, *Chem. Mater.* **11** (1999) pp. 3302-3308.
125. Asefa T., MacLachlan M.J., Coombs N. and Ozin G.A., Periodic mesoporous organosilicas with organic groups inside the channel walls, *Nature* **402** (1999) pp. 867-871.
126. Muth O., Schellbach C. and Fröba M., Triblock copolymer assisted synthesis of periodic mesoporous organosilicas (PMOs) with large pores, *Chem. Commun.* (2001) pp. 2032-2033.
127. Burkett S.L., Sims D. and Mann S., Synthesis of hybrid inorganic-organic mesoporous silica by co-condensation of siloxane and organosiloxane precursors, *Chem. Commun.* (1996) pp. 1367-1368.
128. Macquarrie, Direct preparation of organically modified MCM-type materials. Preparation and characterization of aminopropyl-MCM and 2-cyanoethyl-MCM, *Chem. Commun.* (1996) pp. 1961-1962.
129. Wen J. and Wilkes G.L., Organic/inorganic hybrid network materials by the sol-gel approach, *Chem. Mater.* **8** (1996) pp. 1667-1681.
130. Mercier L. and Pinnavaia T.J., Access in mesoporous materials: advantages of a uniform pore structure in the design of a heavy metal ion adsorbent for environmental remediation, *Adv. Mater.* **9** (1997) pp. 500-503.
131. Feng X., Fryxell G.E., Wang L.Q., Kim A.Y., Liu J. and Kemner K.M., Functionalized monolayers on ordered mesoporous supports, *Science* **276** (1997) pp. 923-926.
132. Fryxell G.E., Liu J., Hauser T.A., Nie Z., Ferris K.F., Mattigod S., Gong M. and Hallen R.T., Design and synthesis of selective mesoporous anion traps, *Chem. Mater.* **11** (1999) pp. 2148-2154.
133. Inumaru K., Kiyoto J. and Yamanaka S., Molecular selective adsorption of nonylphenol in aqueous solution by organo-functionalized mesoporous silica *Chem. Commun.* (2000) pp. 903-904.
134. Huo Q., Margolese D.I., Ciesla U., Feng P., Gier T.E., Sieger P., Leon R., Petroff P.M., Schüth F. and Stucky G.D., Generalized synthesis of periodic surfactant/ inorganic composite materials, *Nature* **368** (1994) pp. 317-321.

135. Braun P.V., Osenar P. and Stupp S.I., Semiconducting superlattices templated by molecular assemblies, *Nature* **380** (1996) pp. 325-328.
136. Attard, G.S., Göltner C.G., Corken J.M., Henke S. and Templer R.H., Liquid-crystal templates for nanostructured metals, *Angew. Chem. Int. Ed.* **36** (1997) pp. 1315-1317.
137. Yang P., Zhao D., Margolese D.I., Chmelka B.F. and Stucky, G.D., Block copolymer templating syntheses of mesoporous metal oxides with large ordering lengths and semicrystalline framework, *Chem. Mater.* **11** (1999) pp. 2813-2826.
138. MacLachlan M.J., Coombs N., Bedard R.L., White S., Thompson L.K. and Ozin G.A., Mesostructured metal germanium sulfides, *J. Am. Chem. Soc.* **121** (1999) pp. 12005-12017.
139. Trikalitis P.N., Rangan K. and Kanatzidis M.G., Platinum chalcogenido MCM-41 analogues. High hexagonal order in mesostructured semiconductors based on Pt^{2+} and $[Ge_4Q_{10}]^{4-}$ (Q = S, Se) and $[Sn_4Se_{10}]^{4-}$ adamantane clusters, *J. Am. Chem. Soc.* **124** (2002) pp. 2604-2613.
140. Rangan K.K., Trikalitis P.N., Bakas T. and Kanatzidis M.G., Hexagonal mesostructured chalcogenide frameworks formed by linking $[Ge_4Q_{10}]^{4-}$ (Q = S, Se) clusters with Sb^{3+} and Sn^{4+}, *Chem. Commun.* (2001) pp. 809-810.
141. Trikalitis P.N., Rangan K., Bakas T. and Kanatzidis M.G., Varied pore organization in mesostructured semiconductors based on the $[SnSe_4]^{4-}$ anion, *Nature* **410** (2001) pp. 671-675.
142. Ryoo R., Joo S.H. and Jun S., Synthesis of highly ordered carbon molecular sieves via template-mediated structural transformation, *J. Phys. Chem. B.* **103** (1999) pp. 7743-7746.
143. Kim J.Y., Yoon S.B., Kooli F. and Yu J.-S., Synthesis of highly ordered mesoporous polymer networks, *J. Mater. Chem.* (2001) pp. 2912-2914.
144. Kang M., Yi S.H., Lee H.I., Yie J.E. and Kim J.M., Reversible replication between ordered mesoporous silica and mesoporous carbon, *Chem. Commun.* (2002) pp. 1944-1945.

MACROPOROUS MATERIALS CONTAINING THREE-DIMENSIONALLY PERIODIC STRUCTURES

YOUNAN XIA,* YU LU, KAORI KAMATA, BYRON GATES, YADONG YIN

Department of Chemistry, University of Washington, Seattle, WA 98195-1700, USA

1   Introduction

Porous materials have been extensively exploited for use in a broad range of applications: for example, as membranes for separation and purification [1], as high-surface-area adsorbants [2], as solid supports for sensors [3] and catalysts [4], as materials with low dielectric constants in the fabrication of microelectronic devices [5], and as scaffolds to guide the growth of tissues in bioengineering [6]. According to the nomenclature suggested by the International Union of Pure and Applied Chemists (IUPAC), porous materials are usually classified into three different categories depending on the lateral dimensions of their pores: microporous (<2 nm); mesoporous (between 2 and 50 nm); and macroporous (>50 nm) [7]. Liquid and gaseous molecules have been known to exhibit characteristic transport behaviors in each type of porous material. For instance, mass transport is achieved via viscous flow and molecular diffusion in a macroporous material; through surface diffusion and capillary flow in a mesoporous material; and by activated diffusion in a microporous material. The focus of this chapter is concentrated on macroporous materials that contain pores of at least 50 nm in size. Due to their technological importance, a rich variety of methods have already been developed for generating such porous materials with relatively large quantities. Conventional approaches include those based on electrochemical etching of aluminum [8] or silicon substrates [9], chemical etching of glasses [10], ion-track etching of polymers [11], excimer laser micromachining [12], and microlithography [13]. Most of these methods are only suitable for generating macroporous materials with essentially one-dimensional (1D) pores. On the other hand, methods based on microemulsion [14] and sintering of colloidal particles [15] have enabled the fabrication of three-dimensional (3D) macroporous materials, albeit these methods are limited in controlling the pore structure and size distribution. More recently, templating against crystalline lattices of various building blocks has been demonstrated as a simple, effective, and versatile approach to 3D macroporous materials with well-defined pore sizes and structures [16-22]. In this approach, the building blocks simply serve as sacrificial scaffolds, against which another material is deposited to a generate structure complementary to that exhibited by the template. As long as the building blocks in the template are in physical contact, the pores in the resultant structure are always interconnected to ensure the formation of a 3D network.

This chapter intends to provide an overview of current research activities that center on 3D macroporous materials fabricated by templating against crystalline lattices self-assembled from spherical colloids of monodispersed in size [19-22]. A unique feature of these materials is their three-dimensionally interconnected and highly ordered lattices of spherical pores (or air balls). The main text of this chapter is organized into five sections: We first explicitly discusses the concept and advantages of template-directed synthesis, with 3D crystalline lattices of spherical colloids as an example. We then briefly survey various methods that have been demonstrated for infiltrating the voids within 3D crystalline lattices of spherical colloids with a diversity of materials. We also evaluate the potential of hierarchical self-assembly as a generic approach to the generation of 3D macroporous materials. In the following sections, we highlight the use of these macroporous materials as photonic bandgap crystals, as well as their liquid permeation and mechanical properties. We conclude the last section with personal perspectives on the directions towards which future work on this new class of materials might be directed.

The objectives of this chapter are the following: i) to fully demonstrate the use of spherical colloids as a class of versatile templates to generate macroporous materials with well-controlled pores and highly ordered structures; ii) to address experimental issues related to the infiltration of voids in a colloidal crystal with a solid material (or a liquid or gas precursor to this material); iii) to evaluate the potential of hierarchical self-assembly in solving various problems associated with conventional infiltration methods; and iv) to assess the use of these macroporous materials as photonic crystals with interesting bandgap properties, as well as their intriguing applications in areas such as filtration-based separation and purification.

2 Template-Directed Synthesis of 3D Macroporous Materials

Template-directed synthesis provides a convenient and versatile procedure for processing a solid material into structures having well-defined morphologies. In this technique, the template simply serves as a scaffold around which other kinds of materials are generated. Like replica molding, the complex topology present on the surface of a template can be faithfully duplicated in a single step [23]. This approach has been extensively explored as a powerful tool for cost-effective and high-throughput fabrication of many types of porous materials. For example, by templating against supramolecular architectures self-assembled from surfactant molecules or organic block co-polymers, it is possible to generate mesoporous materials with pore sizes well-controlled in the range of 0.3-10 nm [24-28]. With the introduction of mesoscale objects as templates, the diameter of these pores has been further extended to cover a broad range spanning from tens of nm to hundreds of μm. In particular, templating against 3D crystalline lattices of monodispersed spherical colloids provides a generic route to macroporous materials that are characterized by well-controlled pore sizes and highly ordered, interconnected porous structures. Figure 1 shows a schematic procedure for this approach. After a cubic-close-packed lattice of spherical colloids has been dried, there is 26% (by volume) of void spaces within the lattice that can be infiltrated (partially or completely) with another material (in the form of a gaseous or liquid precursor, as well as a colloidal suspension) to form a solid matrix around the spherical colloids. Selective removal of the template (e.g., through dissolution or calcination) will lead to the formation of a 3D macroporous material containing a highly ordered architecture of uniform air balls interconnected to each other through small "windows".

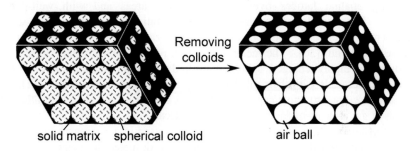

Figure 1. A schematic illustration of the procedure that generates 3D macroporous materials by templating against crystalline lattices of spherical colloids, followed by selective removal of the colloidal templates.

The 3D porous materials prepared using this method have also been referred to as *inverse opals* or *inverted opals* because they exhibit a periodic structure complementary to that of an opal [29]. This method offers a number of advantages for generating macroporous materials: i) It provides a tight control over the size, structure, and surface density of the pores. For example, the diameter of the spherical pores in the bulk could be conveniently controlled in the range of ~30 nm to ~10 μm by varying the diameter of spherical colloids. Surface pore densities up to ~4.5x10^{10} per cm^2 could be achieved routinely. ii) It offers a simple and reliable route to porous materials with three-dimensionally periodic structures. One of the attractive features of these porous materials is their interconnected network of uniform air balls that inherits all ordering and symmetry features from the original template. iii) In principle, such a template-directed method could be extended to any particular material of interest. The only requirement seems to be the availability of a gaseous or liquid precursor that can fill the voids in the crystalline lattice of spherical colloids without significantly swelling or dissolving the template (usually made of polymers or silica). When a liquid precursor is involved, the fidelity of this procedure is mainly determined by van der Waals interactions, wetting of the template's surface by the precursor, kinetic factors such as capillary filling of the small voids among colloidal particles, and volume shrinkage of a precursor during the solidification process.

2.1 Crystalline Lattices of Spherical Colloids

The key component of this synthetic method is the template -- a 3D crystalline lattice self-assembled from spherical colloids such as polymer latexes or silica spheres [19]. A large number of different approaches have already been demonstrated for crystallizing monodispersed spherical colloids into 3D lattices that are sometimes referred to as opals. Several of them are, in particular, useful in generating opaline lattices over relatively large areas and with well-defined structures. For example, sedimentation in the gravitational field represents a simple, convenient, and effective method for the crystallization of spherical colloids with diameters >500 nm [30]. The method based on repulsive electrostatic interactions can organize highly charged spherical colloids into body-center-cubic or face-center-cubic crystals with their thicknesses up to several hundred layers [31]. The approach based on attractive capillary forces (as caused by solvent evaporation) has been used in a layer-by-layer fashion to fabricate opals having well-controlled numbers of layers [32]. The procedure commonly used in our research involved the use of a microfluidic cell that was specifically designed to combine physical confinement and shear flow to organize spherical colloids into 3D crystalline lattices over areas as large as several square centimeters and with thicknesses up to several hundred layers [33, 34].

In our original demonstration [33], the microfluidic cell was constructed by sandwiching a photoresist frame of uniform thickness (<12 μm) between two glass slides and then held together using binder clips. The surface of this photoresist frame could be patterned with parallel arrays of shallow channels (their depth had to be controlled below the diameter of spherical colloids to be assembled) using a double-exposure procedure that involved overlapping of two photomasks. All components of each cell should be assembled in a cleanroom or under a relatively dust-free environment (e.g., a hood equipped with laminar flow and intake filtration). Most recently, the design of fluidic cell was extended to systems foregoing the requirement of cleanroom facilities. We found that the fluidic cells could be fabricated rapidly by sandwiching a square frame (cut from the commercial Mylar film) between two glass substrates [35]. Uniform Mylar films with thickness in the range of 20-100 μm could be obtained from Fralock (Canoga, CA) or Dupont.

Small channels between the Mylar film and glass substrates could be easily fabricated using a number of non-photolithographic methods: for example, by scraping both sides of the Mylar film with a piece of soft paper to create channels via abrasion; by coating both surfaces of the Mylar film with colloids smaller than those to be crystallized in the cell; and by patterning the surface of the glass substrate with an array of channels in thin film of gold by microcontact printing (μCP) and selective wet etching [36]. Although these methods were not as reproducible as the conventional technique based on photolithography, a yield of greater than ~90% could still be routinely accomplished.

The aqueous dispersion of spherical colloids was often injected into the cell through a glass tube glued to the small hole (~3 mm in diameter) drilled in the top glass substrate. Colloids with a diameter larger than the depth of channels in the gasket surfaces could be concentrated at the bottom of the microfluidic cell and crystallized into a long-range ordered 3D lattice. The rate of crystallization (including both nucleation and growth) near the edge of this fluidic cell could be increased by applying a slightly positive pressure of N_2 gas through the glass tube. Constant vibration provided by a Branson 1510 sonicator (Danbury, CT) was used to mechanically agitate the system and thus to ensure that each spherical particle will rest at the lattice site represented as a thermodynamic minimum and no random packing would occur in the sample. Using this procedure, we were able to routinely assemble monodispersed spherical colloids (both polymer latexes and silica beads) into opaline lattices over areas of several square centimeters that were characterized by a homogeneous structure, well-controlled thickness, and long-ranged ordering. Both scanning electron microscopy and optical diffraction studies indicate that the opaline lattice fabricated using this approach exhibited a cubic-close-packing, with its (111) crystallographic planes oriented parallel to the surface of the glass substrate. Figure 2A shows the oblique SEM image of a typical sample that was crystallized from 220-nm diameter polystyrene (PS) beads. The cross-sections of this crystal also confirmed that it had an "ABC" stacking along the direction perpendicular to the surface of the substrate. Most recently, we and other research groups also demonstrated the fabrication of opaline lattices with the (100) planes oriented parallel to the surfaces of substrates by templating against relief structures [37-40]. Figure 2B shows the SEM image of such an example that was fabricated by templating PS beads against a 2D array of pyramidal pits anisotropically etched in the surface of a Si(100) substrate. The key to the success of this approach was to precisely control the lateral dimensions of and separations between the square pyramidal pits to match the diameter of the spherical colloids.

2.2 Infiltration with Liquid Prepolymers

Liquid prepolymers that can be UV- or thermally cured have been widely used in replica molding with resolutions better than a few nanometers [41]. By judicially selecting a monomer, the volume shrinkage involved in a typical process could be controlled well below 1%. Such a feature also makes these materials particularly attractive in the fabrication of 3D macroporous structures by templating against crystalline lattices of spherical colloids. When several drops of this liquid is applied along the edges of this crystalline lattice, the voids within a 3D lattice can be spontaneously filled by a liquid prepolymer (as driven by capillary action). The prepolymer can be solidified by exposure to a broadband UV light source ($\lambda_{max} \approx 365$ nm), or by subjection to thermal treatment at a relatively low temperature (usually <100 °C). Afterwards, the polymer beads can be selectively dissolved by immersing the sample in an organic solvent for ~2 hours. For silica spheres, they can be selectively removed using hydrofluoric acid (49% in water), followed by rinsing with water. The spherical pores left behind in the polymer matrix should have essentially the same dimension as that of the particles contained in the template.

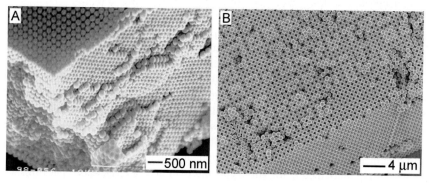

Figure 2. Scanning electron microscopic images of two opaline lattices that were crystallized from polystyrene beads, with their (111) and (100) crystallographic planes oriented parallel to the surfaces of the substrates, respectively.

We have demonstrated the fabrication of macroporous materials using liquid precursors to polymers such as polyurethanes (PU) and poly(acrylate methacrylate) copolymers (PAMC) [42-44]. Both polymer beads and silica colloids have been exploited as templates. After solidifying the prepolymer, the colloidal particles were removed by selective wet chemical etching, leaving behind a porous material characterized by a highly ordered, 3D architecture of interconnected spherical pores. Figure 3 shows SEM images of two typical examples that were fabricated by templating PAMC against (111)- and (100)-oriented crystalline lattices of PS beads (see Figure 2 for their SEM images). The long range ordering and interconnection between adjacent spherical pores could be obviously observed in these SEM images. Based on the cubic-close-packing model, the porous material generated by templating against a 3D crystalline lattice should have an internal porosity greater than 74%. As one of the major advantages over other methods, this approach is also able to generate a highly ordered 2D array of small "windows" on each crystallographic plane of this material. Using this approach, membranes with a uniform pore structure have been readily fabricated over areas ≥ 1 cm^2. Furthermore, the polymer membranes were sufficiently flexible to withstand mechanical deformations and thus could be obtained as freestanding films. Such porous, thin membranes could be subsequently incorporated into flow systems to measure and study the permeabilities of various kinds of gases and liquids [44].

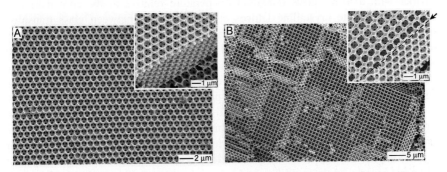

Figure 3. Scanning electron microscopic images of two 3D macroporous membranes that were fabricated by templating a UV-curable prepolymer to poly(acrylate methacrylate) copolymer against the opaline lattices of polystyrene beads shown in Figure 2.

A large number of different prepolymers have been successfully demonstrated for use with this templating procedure. For example, Zakhidov and Baugman et al. have fabricated inverse opals of glassy carbon by infiltrating silica opals with a phenolic resin, followed by thermal curing and pyrolyzing at elevated temperatures [45]. Colvin et al. have synthesized inverse opals made of PS, poly(methyl methacrylate) (PMMA), and PU by directly infiltrating and polymerizing the monomers of these macromolecules [46]. Mallouk et al. have fabricated mesoporous polymers by filling opals of silica spheres (35 nm in diameter) with a mixture of divinylbenzene (DVB) and ethyleneglycol dimethacrylate (EDMA), followed by polymerization [47]. In this system, DVB provided a rigid backbone and EDMA served as a flexible component to control the degree of shrinkage. Norris et al. have synthesized inverse opals made of poly(p-phenylene vinylene) (PPV, a π-conjugated polymer) by filling the voids in silica opals with p-xylylenebis-(tetrathiophenium chloride) monomer, followed by base-induced polymerization and thermal conversion [48]. They have also investigated the photoluminescence properties of these inverse opals, and observed that the emission from a PPV film was altered by the bandgap of the 3D periodic lattice. In a related demonstration, Caruso et al. have synthesized inverse opals of polyaniline by infiltrating the voids in PS opals with aniline, followed by $K_2S_2O_8$-initiated radical polymerization [49]. Braun et al. have recently fabricated waveguiding structures within inverse opals by filling the voids of an opal with a two-photon active dye, followed by writing (i.e., localized polymerization) with a focused laser beam [50].

It is worth mentioning that several new procedures have also been demonstrated to generate polymeric inverse opals more effectively and rapidly. In one approach, Kumachema et al. have synthesized polymeric inverse opals by constructing 3D crystalline lattices from core-shell colloidal particles (with the shell made of a polymer with a lower glass transition temperature relative to the core material). After annealing the crystalline lattices at a temperature between glass transition temperatures of the shell and core materials, inverse opals could be obtained by etching away the cores [51]. In a related study, Ford et al. started with core-shell colloids whose cores were rich in PS and shells rich in 2-hydroxyl methacrylate (HEMA) to fabricate the 3D opaline lattices [52]. Upon exposure to the vapor of a organic solvent good for PS but poor for polyHEMA (e.g., toluene or hexane), the glass transition temperature of the PS-rich cores could be reduced below ambient temperature to allow the migration of this phase to the void spaces within the opaline lattices, and finally led to the formation of inverse opals with PS as their major component. In a recent publication, Kim et al. also demonstrated that polymeric inverse opals could be obtained by sintering opaline lattices of PS beads that had been partially cross-linked with divinylbenzene [53]. Although the exact mechanism of this process remained to be understood, it was believed that thermal treatment might have caused the surface layers of the PS beads to become heavily cross-linked, and thus insoluble in the organic solvent. When 2D arrays of such PS beads were used as templates, they also obtained arrays of polymer rings, as well as honeycomb structures containing highly ordered lattices of openings.

2.3 Infiltration with Sol-Gel Precursors

The void spaces within a crystalline lattice of spherical colloids can also be filled with sol-gel precursors to generate inverse opals made of various ceramic materials. The precursors are often applied as dilute solutions in alcohols, with the reactivity toward atmospheric moisture adjusted by using an acidic or basic catalyst [54]. To facilitate the selective removal of colloidal templates, polymer beads rather than silica spheres are commonly used to construct the 3D opaline lattices. Similar to a liquid prepolymer, the sol-gel precursor solution spontaneously fills the void spaces among polymer beads through capillary action and a ceramic material with the desired composition is obtained upon hydrolysis and condensation of the sol-gel precursor. Even though the voids among polymer beads are completely filled with the precursor solution, the amount of ceramics produced in a sol-gel process might only be enough to form thin coatings on the surfaces of polymer beads after the solvent has been evaporated. In this regard, multiple insertions of the precursor solution are necessary in order to fill completely the voids in the 3D crystalline lattice. Otherwise, ceramic shells rather than an inverse opal will be likely obtained as the final product [55]. The polymer beads can be selectively removed by calcination in air at elevated temperatures. At the same time, such a thermal process may also convert the ceramic material from an amorphous into a polycrystalline phase. When non-cross-linked polymer beads are used as the templates, they can be conveniently dissolved in an organic solvent (e.g., toluene or hexane for PS) to preserve the amorphous structure of the ceramic material.

This templating procedure was first demonstrated by Velev et al. with the fabrication of silica inverse opals as an example [56, 57]. In follow-up studies, it has been further extended by a number of research groups to essentially all types of binary and ternary ceramic oxides. Figure 4 shows electron microscopic images of several typical examples of 3D macroporous materials that we have fabricated as thin films from SiO_2, SnO_2, and TiO_2 [44]. The alkoxide precursors to SnO_2 and TiO_2 hydrolyzed in the presence of atmospheric moisture, while the precursor to SiO_2 only polymerized upon immersion of the sample filled with the sol-gel precursor into a water bath. The SEM images clearly indicate that these porous materials exhibit a long-range ordered structure complementary to that of an opaline lattice: each of them consists of a periodic lattice of uniform air balls that are interconnected to each other. Circular windows into adjacent air balls can be observed where the spherical colloids are in physical contact in the cubic-close-packed lattice. Figure 4B shows a typical TEM image that was recorded from a macroporous material fabricated by templating the silica precursor against a crystalline lattice of 230-nm PS beads. This image is consistent with what has been observed using SEM, and it also clearly shows the interconnected framework between air balls. It is worth mentioning that the relatively high percentage of volume shrinkage associated with the solidification of sol-gel materials often leads to the formation of random cracks in their thin films. In comparison with polymers, most of the ceramic macroporous structures are too fragile to be released from their supports.

This templating procedure has also been modified by many research groups to incorporate new functionalities into the macroporous materials. For example, Stein et al. have successfully prepared inverse opals made of TiO_2, ZrO_2, Al_2O_3, and many other binary oxides by templating their appropriate precursors against crystalline lattices of PS beads [58, 59]. Vos et al. have fabricated inverse opals of anatase TiO_2 and measured their photonic bandgap properties [60]. Fujishima et al. have prepared macroporous films of anatase by templating against 2D arrays of silica spheres and evaluated the use of these porous films in photocatalytic applications [61]. Stucky et al. have demonstrated a simple and convenient method for fabricating hierarchically porous materials by combining spherical colloids and block co-polymers into a single system [62]. Wang et al. have also prepared ordered macroporous structures of anatase TiO_2 that were doped with cobalt [63]. Most recently, this procedure was further extended to cover functional ceramics such as $BaTiO_3$ and PZT [64]. Furthermore, metals (Bi, Te) and semiconductors

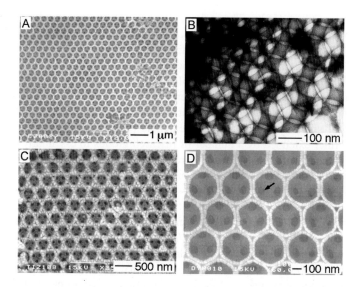

Figure 4. Scanning (A, C, D) and transmission (B) electron microscopic images of 3D macroporous membranes fabricated by templating sol-gel precursors against opaline lattices of polystyrene beads. The membranes were made of amorphous SiO_2 (A, B), SnO_2 (C), and TiO_2 (D), respectively.

(Se) with relatively low melting points have been directed injected (under pressure) into the voids of opaline lattices to generate inverse opals made of these materials [65, 66].

2.4 Infiltration with Solutions

In addition to sol-gel precursor solutions, Stein et al. demonstrated that the void spaces within 3D opaline lattices could be directly filled with salt solutions to generate macroporous materials made of metal oxides and metals. Taking nickel as an example, it was possible to precipitate nickel oxalate in the voids of a crystalline lattice of PS beads by infiltrating these voids with nickel acetate and oxalic acid [67]. Nickel oxide and nickel inverse opal could be obtained by heating the nickel oxalate sample in air or under hydrogen gas. This procedure was also further extended by Stein et al. to fabricate macroporous materials made of cobalt, iron, and nickel/cobalt alloys by templating against opaline lattices assembled from PMMA beads [68]. The use of PMMA instead of PS beads allowed them to achieve better infiltration (as a result of better wettability by polar solvents such as water and alcohols) and easier removal of the templates under milder conditions. We have also demonstrated the fabrication of inverse opals made of buckyballs (C_{60}) by filling the void spaces of silica opals with a C_{60} solution in hexane. Figure 5 shows the SEM image of a C_{60} inverse opal that was fabricated using this procedure. Because the solution was relatively dilute in concentration, at least five times of insertion was needed in order to generate the 3D porous structure shown in this SEM image. Otherwise, the porous structure would be too fragile to sustain its morphology as the colloidal template was removed. In particular, when a wet process is used to remove the template, the capillary force involved in the step of solvent evaporation might be sufficiently strong to collapse the highly porous structure.

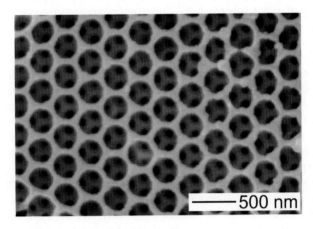

Figure 5. The scanning electron microscopic image of a 3D macroporous material that was fabricated by infiltrating a 3D opaline lattice of silica spheres with a hexane solution containing buckyballs. Note that the porous features underneath were also visible through the pores on the surface of the membrane.

In addition to solutions, colloidal suspensions have also been explored to fill the void spaces within opaline lattices of spherical colloids with the desired solid materials. For example, Norris et al. have synthesized inverse opals of CdSe by filling the voids of silica opals with anhydrous decane that contained CdSe quantum dots [69]. The surfaces of the silica spheres had been pre-derivatized with stearyl alcohol to facilitate the capillary flow of decane. In a related study, Velev et al. have successfully fabricated inverse opals of gold by filling the voids of PS opals with an aqueous dispersion containing gold nanoparticles of 15-25 nm in diameter [70]. They also found that these macroporous materials made of gold could serve as good substrates for surface-enhanced Raman scattering detection [71]. In principle, this procedure can be further extended to many other colloidal suspensions, as long as the nanoparticles contained in the suspensions are much smaller (about one tenth) than the spherical colloids so that they can readily penetrate the interstitial voids. Similar to the cases with sol-gel precursors and solutions, multiple insertions are required to completely fill the voids. As the necks connecting the voids become smaller and smaller, the transport of nanoparticles (contained in the suspension) might eventually be stopped.

2.5 Filling through Electrochemical Deposition

Electrochemical methods have also been explored to deposit metals and semiconductors within the voids of opaline lattices. For example, Colvin et al. have demonstrated the use of electroless deposition to generate macroporous materials of nickel, copper, silver, gold, and platinum [72]. In a typical procedure, they coated the surfaces of silica spheres with 3-mercaptopropyltrimethoxysilane (3-MPTMS) before they were assembled into opaline lattices. Gold nanoparticles (of a few nm in diameter) were then attached to the surfaces of these silica colloids through coupling with the thiol groups. Finally, the template was immersed in an electroless deposition bath to generate a metal within the voids of the 3D opaline lattice. One of the major advantages associated with this new approach is that there exists standard recipes for the electroless deposition of essentially any metal. In a second procedure, Braun and Wiltzius demonstrated the fabrication of inverse opals from CdS and CdSe by electrochemically depositing these materials directly into the voids of opaline lattices [73, 74]. In two related studies, electrochemical deposition was also exploited to generate inverse opals of gold [75, 76]. Other than inorganic materials, Caruso et al. have also fabricated inverse opals of conjugated polymers by electrochemically polymerizing organic monomers such as thiophene or pyrrole within the voids of 3D opaline lattices [77]. Different from the chemical methods described in previous two sections, the electrochemical technique allows one to tune the thickness of inverse opals by controlling the amount of polymer deposited.

2.6 Filling through Chemical Vapor Deposition

Recent studies indicate that the voids within an opaline lattice can also be readily filled with a solid material decomposed from a vapor-phase precursor. For example, CdSe inverse opals have been synthesized by Vlasov et al. using methods based on chemical vapor deposition [78, 79]. These early demonstrations opened the door to the fabrication of inverse opals from a variety of element or compound semiconductors. In the follow-up studies, Blanco et al. successfully synthesized silicon inverse opals and observed the signature of a complete bandgap in this system [80]. This result was later confirmed by Norris et al., and they also demonstrated that it was possible to further pattern these inverse opals into as well-defined microstructures by reactive ion etching through physical masks [81]. Electron microscopic studies revealed that the semiconductor was deposited around each spherical colloid in a layer-by-layer fashion until the voids were completely filled [82]. One of the potential problems associated with this method is that the material deposited on the exterior surface of a opaline sample may block the entrance of precursor molecules in the subsequent step and lead to the formation of an inhomogeneous film (in particular, when the film is relatively thick). In general, this is a method suitable for use with many vapor-phase precursors that have already been developed for the microelectronics [83].

3 Hierarchical Self-Assembly Approaches

As illustrated in the previous sections, templating against 3D opaline lattices of spherical colloids provides a generic route to the fabrication of macroporous materials with highly ordered and three-dimensionally interconnected pore structures. One of the major problems associated with the infiltration scheme shown in Figure 1 is reflected by the difficulty in accomplishing a dense, complete filling of the void spaces among spherical colloids with a solid material. In addition, the infiltration process, itself, might be too slow to be useful because the capillary flow rate of a liquid is inversely proportional to the cross-section of necks connecting adjacent voids [84]. In this regard, hierarchical self-assembly has been demonstrated to provide a potential solution to this problem. In this new approach, spherical colloids are crystallized into a 3D lattice within a liquid medium containing the desired material in the form of nanoparticles. Figure 6 shows a schematic illustration of this process, in which magnetic nanoparticles (see Figure 6A for a TEM image) are packed into a dense solid in the void spaces among spherical colloids (see Figure 6B for an SEM image) when the big, spherical colloids are organized into a closely packed lattice. The volume fraction can be easily varied by controlling the ratio between the nanoparticles and the spherical colloids. In principle, this procedure enables the formation of more uniform and better-controlled 3D macroporous materials over relatively larger areas than those fabricated by infiltration with liquid or gaseous precursors. Furthermore, the hierarchical self-assembly approach is potentially extendable to a broader range of materials, the only requirement seems to be that these materials can be supplied as suspensions of nanometer-sized particles (the sizes of these nanoparticles have to be controlled below one tenth of the diameter of spherical colloids used to construct the 3D opaline lattice). Thanks to many years of continuous efforts, all major classes of inorganic materials (including metals, semiconductors, and dielectrics) can now be readily processed as nanoparticles using various chemical or physical methods [85-87]. In this regard, the hierarchical self-assembly approach described in this section may provide one of the most promising and versatile routes to the fabrication of macroporous materials with diversified compositions.

The capability and feasibility of the hierarchical self-assembly approach have successfully been demonstrated by templating nanoparticles of silica [88], titania [89], and gold [90] against 3D opaline lattices made of PS beads or silica spheres. We have further extended the scope of this procedure to generate 3D macroporous

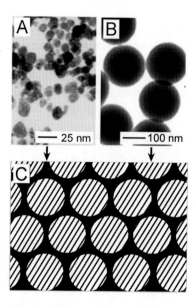

Figure 6. Schematic illustration of the procedure that generates macroporous materials through hierarchical self-assembly. In this route, spherical colloids (B) were assembled into a crystalline lattice in a medium containing an appropriate amount of nanoparticles (A). As the solvent was evaporated, the nanoparticles were fused into a dense, solid matrix around the spherical colloids. The transmission electron microscopic images of the nanoparticles (made of magnetite) and spherical colloids (made of polystyrene) were shown in (A) and (B) respectively.

materials made of magnetite nanoparticles (Figure 7) whose photonic bandgap properties could be addressed using an external magnetic field [91]. In this demonstration, monodispersed PS beads were assembled into a face-center-cubic lattice in an aqueous ferrofluid (EMG 308, Ferrofluidics, Nashua, NH) that contained superparamagnetic particles of magnetite with sizes <15 nm. As the solvent evaporated slowly, this crystalline structure was concentrated into a cubic-close-packed lattice, with the magnetite nanoparticles precipitating out as a dense, continuous solid matrix around the PS beads. The volume fraction of magnetite could be continuously varied in the range of 0-26% by controlling the ratio between the magnetite nanoparticles and the PS beads. Since the magnetite nanoparticles had already been placed around the PS beads, no transport of this material through the small capillaries was necessary and the entire process could be completed within arelatively short period of time. Macroporous materials made of magnetite nanoparticles were obtained after the PS colloids had been selectively removed through calcination at elevated temperatures or by etching in an organic solvent such as toluene or hexane.

Figure 7(A-B) shows the SEM images of an inverse opal that was fabricated by assembling 480-nm PS beads in the presence of an aqueous suspension of magnetite nanoparticles. The PS templates had been removed through calcination in air at ~500 °C. These images indicate the existence of long-range ordering in all three dimensions of space. It was also possible to form such highly ordered, 3D macroporous structures over areas as large as several square centimeters. Although

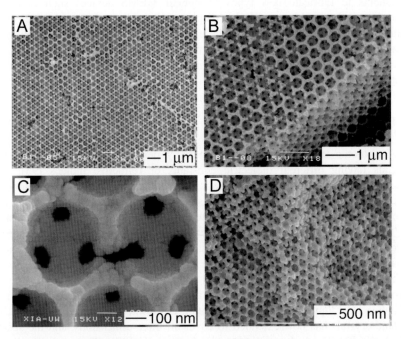

Figure 7. Scanning electron microscopic images of macroporous membranes fabricated through hierarchical self-assembly by templating magnetite nanoparticles against 3D crystalline lattices of polystyrene beads that were 480 nm (A-C) and 270 nm (D) in diameter.

cracks were often formed during the sample drying or calcination process as a result of volume shrinkage, single crystalline domains with lateral dimensions up to several hundred µm^2 could be routinely obtained using this simple method. Figure 7C shows a blow-up view of this sample, suggesting that the void spaces among PS beads had been completely filled to form a dense matrix of magnetite nanoparticles. It was also clear that the PS beads were in a closely packed 3D lattice, since each spherical pore was connected to adjacent ones underneath it through three small "windows". Compared to the magnetite nanoparticles (Fig. 6A) contained in the original ferrofluid, these nanoparticles have been sintered into a dense structure without changing their sizes. Figure 7D shows the cross-sectional SEM of another sample of magnetite inverse opal that was fabricated by templating against PS beads of 270 nm in diameter. Again, a highly ordered, 3D macroporous structure was also formed on this length scale. It is believed that the light weight and high surface area associated with these macroporous materials should make them attractive candidates for the fabrication of magnetic membranes to be incorporated into filtration and separation systems [92]. These field-responsive materials can be explored as active components to fabricate new types of optical MEMS devices such as tunable mirrors, optical switches, and color display units, where the mechanical forces (including torsion) can be applied to these devices by changing the distribution of a magnetic gradient. One of the major advantages to operate with a magnetic field (rather than an electric field) is that very little or no power (or energy) is required to maintain a static magnetic field.

4 Photonic Bandgap Properties

A photonic bandgap crystal is a spatially periodic lattice constructed from materials having different dielectric constants or refractive indices [93]. The concept of this new class of material was independently proposed by Yablonovich [94] and John [95] in 1987, and ever since a variety of applications has been demonstrated (or proposed) for these materials. For example, a photonic crystal is able to influence the propagation of electromagnetic waves in a similar way as a semiconductor does for electrons -- that is, its band gap can exclude the existence or propagation of photons of a chosen range of frequencies. As a result, the adoption of photonic crystals provides a powerful tool to confine, control, and manipulate photons in all three dimensions of space, and should enable the fabrication of more efficient optical, optoelectronic, and quantum electronic devices [96]. Research on photonic crystals has now extended to cover all three dimensions, and the spectral region extending from ultraviolet to radio frequencies [97]. The motivation in this area, however, has been the strong desire to obtain a complete bandgap around 1.55 μm -- the wavelength now used in optical fiber communications. Although many advances have been made, it is still a grand challenge to apply conventional microlithographic techniques to the fabrication of 3D periodic lattices when the feature size has to be precisely controlled on the same scale as the wavelength of near-infrared or visible light [98].

In principle, the band structure of a photonic crystal can be calculated in advance by solving the Maxwell's equations, with the dielectric constant being expressed as a spatially periodic function [99]. Since the Maxwell's equations enjoy scale invariance, it is also possible to shift a bandgap theoretically to any frequency range simply by scaling the feature sizes of a periodic structure. Several numerical recipes have been demonstrated to calculate the photonic band structure of a periodic lattice. The plane wave expansion method (PWEM) seems to be the most commonly used tool for 3D systems, albeit it cannot be applied to those systems whose dielectric constants contain large imaginary parts (as a result of absorption) [100]. The photonic band structure calculated for an cubic-close-packed lattice of spherical colloids indicate that this simple system does not possess a full bandgap; it only has a number of pseudo bandgaps (or stop bands). In this case, the existence of a complete gap is largely inhibited by the degeneracy at W- or U-point that is induced by the spherical symmetry of the building blocks. Computational studies also indicate that it is impossible to eliminate the degeneracy by increasing the refractive index contrast, or by changing the filling ratio of spheres from 74% (closely packed) to any other values.

Although face-center-cubic lattices assembled from spherical colloids do not have complete bandgaps, they offer a simple and easily prepared model system to experimentally probe the photonic band diagrams of certain types of 3D periodic lattices. They also serve as sacrificial templates to prepare inverse opals, which are expected to exhibit complete bandgaps when the contrast in dielectric constant is high enough [101]. Computational simulations also indicate that the minimum contrast in the refractive index at which a complete bandgap will be observed (between the 8th and 9th bands) in an inverse opal is around 2.8. It is possible to accomplish this using a number of inorganic materials such as group IV or II-VI semiconductors (Si, Ge, or CdSe), rutile, and iron oxides. Figure 8 shows the photonic band structure of an inverse opal made of ferric oxide whose refractive index is around 3.0 [102]. This diagram suggests the existence of a complete bandgap (the hatched region, between the 8th and 9th bands) that extends over the entire Brillouin zone. Figure 9 shows the transmission and reflectance spectra taken from an inverse opal of magnetite. For this sample, there only existed a stop band (although its bandwidth is broader than that of an opal). Part of the reason lies in the fact that the filling of void spaces among spherical colloids was not complete and the resultant materials might not be sufficiently dense to acquire a refractive index close to that of the bulk solid. Through the use of semiconductors with higher refractive indices (such as Si or Ge), the definitive existence of the signature of a complete photonic bandgap has been experimentally observed by a number of research groups [80-82]. More recently, Ozin et al. have also processed inverse opals of silicon into the form of nanofibers that might find use as optical components for microphotonics [103].

The photonic bandgap structure of an inverse opal can be tuned using a number of different methods. For example, computational studies by John et al. have demonstrated that an inverse opal could exhibit a fully tunable gap if the surface of the 3D porous structure would be coated with a few layers of an optically birefringent material such as a nematic liquid crystal [104]. In this case, the bandgap could be readily opened or closed by applying an external electric field to rotate the liquid crystal molecules with respect to the normal to the surface of the inverse opal. With a polymeric inverse opal as the example, Ozaki et al. recently demonstrated that the stop band could be shifted to different positions by infiltrating the inverse opal with a nematic liquid crystal, followed by the application of a direct electrical field [105]. In another demonstration, Stein et al. demonstrated the stop band position of a ceramic inverse opal could be tuned to various positions by filling the interconnected pores with solvents having different refractive indices [106]. Most recently, Tsutsui et al. demonstrated that polymer inverse opals could serve as sensors to measure mechanical strains applied by stretching the samples [107]. In this case, the position of stop band was found to be directly proportional to the stretching ratios. In several recent publications, inverse opals have also been exploited as substrates to fabricate diffractive optical sensors by monitoring the change in stop band position [108-110].

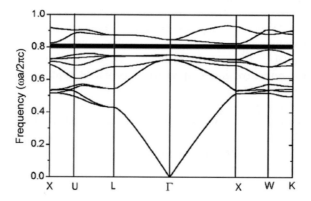

Figure 8. The photonic band structure calculated for an inverse opal having a refractive index contrast of 3. Note that there exists a complete bandgap between the 8th and 9th bands when the contrast in refractive index is higher than 2.8.

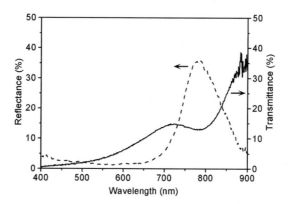

Figure 9. Transmission and reflectance spectra of an inverse opal that was fabricated through hierarchical self-assembly by templating magnetite nanoparticles against a 3D crystalline lattice of polystyrene beads that were 480 nm in diameter.

5 Mechanical and Liquid Permeation Properties

In addition to their photonic bandgap properties, the 3D macroporous materials described in this chapter may also find use in a range of other applications such as structural mechanics and filtration-based separation or purification. To these ends, we have quantitatively evaluated the mechanical strength of these macroporous membranes using an approach similar to the Whilhemy plate technique that is widely used to measure the surface tension of a liquid [44]. It was found that a macroporous membrane (made of UV-curable PMAC and with a pore diameter of ~1 μm) was able to hold a pressure of ~0.25 atm without being fractured. In other words, such a highly porous membrane (with only ~26% of solid content) was sufficiently strong to support an object weighing ~10^5 times heavier than the membrane itself. The permeability of air through such a 3D macroporous membrane was measured as high as 1.24×10^{-3} mLs^{-1}cm^{-1}atm^{-1} at room temperature. In comparison, the permeability of the smallest gas molecule (H$_2$) through low-density polyethylene films is much lower, on the order of 5×10^{-8} mLs^{-1}cm^{-1}atm^{-1}.

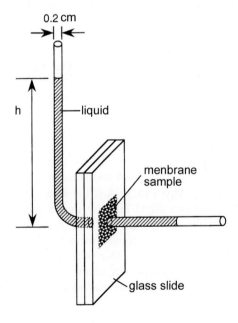

Figure 10. A schematic illustration of the flow cell used for measuring the permeability of a liquid through a 3D macroporous material. The macroporous membrane was held perpendicular to the flow direction of the solvent. A solvent flow could be initiated by creating a height difference for the liquid columns contained in the two arms.

We have also measured the liquid permeability of these 3D macroporous membranes using a homemade flow cell that is schematically shown in Figure 10 [44]. This cell was designed with two glass tubes as the side arms, and each one of them was attached to a glass slide over an ~1-mm hole. The sample (a piece of the freestanding porous membrane) was held between two Teflon™ spacers and centered over the two holes. This apparatus was then tightened with several binder clips around the edges of the glass slides. The Teflon™ tape around the membrane was able to assure a tight sealing for this flow cell. The difference in initial heights for the liquid in the two arms allows a spontaneous, continuous flow of the liquid through the membrane from one side to the other, and a set of heights (h) versus time (t) can be established. The permeation constant (τ) for each individual liquid could be calculated by exponentially fitting such a curve with the following equation: $h = h_0 \exp(-t/\tau)$. According to previous studies, the permeation rate of a liquid through a macroporous membrane is expected to depend on the exposed area of the membrane, the pressure difference across the membrane barrier, the thickness of the membrane, the pore size, and the total volume of membrane traveled by the solvent [1]. Holding these parameters constant and only varying the solvent, the permeation constant of a liquid through a 3D porous membrane should be solely dependent on the viscosity of the liquid, as well as the interactions between the liquid and the surface of the membrane.

Figure 11 shows a typical permeation curve that was measured using ethanol as the example. Table 1 summarizes the permeabilities calculated for four alcohols, together with the value measured for water. A comparison between the viscosities (η) of these liquids and their permeabilities suggests that the permeability is inversely proportional to the viscosity, indicating the predominance of capillary flow in the permeation of these liquids through the 3D macroporous membrane. Another parameter that is commonly used to describe the permeability of a liquid through a porous material is the so-called transport resistance or R_m [1], which is related to the permeability through: $R_m = A\tau/\eta$ (where A is the area of the membrane). The data exhibited in Table 1 shows a slight increase in the transport resistance when the liquid was changed in the order from methanol to butanol. Such an increase in transport resistance might reflect a stronger interaction between the liquid molecules and the membrane surface as the alkyl chain of the alcohol became bulkier and more hydrophobic. It is worth noting that comparative studies should also be possible by controlling the pore size, surface porosity, and wettability of the membrane surface.

6 Concluding Remarks

In summary, templating against opaline lattices assembled from spherical colloids provides a convenient and versatile method for generating 3D macroporous materials that contain long-range ordered structures. The templating procedure has been successfully applied to a variety of materials that include organic polymers, ceramics, semiconductors, and metals. Table 2 gives a partial list of such materials that have been fabricated and characterized. Fabrication based on this approach is remarkable for its simplicity, and for its fidelity in transferring the structure from the colloidal template to the replica. By changing the diameter of spherical colloids, the pore size of these 3D macroporous materials could be easily tuned to cover a broad range that varies from tens of nanometers to hundreds of micrometers. At present, the smallest spherical colloids that have been successfully adopted for use with this method were ~35 nm in diameter; the lower limit to the size of particles that can be incorporated into this technique is yet to be completely established [47].

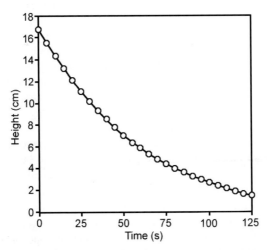

Figure 11. The permeation curve recorded for ethanol through a macroporous membrane of PAMC that was fabricated by templating against an opaline lattice of 1.05-μm PS beads. Here the height of liquid column in the left arm was measured as a function of time.

Table 1. Permeation properties measured for various solvents

Liquid	Viscosity η (cP)	Permeability τ (s)	Transport resistance R_m (cm^2/cP)
water	0.890	283	2.50
methanol	0.554	180	2.60
ethanol	1.074	440	3.22
2-popanol	2.083	787	2.97
1-butanol	2.544	1220	3.77

Although this template-based procedure may lack the characteristics required for high volume production, it does provide a simple and effective route to 3D porous materials having tightly controlled pore sizes and well-defined periodic structures. These macroporous materials can serve as a unique model system to investigate many interesting subjects such as adsorption, condensation, transport (diffusion and flow), and separation of molecules in porous structures, as well as thermal, optical, and mechanical properties associated with highly porous materials. With appropriate modification to the surfaces, these three-dimensionally open structures (with porosity >74%) should also find use as substrates in fabricating prototype sensors with enhanced sensitivities, or as templates to generate complex 3D structures of various functional materials that cannot be produced using conventional lithographic techniques. Because of their long-ranged ordering, these 3D periodic structures should particularly hold the promise as photonic crystals that may exhibit complete bandgaps in the spectral regime ranging from ultraviolet, through visible, to near-infrared. Other types of applications that have been proposed for these 3D macroporous materials include their potential usage as collectors of solar energies, as model systems for studying quantum confinement, as electrodes for fuel cells and many other types of electrochemical processes, as supports for catalysts, as scaffolds in tissue engineering, and as low-dielectric materials in fabricating capacitors.

Plain polymer latexes and silica spheres will, of course, continue to serve as the dominant building blocks in generating opaline lattices to be used as templates for the fabrication of 3D macroporous materials. New types of colloids are also being synthesized and evaluated as alternatives. Some of them might render the macroporous materials with attractive functionality not offered by the plain polymer latexes and silica beads. For example, liquid droplets form by microemulsion have been exploited by Pine et al. as templates to generate 3D macroporous materials [111]. Organic block co-polymers have also been demonstrated by several groups as the building blocks to form macroporous materials [112-114]. The major advantage associated with these two methods is that there is no need to remove the templates since they are often made of liquids or air. Most recently, we also demonstrated that Au@SiO_2 core-shell spherical colloids with well-controlled shell thicknesses could be synthesized by directly coating gold sols with a sol-gel precursor to amorphous silica. These core-shell colloids could also be assembled into 3D crystalline lattices (Figure 12A) over areas as large as several square centimeters [115]. Subsequent infiltration of the voids with a liquid prepolymer, followed by curing of the polymer matrix and selective removal of the silica shells, eventually led to the formation of a polymer macroporous membrane, with each of its spherical pores containing one gold nanoparticle. Figure 12B show the SEM image of such a macroporous membrane, with gold nanoparticles indicated by arrows. These nanoparticles can serve as catalytic sites for the conversion (e.g., decomposition) of chemical reagents as they are passing through the interconnected

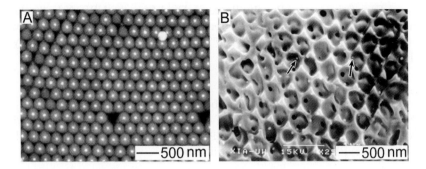

Figure 12. (A) The scanning electron microscopic image of a 3D crystalline lattice assembled from Au@SiO$_2$ core-shell colloids. The gold cores were ~50 nm in diameter and the silica shells were ~120 nm in thickness. (B) The scanning electron microscopic image of a macroporous membrane that was fabricated by templating PMAC against an opaline lattice of Au@SiO$_2$ core-shells colloids. The gold cores (as indicated by arrows) were left behind in the spherical pores, after the silica shells had been selectively removed by etching with HF solution.

pores. The existence of a polymeric matrix around the gold nanoparticles should provide a physical barrier to hinder the sintering of these nanoparticles during chemical reactions.

There are a number of issues that remain to be addressed before these macroporous materials will become widely used in various applications. As one of the major drawbacks, the technique based on infiltration is relatively slow. Although it has been shown that spherical colloids could be assembled into 3D crystalline lattices over areas of ~1 cm^2 in 1-2 days, subsequent infiltration of the small voids may take days to weeks to complete. Another drawback is related to the formation of cracks in the colloidal templates. Due to volume shrinkage involved in the sample drying process, the resultant opaline lattices are often decorated with cracks that divide the crystals into many small domains. These cracks may cause diffusive scattering of light and thus greatly limit the use of these inverse opals as photonic bandgap crystals or in many other types of diffractive optical devices. The last but not least, it is still hard to control the type and density of defects that might be trapped in the crystalline lattice of an inverse opal. These defects may play the most important role in determining the performance of an inverse opal in photonics-related applications.

Acknowledgments

This work has been supported in part by AFOSR, ONR, NSF, the Camille and Henry Dreyfus Foundation, the Alfred P. Sloan Foundation, the David and Lucile

Packard Foundation, and a Royalty Research Fund from the UW. Y.L. thanks the Center for Nanotechnology at the UW for a student fellowship award. B.G. thanks the Center for Nanotechnology at the UW for an IGERT fellowship supported by the NSF.

Table 2. Materials that have been processed as inverse opals
Abbreviations: PU: poly(urethane); PMAC : poly(acrylate methacrylate) copolymers; PS: poly(styrene); DVB: dininylbezne; EDMA: ethyleneglycol dimethacrylate; PVP: poly(p-phenylene vinylene).

Materials		Methods	Comments	Refs.
Polymers	PU, PAMC	Infiltration with liquid prepolymers, UV-curing	Extraction with solvents, highly ordered, large free-standing membranes	[42-44]
	Glassy carbon	Infiltration with phenolic resin, thermal curing and pyrolysis	HF etch, highly ordered	[45]
	PS, PMMA	Infiltration with liquid monomers, and *in situ* polymerization	HF etch, highly ordered, UV-Vis spectra measured	[46]
	Copolymers of DVB and EDMA	Infiltration with liquid monomers, followed with polymerization	HF etch, partially ordered the smallest beads: 35 nm	[47]
	PVP	Infiltration with monomer solutions, polymerization, and thermal conversion	HF etch, highly ordered, UV-Vis and fluorescence spectra measured	[48]
	Polyaniline	Infiltration with liquid monomer, radical polymerization	Extraction with solvents, ordered, UV-vis spectra	[49]
	PS	Solvent vapor treatment of poly(HEMA-*co*-styrene) spheres	Partially ordered	[52]
	Poly(divinyl benzene)	Polymer sintering of poly(styrene-*co*-divinylbenzene)	Extraction with solvents, ordered	[53]
	Polythiophe, polypyrrole	Electropolymerization	Extraction with solvents, ordered, UV-vis spectra	[77]
Ceramics	SiO_2, SnO_2, TiO_2, ZrO_2, Al_2O_3	Sol-gel, hydrolysis of metal alkoxides	Calcination, HF etch, or extraction with solvents	[44, 56-61]

Table 2. (*Continued*)

	SiO_2, Nb_2O_5, TiO_2	Sol-gel with amphiphilic triblock copolymers and beads as templates	Calcination, ordered, and with hierarchical pores	[62]
	$BaTiO_3$, PZT	Sol-gel		[64]
	SiO_2, TiO_2, Fe_3O_4	Hierarchical self-assembly with nanoparticles and spherical beads	Calcination, ordered, UV-vis spectra measured	[88-91]
Metals	Bi, Te	Infiltration with melts	highly ordered, spectra	[65, 66]
	Ni, Co, Fe, and Ni-Co alloy	Infiltration with solutions, thermal conversion of the metal salts	Calcination, ordered	[67, 68]
	Au	Infiltration with suspensions of gold colloids	Extraction with solvents, partially ordered	[70, 71]
	Ni, Co, Ag, Au, Pt	Electroless deposition or electrodeposition	HF etch, highly ordered, spectra measured	[72, 75, 76]
Semi-conductors	C_{60}	Infiltration with solutions in hexane	HF etch, highly ordered	
	Se	Infiltration with melt		[66]
	CdSe	Infiltration with suspension of CdSe quantum dots	HF etch, partially ordered	[69]
	CdSe	Chemical vapor deposition	HF etch, highly ordered, spectra measured	[78, 79]
	CdS, CdSe	Electrodeposition	HF etch, highly ordered	[73, 74]
	Si, Ge	Chemical vapor deposition	HF etch, complete band gaps observed	[80-82]

References

1. Bhave R. R., Inorganic Membranes: Synthesis, Characteristics and Applications, Van Nostrand Reinhold: New York, N.Y., 1991.
2. Fain D. E., Membrane gas separation principles *MRS Bulletin* **19** (1994) pp. 40-43.

3. Lin V. S.-Y., Motesharei K., Dancil K. P. S., Sailor M. J. and Ghadiri M. R., A porous silicon-based optical interferometric biosensor *Science* **278** (1997) pp. 840-843.
4. Keizer K. and Verweij H., Process in inorganic membranes *CHEMTECH* (1996) pp. 37-41.
5. Hedrick J. L., Miller R. D., Hawker C. J., Carter K. R., Volksen W., Yoon D. Y. and Trollsås M., Templating nanoporosity in organosilicates using dendritic macromolecules *Adv. Mater.* 10 (1998) pp. 1049-1053.
6. Hubbell J. A. and Langer R., Tissue engineering *Chem. Eng. News* **13** (1995) pp. 42-45.
7. Schaefer D. W., Engineered porous materials *MRS Bulletin* **19** (1994) pp. 14-17.
8. Furneaux, R. C.; Rigby, W. R.; Davidson, A. P. The formation of controlled-porosity membranes from anodically oxidized aluminum *Nature* **337** (1989) pp. 147-149.
9. Kendall, D. L. Vertical etching of silicon ay very high aspect ratios *Annu. Rev. Mater. Sci.* **9** (1979) pp. 373-403.
10. Tonucci R. J., Justus B. L., Campillo A. J. and Ford C. E., Nanochannel array glass *Science* **258** (1992) pp. 783-785.
11. Yoshida M., Asano M., Suwa T., Reber N., Spohr R. and Katakai R., Creation of thermo-responsive ion-track membranes *Adv. Mater.* **1997**, 9, 757-758.
12. Rebhan U., Endert H. and Zaal G., Micromanufacturing benefits from excimer-laser development *Laser Focus World* **11** (1994) pp. 91-96.
13. Madou M., Fundamentals of Microfabrication, CRC Press: Boca Raton, FL, 1997.
14. Even Jr. W. R. and Gregory D. P., Emulsion-derived foams: preparation, properties, and application *MRS Bulletin* **19** (1994) pp. 29-33.
15. Leenaars A. F. M., Keizer K. and Burggraaf A. J., Porous alumina membranes *CHEMTECH* **2** (1986) pp. 560-564.
16. Walsh D., Hopwood J. D. and Mann S., Crystal tectonics: construction of reticulated calcium phosphate frameworks in bicontinuous reverse micromulsions *Science* **264** (1994) pp. 1576-1578.
17. Imhf A. and Pine D. J., Ordered macroporous materials by emulsion templating *Nature* **389** (1997) pp. 948-951.
18. Davis S. A., Burkett S. L., Mendelson N. H. and Mann S., Baterial templating of ordered macrostructures in silica and silica-surfactant mesophases *Nature* **385** (1997) pp. 420-423.
19. Xia Y., Gates B., Yin Y. and Lu Y., Monodispersed colloidal spheres: old materials with new applications *Adv. Mater.* **12** (2000) pp. 693-713.
20. Orlin D. V. and Abraham M. L., Colloidal crystals as templates for porous materials, *Current Opin. Colloid Interf. Sci.* **5** (2000) pp. 56-63.

21. Kulinowski K. M., Jiang P., Vaswani H. and Colvin V. L., Porous metals from colloidal templates *Adv. Mater.* **12** (2000) pp. 833-838.
22. Schroden R. C., Al-Daous M., Blanford C. F. and Stein A., Optical Properties of Inverse Opal Photonic Crystals *Chem. Mate.* **14** (2002) pp. 305-3315.
23. Xia Y. Kim E., Zhao X.-M., Rogers J., Prentiss M. and Whitesides G. M. Complex optical surfaces formed by replica molding against elastomeric master *Science* **273** (1996) pp. 347-349.
24. Kresge C. T., Leonowicz M. E., Roth W. J., Vartuli J. C. and Beck J. S. Ordered mesoporous molecular sieves synthesized by a liquid-crystal template mechanism *Nature* **359** (1992) pp. 710-712.
25. Davis M. E. Ordered porous materials for emerging applications *Nature* **417** (2002) pp. 813-821.
26. Brinker C. J., Lu Y., Sellinger A. and Fan H. Evaporation-induced self-assembly: nanostructures made easy *Adv. Mater.* **11** (1999) pp. 579-585.
27. Behrens P. and Stucky G. *Angew Chem. Int. Ed. Engl.* **32** (1993) pp. 696-699.
28. Braun P., Osenar P. and Stupp S. I. Semiconducting superlattices templated by molecular assemblies *Nature* **380** (1996) pp. 325-328.
29. Sanders J. V. Colour of precious opal *Nature* **204** (1964) pp. 1151-1152.
30. Mayoral R., Requena J., Moya J. S., Lopez C., Cintas A., Miguez H., Meseguer F., Vazquez L., Holgado M. and Blanco A., 3D long-range ordering in an SiO2 submicrometer-sphere sintered superstructure. *Adv. Mater.* **9** (1997) pp. 257-260.
31. Flaugh P. L., O'Donnell S. E. and Asher S. A., Development of a new optical wavelength rejection filter: demonstration of its utility in Raman spectroscopy. *Appl. Spec.* **38** (1984) pp. 847-850.
32. Jiang P., Bertone J. F., Hwang K. S. and Colvin V. L. Single-crystal colloidal multilayers of controlled thickness *Chem. Mater.* **11** (1999) pp. 2132-2140.
33. Park S. H., Gates B. and Xia Y., A three-dimensional photonic crystal operating in the visible *Adv. Mater.* **11** (1999) pp. 462-466.
34. Park S. H. and Xia Y., Crystallization of meso-scale particles over large areas and its application in fabricating tunable optical filters *Langmuir* **15** (1999) pp. 266-273.
35. Lu Y., Yin Y., Gates B. and Xia Y., Growth of large crystals of monodispersed spherical colloids in fluidic cells fabricated using non-photolithographic methods *Langmuir* **17** (2001) pp. 6344-6350.
36. Xia Y. and Whitesides G. M. Soft lithography *Angew Chem. Int. Ed. Engl.* **37** (1998) pp. 550-575.
37. Yin Y. and Xia Y., Growth of large colloidal crystals with their (100) planes oriented parallel to the surfaces of supporting substrates *Adv. Mater.* **14** (2002) pp. 605-608.
38. Yin Y., Li Z.-Y. and Xia Y., Template-directed growth of (100)-oriented colloidal crystals *Langmuir* **19** (2003) pp. 622-631.

39. van Blaaderen A., Rue R. and Wiltzius P., Template-directed colloidal crystallization *Nature* **385** (1997) pp. 321-324.
40. Lin K.-H., Crocker J. C., Prasad V., Schofield A., Weitz D. A., Lubensky T. C. and Yodh A. G., Entropically driven colloidal crystallization on patterned surfaces *Phys. Rev. Lett.* **85** (2000) pp. 1770-1773.
41. Xia Y., McClelland J. J., Gupta R., Qin D., Zhao X.-M., Sohn L. S., Celotta R. J. and Whitesides G. M. Replica molding using polymeric materials: a practical step toward nanomanufacturing *Adv. Mater.* **9** (1997) pp. 147-150.
42. Park S. H. and Xia Y., Fabrication of three-dimensional macroporous membranes with assemblies of microspheres as templates *Chem. Mater.* **10** (1998) pp. 1745-1747.
43. Park S. H. and Xia Y., Macroporous membranes with highly ordered and three-dimensionally interconnected spherical pores *Adv. Mater.* **10** (1998) pp. 1045-1048.
44. Gates B., Yin Y. and Xia Y., Fabrication and characterization of porous membranes with highly ordered three-dimensional periodic structures *Chem. Mater.* **11** (1999) pp. 2827-2836.
45. Zakhidov A. A., Baughman R. H., Iqbal Z., Cui C., Khayyrullin I., Dantas O., Marti J. and Ralchenko V. G., Carbon structures with three-dimensional periodicity at optical wavelengths. *Science* **282** (1998) pp. 897-901.
46. Jiang P., Hwang K. S., Mittleman D. M., Bertone J. F. and Colvin V. L., Template-directed preparation of macroporous polymers with oriented and crystalline arrays of voids *J. Am. Chem. Soc.* **121** (1999) pp. 11630-11637.
47. Johnson S. A., Ollivier P. J. and Mallouk T. E., Ordered mesoporous polymers of tunable pore size from colloidal silica templates *Science* **283** (1999) pp. 963-965.
48. Deutsch M., Vlasov Y. A. and Norris D. J. Conjugated-polymer photonic crystals *Adv. Mater.* **12** (2000) pp. 1176-1180.
49. Wang W. and Caruso F., Fabrication of polyaniline inverse opals via templating ordered colloidal assemblies *Adv. Mater.* **13** (2001) pp. 350-353.
50. Lee W., Pruzinsky S. A. and Braun P. V., Multi-photon polymerization of waveguide structures within three-dimensional photonic crystals. *Adv. Mater.* **14** (2002) pp. 271-274.
51. Kumacheva E., Kalinina O. and Lilge L., Three-dimensional arrays in polymer nanocomposites *Adv. Mater.* **11** (1999) pp. 231-234.
52. Chen Y., Ford W. T., Materer N. F. and Teeters D., Facile conversion of colloidal crystals to ordered porous polymer nets *J. Am. Chem. Soc.* **122** (2000) pp. 10472-10473.
53. Yi D. K. and Kim D.-Y., Novel approach to the fabrication of macroporous polymers and their use as a template for crystalline titania nanorings *NanoLett.* **3** (2003) ASAP.

54. Hench L. L. and West J. K. The sol-gel process. *Chem. Rev.* **90** (1990) pp. 33-72.
55. Zhong Z., Yin Y., Gates B. and Xia Y., Preparation of mesoscale hollow spheres of TiO$_2$ and SnO$_2$ by templating against crystalline arrays of polystyrene particles *Adv. Mater.* **12** (2000) pp. 206-209.
56. Velev O. D., Jede T. A., Lobo R. F. and Lenhoff A. M., Porous silica via colloidal crystallization *Nature* **389** (1997) pp. 447-448.
57. Velev O. D., Jede T. A., Lobo R. F. and Lenhiff A. M., Microstructured porous silica obtained via colloidal crystal templates *Chem. Mater.* **10** (1998) pp. 3597-3602.
58. Holland B. T., Blanford C. F. and Stein A., Synthesis of macroporous minerals with highly ordered three-dimensional arrays of spheroidal voids *Science* **281** (1998) pp. 538-540.
59. Hollord B. T., Blanford C. F., Do T. and Stein A., Synthesis of highly ordered, three-dimensional, macroporous structures of amorphous or crystalline inorganic oxides, phosphates, and hybrid composites *Chem. Mater.* **11** (1999) pp. 795-805.
60. Wijinhoven J. E. G. J. and Vos W. L., Preparation of photonic crystals made of air spheres in titania *Science* **281** (1998) pp. 802-804.
61. Matsushita S. I., Miwa T., Tryk D. A. and Fujishima A., New mesostructured porous TiO$_2$ surface prepared using a two-dimensional array-based template of silica particles *Langmuir* **14** (1998) pp. 6441-6447.
62. Yang P., Deng T., Zhao D., Feng P., Pine D., Chmelka B. F., Whitesides G. M. and Sturcky G. D., Hierarchically ordered oxides *Science* **282** (1998) pp. 2244-2246.
63. Yin J. S. and Wang Z. L., Template-assisted self-assembly and cobalt doping of ordered mesoporous titania nanostructures *Adv. Mater.* **11** (1999) pp. 469-472.
64. Kulinowski K. M., Jiang P., Vaswani H., Colvin V. L., Porous metals from colloidal templates. *Adv. Mater.* **12** (2000) pp. 833-838.
65. Bogomolov V. N., Sorokin L. M., Kurdyukov D. A. and Pavlova T. M., A comparative TEM study of 3D lattice of tellurium nanoclusters fabricated by different techniques in an opal host *Phys. Solid State* **9** (1997) pp. 1869-1874.
66. Braun P. V., Zehner, R. W., White C. A., Weldon M. K., Kloc C., Patel S. S., Wiltzius P., Epitaxial growth of high dielectric contrast three-dimensional photonic crystals. *Adv. Mater.* **13** (2001) pp. 721-724.
67. Yan H., Blanford C. F., Hollord B. T., Parent M., Smyrl W. H. and Stein A., A chemical synthesis of periodic macroporous NiO and metallic Ni *Adv. Mater.* **11** (1999) pp. 1003-1006.
68. Yan H., Blanford C. F., Lytle J. C., Carter C. B., Smyrl W. H. and Stein A., Influence of processing conditions on structures of 3D ordered macroporous metals prepared by colloidal crystal templating *Chem. Mater.* **13** (2001) pp. 4314-4321.

69. Vlasov Y. A., Yao N. and Norris D. J., Synthesis of photonic crystals for optical wavelengths from semiconductor quantum dots *Adv. Mater.* **11** (1999) pp. 165-169.
70. Velev O. V., Tessier P. M., Lenhoff A. M. and Kaler E. W., A class of porous metallic nanostructures *Nature* **401** (1999) pp. 548-548.
71. Tessier P. M., Velev O. D., Kalambur A. T., Rabolt J. F., Lenhoff A. M. and Kaler E. W., Assembly of gold nanostructured films templated by colloidal crystals and use in surface-enhanced raman spectroscopy *J. Am. Chem. Soc.* **122** (2000) pp. 9554-9555.
72. Jiang P., Cizeron J., Bertone J. F. and Colvin V. L., Preparation of macroporous metal films from colloidal crystals *J. Am. Chem. Soc.* **121** (1999) pp. 7957-7958.
73. Braun P. V. and Wiltzius P. Microporous materials: Electrochemically grown photonic crystals. *Nature* **402** (1999) pp. 603-604.
74. Braun P. V. and Wiltzius P., Electrochemical fabrication of 3D microperiodic porous maters *Adv. Mater.* **13** (2001) pp. 482-485.
75. Netti M. C., Coyle S., Baumberg J. J., Ghanem M. A., Birkin P. R., Bartlett P. N. and Whittaker D. M., Confined surface plasmons in gold photonic nanocavities *Adv. Mater.* **13** (2001) pp. 1368-1370.
76. Wijinhoven J. E. G. J., Zevenhuizen S. J. M., Hendricks M. A., Vanmaekelbergh D., Kelly J. J. and Vos W. L., Electrochemical assembly of ordered macropores in gold *Adv. Mater.* **12** (2000) pp. 888-890.
77. Cassagneau T. and Caruso F., Semiconducting polymer inverse opals prepared by electropolymerization *Adv. Mater.* **14** (2002) pp. 34-38.
78. Astratov V. N., Vlasov Y. A., Karimov O. Z., Kaplyanskii A. A., Musikhin Y. G., Bert N. A., Bogomolov V. N. and Prokofiev A. V., Photonic band gapsin 3D ordered fcc silica matrices *Phys. Lett. A* **222** (1996) pp. 349-353.
79. Romanov S. G., Fokin A. V., Tretijakov V. V., Butko V. Y., Alperovich V. I., Johnson N. P. and Sotomayor Torres C. M., Optical properties of ordered three-dimensional arrays of structurally confined semiconductors *J. Crystal Growth* **159** (1996) pp. 857-860. l
80. Blanco A., Chomski E., Grabtchak S., Ibisate M. John S., Leonard S. W., Lopez C., Meseguer F., Miguez H., Mondia J. P., Ozin G. A., Toader O. and van Driel H. M., Large-scale synthesis of a silicon photonic crystal with a complete three-dimensional bandgap near 1.5 micrometers *Nature* **405** (2000) pp. 437-440.
81. Vlasov Y. A., Bo X.-Z., Sturm J. C. and Norris D. J., On-chip natural assembly of silicon photonic bandgap crystals *Nature* **414** (2001) pp. 289-293.
82. Miguez H., Chomski E., Garcia-Santamaria F., Ibisate M., John S., Lopez C., Meseguer F., Mondia J. P., Ozin G. A., Toader O. and Van Driel H. M., Photonic bandgap engineering in germanium inverse opals by chemical vapor deposition. *Adv. Mater.* **13** (2001) pp. 1634-1637.

83. Meseguer F., Blanco A., Miguez H., Garcia-Santamaria F., Ibisate M. and Lopez C., Synthesis of inverse opals. *Colloids Surf. A:* **202** (2002) pp. 281-290.
84. Kim E., Xia Y. and Whitesides G. M. Polymer microstructures formed by moulding in capillaries *Nature* **376** (1995) 581-584.
85. Alivisatos A. P. Semiconductor nanocrystals *MRS Bull.* **August** (1995) pp. 23-32.
86. Henglein A. Small-particle research: physicochemical properties of extremely small colloidal metal and semiconductor particles *Chem. Rev.* **89** (1989) pp. 1861-1873.
87. Matijevic E. Uniform inorganic colloid dispersions. achievements and challenges *Langmuir* **10** (1994) pp. 8-16.
88. Subramanian G., Manoharan V. N., Thorne J. D. and Pine D. J., Ordered macroporous Materials by colloidal assembly: A possible route to photonic bandgap materials *Adv. Mater.* **11** (1999) pp. 1261-1265.
89. Subramanian G., Constant K., Biswas R., Sigalas M. M. and Ho K.-M., Optical photonic crystals fabricated from colloidal systems *Appl. Phys. Lett.* **74** (1999) pp. 3933-3935.
90. Tessier P. M., Velev O. D. Kalambur A. T., Lenhoff A. M., Rabolt J. F. and Kaler E. W. Structured metallic films for optical and spectroscopic applications via colloidal crystal templating *Adv. Mater.* **13** (2001) pp. 396-400.
91. Gates B. And Xia Y., Photonic crystals that can be addressed with an extendal magnetic field *Adv. Mater.* **13** (2001) pp. 1605-1608.
92. Watson J. H. P. and Beharrell P. A., Magnetic separation using a switchable system of permanent magnets. *J. Appl. Phys.* **81** (1997) pp. 4260-4262.
93. Joannopoulos J. D., Meade R. D. and Winn J. N., *Photonic Crystals: Molding the Flow of Light*. Princeton University Press, Princeton, NJ (1995).
94. Yablonovitch E., Inhibited spontaneous emission in solid-state physics and electronics. *Phys. Rev. Lett.* **58** (1987) pp. 2059-2062.
95. John S., Strong localization of photons in certain disordered dielectric superlattices. *Phy. Rev. Lett.* **58** (1987) pp. 2486-2489.
96. Scherer A., Doll T., Yablonovitch E., Everitt H. O. and Higgins J. A., Electromagnetic crystal structures, design, synthesis, and applications. *J. Lightwave Technol.* **17** (1999) pp. 1928-1932.
97. Special issue: photonic crystals *Adv. Mater.* **13** (2001) pp. 369-450.
98. Lin S. Y., Fleming J. G., Hetherington D. L., Smith B. K., Biswas R., Ho K. M., Sigalas M. M., Zubrzycki W., Kurtz S. R. and Bur J., A three-dimensional photonic crystal operating at infrared wavelengths. *Nature* **394** (1998) pp. 251-253.
99. Haus J. W., A brief review of theoretical results for photonic band structures. *J. Mod. Opt.* **41** (1994) pp. 195-207.

100. Moroz A. and Sommers C., Photonic band gaps of three-dimensional face-centered cubic lattices. *J. Phys: Condens. Matter* **11** (1999) pp. 997-1008.
101. Ho K. M., Chan C. T. and Soukoulis C. M., Existence of a photonic gap in periodic dielectric structures. *Phys. Rev. Lett.* **65** (1990) pp. 3152-3155.
102. Xia Y., Gates B. and Li Z.-Y., Self-assembly approaches to three-dimensional photonic crystals *Adv. Mater.* **13** (2001) pp. 409-413.
103. Miguez H., Yang S. M., Tetreault N. and Ozin G. A. Oriented free-standing three-dimensional silicon inversed colloidal photonic crystal microfibers *Adv. Mater.* **14** (2002) pp. 1805-1808.
104. Busch K. and John S. Liquid-crystal photonic-band-gap materials: The tunable electromagnetic vacuum. *Phys. Rev. Lett.* **83** (1999) pp. 967-970.
105. Ozaki M., Shimoda Y., Kasano M. and Yoshino K., Electric field tuning of the stop band in a liquid-crystal-infiltrated polymer inverse opal *Adv. Mater.* **14** (2002) pp. 514-518.
106. Blanford C. F., Schroden R. C., Al-Daous M. and Stein A., Tuning olvent-dependent color changes of three-dimensionally ordered macroporous materials through compositional and geometric modifications *Adv. Mater.* **13** (2001) pp. 26-29.
107. Sumioka K., Kayashima H. and Tsutsui T., Tuning the optical properties of inverse opal photonic crystals by deformation. *Adv. Mater.* **14** (2002) pp. 1284-1286.
108. Gu Z.-Z., Horie R., Kubo S., Yamada Y., Fujishima A. and Sato O., Fabrication of a metal-coated three-dimensionally ordered macroporous film and its application as a refractive index sensor *Angew. Chem.* **41** (2002) pp. 1153-1156.
109. Qian W., Gu Z.-Z., Fujishima A. and Sato O., Three-dimensionally ordered macroporous polymer materials: An approach for biosensor applications *Langmuir* **18** (2002) pp. 4526-4529.
110. Holtz J. H. and Asher S. A., Polymerized colloidal crystal hydrogel films as intelligent chemical sensing materials. *Nature* **389** (1997) pp. 829-832.
111. Imhof A. and Pine D. J., Uniform macroporous ceramics and plastics by emulsion templating *Adv. Mater.* **10** (1998) pp. 697-700.
112. Jenekhe S. A. and Chen X. L., Self-assembly of ordered microporous materials from rod-coil block copolymers. *Science* **283** (1999) pp. 372-375.
113. Srinvasarao M., Collings D., Philips A. and Patel S., Three-dimensionally ordered array of air bubbles in a polymer film *Science* **292** (2001) pp. 79-83.
114. Jiang P., Bertone J. F. and Colvin V. L., A lost-wax approach to monodispersed colloids and their crystals *Science* **291** (2001) pp. 453-457.
115. Lu Y., Yin Y. and Xia Y., Synthesis and self-assembly of Au@SiO$_2$ core-shell colloids *Nano Lett.* **2** (2002) pp. 785-788.

CVD SYNTHESIS OF SINGLE-WALLED CARBON NANOTUBES

BO ZHENG, JIE LIU
Department of Chemistry, Duke University, Durham, NC 27708

Since their discovery, single-walled carbon nanotubes (SWNTs) have been a focus in materials research. However, the fundamental research and application development were greatly hampered by the limited resources of high quality SWNT materials. The aim of this article is to provide an updated review of current progress in chemical vapor deposition (CVD) synthesis of SWNTs. Various CVD methods and related experimental technique issues are discussed. This is followed by a discussion of growth mechanisms of SWNTs. The final part of the article is a summary and an outlook of this exciting new material.

1 Introduction

First discovered in 1993[1, 2], single-walled carbon nanotubes are a new form of carbon materials with many unique electrical, mechanical and chemical properties. They have attracted much attention due to potential applications in high-strength materials, efficient field emitters and nanoscale electronic devices.

A SWNT can be thought of as a graphene sheet rolled-over to form a seamless cylinder with a diameter of about 0.7-5 nm. A SWNT can be as long as several hundred microns. The structure of SWNT is usually denoted by means of two indices, n and m (Figure 1), which define the chiral vector C_h by equation (1). a_1 and a_2 are unit vectors in a 2D graphene lattice. Imagine that the nanotube is unraveled into a planar sheet. The two lines (the dash lines) along the tube axis are where the separation takes place. The chiral vector connects two carbon atoms from the two lines that would be the same carbon atom on the tube. The chiral angle (θ) measures the deviation of the chiral vector from the 'zigzag' configuration and ranges from $0°$ to $30°$. Zigzag and armchair SWNTs correspond to $\theta=0°$ and $\theta=30°$, respectively. On the basis of tight-binding band-structure calculations, an infinite-length SWNT is metallic if n-m=0, has a narrow band gap if n-m is a multiple of 3, and is a moderated-gap semiconductor otherwise[3-7].

$$C_h = n\, a_1 + m\, a_2 \tag{1}$$

The preparation of high-quality SWNTs with high yield has been the goal of many research endeavors. So far arc-discharge[1, 8], laser-ablation[9] and chemical vapor deposition (CVD) are the three main methods for SWNTs production. Arc-discharge and laser-ablation are the first two methods that allow synthesis of

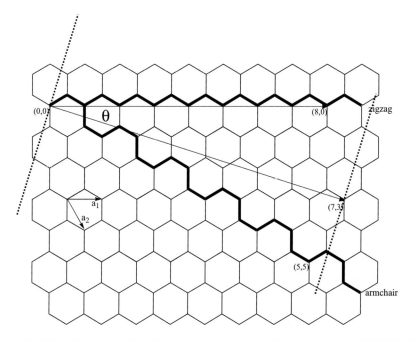

Figure 1. Schematic representation of the vectors representing the circumferences of SWNTs. The SWNTS are named according to the two integers representing the vectors.

SWNTs in relatively large (gram) amounts. Both methods involve the condensation of hot gaseous carbon atoms generated from evaporation of solid carbon. However, the equipment requirement and large energy consumption make them suitable only at a lab scale. It is the CVD method that can be easily scaled up to industry level and has become the most important commercial method for SWNT production.

There have been many review articles[10-16] and special issues of journals[17, 18] in addition to several books[19-22] focusing with carbon nanotubes during the past several years. In this article we will present an overview on the current state of the art of SWNT synthesis by CVD method.

2   Chemical Vapor Deposition

CVD is a term used to represent heterogeneous reactions in which both solid and volatile reaction products are formed from a volatile precursor, and the solid reaction products are deposited on a substrate[23, 24]. It has become a common method for thin film growth on various solid substrates. CVD of carbon has been successful in making carbon films[25], fiber[26, 27], carbon-carbon composites[28] and multiwalled carbon nanotube (MWNT) materials[29] at industry scale for more than 20 years. Only recently, the growth of SWNT using CVD has become possible[30]. In general, the growth process for SWNT synthesis involves a series of steps: (1) heating a catalyst material to high temperature, usually between 700 to 1000°C. The catalysts usually are nanoparticles composed of transition metals supported on either porous or flat supports. It can also be metal nanoparticles formed in gas phase and floating in the flow of feeding gas; (2) introduction of precursor gas containing carbon source into the furnace; (3) diffusion and decomposition of precursor on the surface of catalyst nanoparticles and dissolution of carbon atoms within the metal nanoparticles and (4) nucleation and growth of nanotubes from the metal nanoparticles saturated with carbon atoms. As will be discussed in details in the following sections, the control of experimental conditions are critical for the production of high quality SWNTs.

Comparing with arc and laser methods, the biggest advantage of CVD method is the more straightforward ways to scale up the production to industry level. Indeed, a company in Houston, TX, Carbon Nanotechnology Inc., has already established the capability for the production of SWNTs in pounds/day scale (Figure 2) using a process called HiPCO (High Pressure Carbon Monoxide), which is a floating catalyst CVD method. Another advantage of CVD methods is that we have more control on the morphology and structure of the produced nanotubes. With arc and laser method, only powdered samples with nanotubes tangled into bundles can be produced. With CVD methods, we can produce well-separated individual nanotubes either supported on flat substrates or suspended across trenches. These nanotubes can be directly used to fabricate nanoscale electronics. For such applications, these well-separated nanotubes present a big advantage over bulk samples since no separation of the nanotubes and purifications are needed. It is known that the purification and separation process may create defects on the nanotubes that could alter their electronic properties. In addition, there are also recent reports that the orientation of the nanotubes[31-33] and their diameters[34-36] can be controlled by controlling the experimental parameters. In the following sections, we will discuss in more details about the various CVD methods for the synthesis of SWNTs. We have divided all the methods into two categories, one for bulk synthesis and the other for surface synthesis.

3 Bulk CVD Synthesis of SWNT

3.1 Supported Catalyst CVD Method

A typical CVD experimental setup for bulk synthesis of SWNTs is depicted in Figure 3. It normally is composed of a tube furnace with temperature and flow control. Feeding gas containing carbon source flows through the tube furnace after a catalyst material is heated at elevated temperature. Type of catalysts, type of feeding gas, temperature, mass flow rate and pressure are the main parameters that define the experimental conditions. Depending on the form of catalysts used in the process, there are two main methods for bulk CVD synthesis of SWNTs: floating catalyst method and supported catalyst method. For supported catalyst CVD, the catalysts are typically transition metal of metal oxides dispersed on a porous support material. In the following section, we will discuss the preparation of catalysts and the effect of catalysts on the quality and yield of SWNT products.

Figure 2. The setup at CNI for large scale production of SWNTs. Picture provide by CNI.

3.1.1 Catalysts

Three different methods of catalyst preparation have been used for CVD growth of SWNTs:

(1) Solution impregnation method. This is the method used in several of the earliest reports for the synthesis of SWNTs using CVD[30, 37-39]. It is the easiest method for catalyst preparation but the yield of the SWNTs is limited. Typically, catalysts are prepared by mixing aqueous or alcoholic solution containing soluble

precursors of the desired metallic catalyst with commercially available inert porous support materials, such as Al_2O_3, SiO_2 or MgO, followed by slow evaporation of the solvents. The catalyst precursors are normally transition metal salts that can be decomposed to form metal oxides at higher temperature. In order to decompose the precursors, the as-prepared catalysts need to be calcined at high temperature before used for SWNT growth. It has been shown that the type of precursor and the condition for the calcinations are important factors that can affect the efficiency of the catalysts [37, 38].

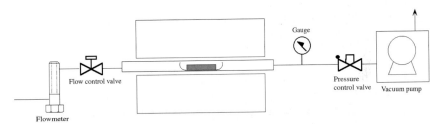

Figure 3. A typical CVD system for SWNT synthesis.

(2) Sol-gel method. The main difference between this method and the solution impregnation method is that the porous catalyst supports are made in situ by sol-gel method. The supports are normally inert metal oxides such as alumina and silica made from alkoxide precursors. To make the catalyst support, the precursors are dissolved in alcohols and reacted with water using either acids or bases as catalyst. Under well-controlled reaction conditions, a wet gel composed of solvents and a net works of metal oxide nanoparticles will be formed. If such wet gels are dried under supercritical conditions, highly porous solid structures with ultrahigh surface area will be formed. These porous structures are called "aerogels". Because of the high surface area and large pore volume these aerogels possess, they are ideal candidates for catalyst supports. The loading of catalysts on to such supports are can be performed during the formation of the gel by adding catalyst precursors into the mixture. Su et al. had used such a method to prepare catalysts that significantly increased the yield of SWNTs [40].

(3) Solid-solution: The third way to make catalysts for SWNT synthesis is to prepare solid solutions containing the catalytic metal nanoparticles. Typically it is prepared by thermal decomposition of metal salts mixtures. For example, Li et al. prepared MgO supported Fe/Mo catalyst by decomposition of Mg nitrates and alkaline magnesium carbonates at 400°C[41]. In particular, combustion synthesis is a convenient way to prepare this kind of catalyst because it is simple, fast and economical [42-44].

Weight gain (also called yield) of the catalyst during a CVD process is an important parameter to indicate the quality of the catalyst. It is defined by weight percentage of carbon relative to the catalyst. Much effort to enhance the production of SWNTs have focused on optimizing the catalyst used in the CVD process. The three magnetic transition metals Fe, Co and Ni are the most popular catalyst. Mo is rarely used as catalyst[30, 39] but it is often added to be a synergetic catalyst[41, 45]. The use of several other bimetallic catalysts such as Fe-Co also greatly increases the yield of SWNTs [44].

3.1.2 Feeding gas

a). Hydrocarbon

Catalytic pyrolysis of hydrocarbons had been practiced in carbon fiber industry long before SWNT was discovered, which laid the foundation for SWNTs growth. For SWNT synthesis, methane is the most often used feeding gas. Ethylene is also reported as another option[39].

Cassell et al. reported the synthesis of bulk amounts of SWNTs by CVD from methane at 900°C[37]. A systematic study of the catalyst leads to the conclusion that the best catalyst is a Fe/Mo bimetallic catalyst supported on Al_2O_3-SiO_2 hybrid material. The Al_2O_3-SiO_2 hybrid support exhibits both strong metal-support interaction from Al_2O_3 and better stability at temperature at temperature as high as 900°C. The catalyst has a surface area of almost 200m^2/g and mesopore volume of 0.8ml/g. Weight gain measurement showed that the yield of SWNT is ~35% with 30min growth time. Su et al. significantly improved the yield of this method to about 100% (30min growth time) using Al_2O_3 aerogel impregnated with Fe/Mo nanoparticles as catalyst[40]. The catalyst was prepared by sol-gel method followed by supercritical drying. Simply evaporating the solvent of the wet gel would cause collapse of the pore-walls inside the wet gel by strong forces from surface tension at the liquid/gas interface within the pores. The shrinkage would significantly reduce the surface area and pore volume of the dried gel. In the reported method, the solvent in the wet gel is brought to supercritical state at high temperature and high pressure where no liquid/gas interface is formed before being removed from wet gel. Therefore the porous structure would survive this procedure. The aerogel-based catalyst showed a surface area of 500-600m^2/g. Figure 4 shows the scanning electron microscopy (SEM) and transmission electron microscopy (TEM) images of the SWNTs in the raw product.

Several groups also used other hydrocarbon and catalysts to prepare SWNTs. For example, Hafner et al. prepared SWNTs using an extremely small amount of C_2H_4 diluted by Ar and Fe/Mo bimetallic catalyst with Al_2O_3 support[39]. Both single- and double- wall nanotubes were observed for reaction temperature from 700°C to 850°C. However, methane is still the most common gas used to prepare SWNTs. It has been shown that if using different catalysts, the optimal reaction conditions are different. Unfortunately, there are no systematic theory that can explain all the experimental observations. These experimental observations tend to be isolated from each other and the observed rules only apply to specific system.

Take methane as an example, when 2.5 wt% Co/MgO catalysts were used, 1000°C is the best reaction temperature as reported by Colomer et al[46]. On the other hand, Li et al. prepared high-quality SWNTs by CVD from methane at 850°C on Fe/Mo catalyst with MgO support [41] and Harutyunyan et al. reported high quality SWNTs growth at low temperature (680°C) and low methane flow rate (40 sccm)[45]. Clearly, more systematic studies are needed in this research field to provide a better understanding on the general growth mechanism and explain all experimental observations.

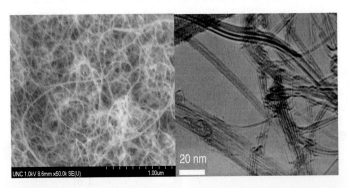

Figure 4. SEM and TEM images of as-grown SWNTS using CVD method. Figures reproduced with permission from reference 40.

b). Carbon monoxide

CO is actually the first feeding gas used for the growth of SWNTs. Dai et al. performed the first CVD synthesis of SWNTs by Mo catalyzed disproportionation of CO at 1200°C in 1996[30]. It was reported that most resulting SWNTs have catalytic particles attached to the ends, indicating that the growth of SWNTs is catalyzed by pre-formed nanoparticles. However, due to the safety reasons related to the use of CO, the reports on the growth of SWNTs using CO is limited comparing with other feeding gases. One exception is the use of CO in HiPCO process which will be discussed in the section on floating catalysts CVD methods. However, the use of CO as feeding gas does posses certain advantages over hydrocarbons. For example, Zheng et al. reported the use of CO and Al_2O_3 aerogel supported catalyst to synthesized SWNTs at 900°C[47]. Comparing with the samples made using the same catalyst and methane, the amount of amorphous carbon is greatly reduced. An acid treatment followed by an oxidation process produced a high purity of SWNTs. Recently Co-Mo catalyst has been found to selectively produce SWNTs at 700°C when CO was used as carbon source[48, 49].

The impact of addition of hydrogen to the CO CVD synthesis was studied by Bladh et al.[50] and Zheng et al.[51] It is found that hydrogen can greatly enhance the SWNTs synthesis by CO disproportionation. The effect of hydrogen can be explained by two possible reasons: (1) hydrogen directly reacts with CO, producing carbon and H_2O (2) hydrogen interacts with catalyst nanoparticles so that the activity of the catalyst towards CO disproportionation is enhanced [51].

c). Alcohols

Recently Maruyarna et al. reported a new CVD method of synthesizing high-purity SWNTs[52] using alcohols such as methanol and ethanol as carbon source. The synthesis temperature is at 700-800°C. The catalyst used in the method is bimetallic (Fe-Co) catalyst supported on zeolite. TEM and SEM showed that the products are very clean SWNTs (Figure 5) without any amorphous carbon coating. It is hypothesized that the OH radical formed at high temperature from alcohols can remove the amorphous carbon efficiently during the growth, leaving only pure SWNTs as product. If proven to be scalable, alcohols may become a better carbon source for industry scale synthesis of SWNTs because of the lower price of the alcohols and their easiness of handling and storage comparing with methane and carbon monoxide.

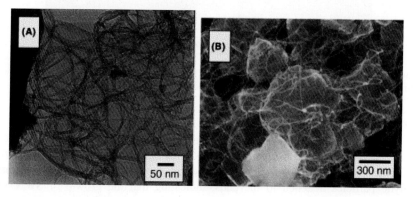

Figure 5. SWNTs produced using alcohol as feeding gas. Figures reproduced with permission from reference 52.

3.1.3 Synthesis temperature

The optimal synthesis temperature is closely related to the nature of feeding gas. Equation (2) and (3) show the thermodynamics data for CH_4 decomposition and CO disproportionation. T is reaction temperature (unit Kelvin).

$CH_4 \rightarrow C + 2 H_2$
$\Delta G / (kJ/mol) = 84.680 - 0.105 * T$

$2 CO \rightarrow C + CO_2$
$\Delta G / (kJ/mol) = -171.88 + 0.18 * T$

The usual temperature in CVD synthesis ranges from 700°C to 1200°C. So far the lowest temperature for CVD synthesis of SWNTs is 550°C[52]. All methane CVD procedures have been done at temperature between 600°C and 1000°C. For methane decomposition, ΔG is zero at 533°C. Deposition of amorphous/graphitic carbon product can be competitive or dominant if temperature is too high. Hornyak et al. reported a temperature window from 680°C to 850°C for methane CVD process[53]. The turn-on at the low-temperature end appears to be controlled by the thermodynamics of SWNTs growth, while the turn-off at high-temperature is due to competition deposition of amorphous and nanocrystalline carbon. In contrast, CO disproportionation is thermodynamically unfavorable at the usual temperature for CVD from CO so amorphous carbon is much less serious a problem for CO as feeding gas at high temperature [47].

3.1.4 Growth kinetics

One common feature of bulk synthesis of SWNT with CVD method is the decrease of growth rate with time [37, 47, 49]. In CVD of CO, it was found that weight gain of the catalyst is proportional to $t^{1/2}$ with t as growth time (Figure 6) [39, 47]. Generally it is believed that the gas-diffusion rate, which is one rate-limiting step in CVD process, becomes lower with more SWNTs grown on the catalyst. This is also consistent with the observation that high porosity of the catalyst can enhance the SWNT yield.

3.2 Floating Catalyst CVD Method

In floating catalyst method, the catalysts are formed in gas phase from volatile organometallic catalyst precursor introduced into the furnace. The organometallic species decompose at high temperature, forming metal clusters on which SWNTs nucleate and grow. This method is more suitable for a large-scale synthesis because the reaction can be operated continuously.

The first SWNT synthesis using floating catalyst approach was carried out by Cheng et al.[54] In their study, ferrocene was vaporized and carried by hydrogen gas into the furnace together with benzene as carbon source. The prepared nanotubes were transported out of the reaction zone by the flowing gas and collected on a graphite plate. Since then this method has been adopted by several other groups. Satishkumar et al. used metallocenes in admixture with C_2H_2 in their floating catalyst synthesis [55]. They also found that $Fe(CO)_5$ can be used as catalyst precursor. Bladh et al. used CO and CO/H_2 mixture as carbon feedstock to prepare SWNTs [50]. Ci et al. prepared high-quality SWNTs by pyrolysis of acetylene at 750°C-1200°C in a float iron catalyst system [56]. However, it is at Rice University that a serious commercial method for SWNT production was developed using the so-called HiPco process [57, 58]. In this procedure, high-pressure (30~100 atm) and high-temperature (1050°C) CO with $Fe(CO)_5$ as a catalyst precursor produce high-quality SWNTs at a rate of approximately 450 mg/h. The product consists of entangled SWNT bundles interspersed with Fe nanoparticles (Figure 7). A company (Carbon Nanotechnology Inc.) was formed in 2001 at Houston working on the commercialization of HiPCO SWNTs.

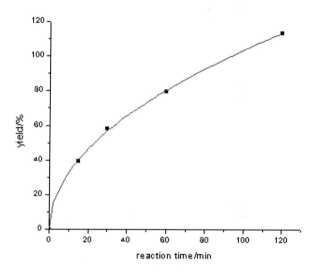

Figure 6. The relationship between the yield and growth time for CO CVD. Figure reproduced with permission from reference 47.

Bronikowski et al investigated the change of the SWNT yield with pressure of CO [58]. Production rate was found to increase with increased pressure up to 50 atm, indicating this is a surface-reaction-limited process. Ci et al. also studied the yield change with the partial pressure of C_2H_2 [56]. When the C_2H_2 partial pressure is below 5.0 Torr, the yield also increased with the pressure. However, more thermally decomposed amorphous carbon was produced when the pressure is above 5 Torr, thus SWNTs yield decreases.

There are several reports that sulfur and sulfur-containing additive (thiophene) could enhance the yield of SWNTs [54, 59]. One possible explanation for the role of sulfur is the change of the surface state of catalyst nanoparticles since sulfur is able to form chemical bond with metal atom and block a catalytically active site on nanoparticles surface. This can change the precipitation process of carbon from carbon solution in a metal nanoparticles. Another possibility is that sulfur lowers the eutectic temperature of the metal-carbon molten mixture. This can also affect the carbon precipitation process. However, recently it is also found that sulfur will promote double-wall nanotube growth [60].

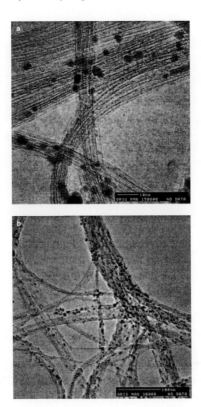

Figure 7. SWNTs produced by HiPCO method at Rice. Figures reproduced with permission from reference 57.

Recently, ultralong SWNT strands were synthesized by Zhu et al.[61] using a floating catalyst method. In the reported method, a hexane solution of ferrocene and thiophene was introduced into the reactor with hydrogen as the carrier gas. The as grown SWNT strands can be as long as 10-20 cm (Figure 8) with Young's modulus ranging from 49 to 77Gpa. The main difference between this new method and previous method[54] is that the furnace used in the CVD system is changed from a horizontal position tom a vertical position.

4   Surface CVD Synthesis

Direct growth of SWNTs on surfaces has several advantages over deposition of SWNTs for device fabrication. No sonication or oxidative purification is involved for surface grown SWNT samples, which greatly reduces the possibility of defect formation on SWNTs. Patterns can be introduced by various lithography techniques

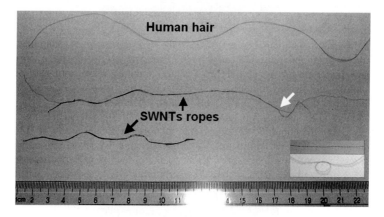

Figure 8. Long Strands of SWNTs produced by floating catalyst CVD method. Figure reproduced with permission from reference 61.

which is especially useful for field emission, sensors and other applications. Also the sample is usually very clean after surface growth, containing only SWNTs and nanoparticles.

Two types of surface growth procedure have been reported. The first one involves transferring catalyst, including both the nanoparticle catalysts and support materials, onto a substrate. SWNTs are grown by CVD from the catalyst. This method is essentially a mini version of bulk CVD synthesis. Kong et al. used electron-beam lithography to prepare catalytic islands on silicon substrates [62]. CVD from methane was then carried out to synthesize SWNTs. From SEM observation, the nanotubes are rooted in the islands, some of which bridge two islands. The as-grown SWNTs are mostly straight, indicating that they are defects-free along the lengths. Conductivity measurement on individual nantoubes by evaporating electrodes on the nanotubes have shown that the CVD nanotubes is of similar quality as laser grown nanotubes [11, 62-66]. Due to its simplicity and the high quality of the produced SWNTs, this procedure was widely adapted to develop electrical interconnects and electronic devices [66, 67]. Gu et al. used microcontact printing method to fabricate catalyst patterns on silicon nitride substrates [68]. A higher yield of SWNTs was obtained by CVD using a mixture of methane and hydrogen instead of pure methane. The ultrathin silicon nitride membrane enables observation of the as-grown nanotubes by TEM. It also provides direct evidence that the nanotubes are single-walled.

The second type of surface growth of nanotubes starts with only the metal or metal oxide nanoparticles directly deposited on substrates. In this way a better control of the density of SWNTs grown on surfaces is achieved. Li et al. prepared monodispersed Fe-Mo nanoparticles by thermal decomposition of metal carbonyl complexes [69]. The size of the nanoparticles can be systematically varied from 3 to 14 nm. The prepared nanoparticles were used as catalyst for SWNTs growth. It is found that there is an upper limit for the size of the particles to nucleate SWNTs. This observation was confirmed later by Li et al.[35] Recently Kim et al. used pre-formed Fe_2O_3 nanoparticles and a mixture carbon source of CH_4 and C_2H_2 to grow SWNT on SiO_2 substrate.[64] The SWNTs can be as long as 600µm, some of which exhibit loop and closed ring structures. Electrical transport measurements over 40 individual SWNTs reveal that their CVD condition produces SWNTs with no preference in chirality. Recent work in our lab has shown that using mixture of CO and H_2 offers a much higher SWNT yield [51]. Very high growth yield was achieved when the reaction temperature was between 800 and 900 °C and H2 concentration between 20% and 80%. This large window of growth temperature and hydrogen concentration for the best yield offers a higher consistency in experimental result.

In addition to using preformed nanoparticles, the catalyst can be generated by physical vapor deposition of ultrathin metal film on substrates. Anantram et al. demonstrated physical sputtering of transition metals on patterns to be a viable technique in preparation of SWNTs [70]. Ultrathin film of few nanometers of transition metal was found to be effective for SWNTs growth. Delzeit et al. used ion beam sputtering for the deposition of metal catalyst to grow SWNTs [71]. No SWNTs were grown with 0.1-1.0 nm pure Fe on substrates. However, a thick Al underlayer greatly increases the production of SWNTs. The density of the SWNTs grown increases with an increase in Al underlayer thickness from 1 to 10 nm. The thickness of Al underlayer beyond this range does not exhibit a significant increase in the density of SWNTs. An extra thin layer of Mo (0.2 nm) as co-catalyst produced a slight increase in SWNTs. It was suggested that the metal underlayer increase the surface roughness and provide more active nucleation sites. Hongo et al. investigated the dependence of the SWNTs growth on the orientation of the sapphire's crystallographic face [72]. Sapphire substrates were coated with a 2- to 5-nm thick Fe film by e-beam evaporation. CVD from methane at 800°C revealed that the largest quantity of SWNTs was produced on the A-face substrate, followed by the R- and C-face substrates (Figure 9). Meanwhile, the amount of amorphous carbon deposit increases as the substrate changes from A-face to R-face then to C-face. This result implies that the interaction between metal catalyst and the substrate plays an important role in SWNTs growth.

Growth Promoters, normally bulk amounts of catalyst materials, placed upstream near the substrate can significantly increase the yield of SWNTs for CVD from methane [33]. It is suggested that the bulk catalyst can activate methane, thus providing an efficient carbon-feedstock for SWNT growth.

5 Discussion

5.1 Diameter Control

Theoretical calculations have shown that the electron structures of SWNTs depend strongly on their diameter and chirality. Therefore the diameter control of SWNT is very important for both research and industry application purposes.

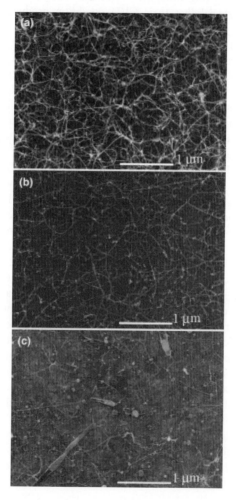

Figure 9. SEM images of 2-nm-thick Fe-coated sapphire substrates after methane CVD at 800 °C. (a) A-face, (b) R-face, and (c) C-face orientations. Figures reproduced with permission from reference 72.

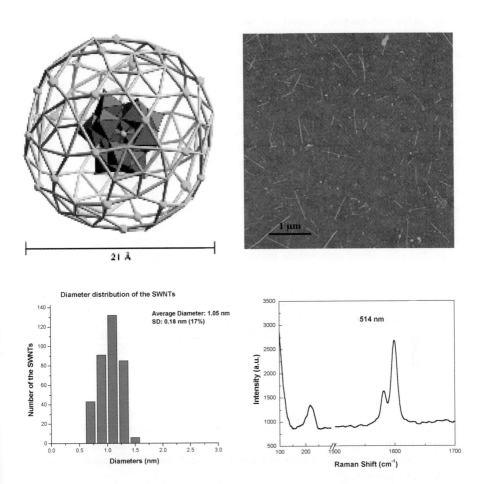

Figure 10. Top left: Structure of the molecular cluster used as catalysts for SWNT growth. Top right: AFM image of SWNTs grown using the molecular cluster as catalyst. Bottom left: the diameter distribution of the SWNTs grown on the chemically attached Fe/Mo nanoclusters on the silicon dioxide surfaces. The sample size is more than 350. The average diameter is 1.0 nm and the standard deviation is 0.18 or 17%. Bottom right: the Raman spectrum of the SWNTs grown upon the chemically attached Fe/Mo nanoclusters on the silicon nitride membranes. The excitation wavelength was 514.5 nm sourced by an argon ion laser at a power of 120 mW. Figures reproduced with permission from reference 83.

The diameter of SWNTs is believed to be closely related with the size of catalyst nanoparticles. In most catalyst used for SWNT synthesis, the size of the metal oxide nanoparticle cannot be well controlled. As a result, the SWNTs produced usually exhibit a broad distribution of diameter. One method to solve this problem is to use preformed monodispersed nanoparticles as catalyst. Dai and coworkers loaded iron atoms into the cores of horse spleen apoferritin, an iron-storage protein [35]. Iron oxide nanoparticles were obtained by heating the artificial ferritin to 800°C in air. The size of the resulting nanoparticles can be controlled by the number of the iron atoms in the protein. In our lab, a metal containing nanocluster molecule was found to be a good catalyst for CVD growth of SWNTs. This cluster has a well-defined chemical composition with 84 Mo atoms and 30 Fe atoms. So the diameter of the nanoparticle is identical, which is about 1.3nm after being reduce in H_2 at 900°C. SWNTs could be grown by CVD from methane at 900°C using this catalyst. The resulting SWNTs have remarkably uniform diameter, as revealed by AFM measurement and Raman spectroscopy (Figure 10) [83].

Another approach toward controlling the diameter of nanotubes is the use of templates. Tang et.al. synthesized mono-sized SWNTs inside 0.73nm sized channels of microporous aluminophosphate crystallites by pyrolysis of tripropylamine [36]. The resulting SWNTs have a diameter of 0.4 nm, which would be unstable without the support from the channels.

Temperature also affects the diameter of the SWNTs. It is generally known that with other conditions unchanged an increase of synthesis temperature results in an increase of the diameter of the nanotubes as well as an increase of the range of the diameter distribution [40, 45, 52]. This temperature dependence of diameter distribution also holds for the laser-ablation method [73]. One possible explanation is that metal catalyst nanoparticles sinter and grow larger at higher temperature, which will grow SWNTs with a larger diameter.

In floating catalyst CVD synthesis, Nikolaev et al. observed that increasing CO pressure leads to smaller SWNT diameter and narrower diameter distribution [57]. It is suggested that higher CO pressure increases CO disproportionation rate, giving more C atoms to each Fe particle and faster (earlier) SWNT nucleation relative to continued Fe particle growth.

5.2 Purity Control

Generally as-grown SWNTs are accompanied by other forms of carbon materials including MWNT and amorphous carbon. It is also a common phenomenon that the yield of other carbon impurities increase along with the yield of SWNT, which makes it a time and labor consuming task to remove these unwanted materials. Therefore maximizing the percentage of SWNTs in the raw product is especially important. Thermodynamically CH_4 tends to decompose at high temperature, which forms amorphous carbon. Therefore it is essential to limit the reactions to the surface of the catalyst. Other than controlling the temperature, hydrogen can be added to curb the self-pyrolysis of CH_4. Peigney et al. studied the influence of the composition of a H_2-CH_4 mixture on the SWNT quality and yield. For their catalyst and procedure the best molar ratio for CH_4 is 9~18% [74].

For CVD from both CH_4 and CO, synergistic effect between Co and Mo has been reported. Using bimetallic Co-Mo catalysts, Resasco and coworkers investigated the selectivity towards SWNTs by changing the Co:Mo ratio in the catalytic CVD from CO [48]. At a Co:Mo molar ratio 2:1, a significant fraction of the product is MWNT, while at a ratio of 1:2, majority of the product is SWNT. Tang et al. reported that the addition of Mo to Co/MgO catalysts promote the formation of SWNTs preferentially and suppresses the formation of amorphous carbon [42].

5.3 Orientation Control

Ordered SWNT architectures are of special interest to both fundamental research and industrial applications. To date, CVD method is the most successful method for synthesis of SWNTs with controlled orientation.

In order to achieve orientation control, an external guidance must be applied for SWNTs growth. This external force could be from electrical, gravity or magnetic field, or from the interaction between the growing SWNT and the feeding gas. The first direct growth of suspended SWNTs is demonstrated by Cassel et al [75]. First lithographically patterned silicon pillars are fabricated on substrate. Contact printing is used to transfer catalyst precursor material onto the tops of pillars. CVD from methane produces suspended SWNTs with the nanotube orientation directed by the pattern of the pillars (Figure 11). Zhang et al. applied a lateral electric field during CVD growth and the SWNTs exhibit alignment guided by the field (Figure 12) [31]. This field-alignment effect originates from the high polarizability of SWNTs. Aligning torques and forces on the nanotubes are generated by induced dipole moment. A second aligning method utilizing electric field is plasma techniques. However, so far all plasma techniques have produced only MWNTs [76].

Another promising way of aligning SWNTs is magnetic orientation. Calculations reveal that metallic SWNTs are paramagnetic along their long axes and tend to align themselves parallel to the ambient magnetic field. Non-metallic SWNTs are diamagnetic, but their diamagnetic susceptibilities are most negative in the direction perpendicular to the tube axis, forcing them to align also parallel to the ambient field. So far a high magnetic field has been applied only on suspensions of SWNTs during a deposition process [77]. The resulting SWNT film showed a preferential orientation and was denser than usual buckypaper [77, 78].

5.4 SWNT Formation Mechanisms

Understanding the formation mechanism of SWNTs is a key issue for the development of both SWNT synthesis and its application. By the observation of the SWNTs in the raw product, it was suggested that two growth modes exist [79, 80], e.g., base growth in which the SWNT grows upwards while the nanoparticle remains attached to the substrate, and tip growth in which the particle detaches and move at the head of the growing SWNT. However, the role of these metal nanoparticles in determining the growth has been inaccessible to direct observation and remains controversial. By examining the states of the SWNT ends, Dai et al. proposed the so-called "yarmulke" mechanism for the first CVD growth of SWNT [30]. In this mechanism, the excess of carbon that is dissolved in the metal nanoparticles assembles a graphene cap on the particle surface, which is called a yarmulke. This cap is strongly chemisorbed to the metal, relieving the high surface energy problem of the metal nanoparticle. Newly arriving carbon continue to assemble on the surface of the metal nanoparticle, adding to the cylindrical section between the cap and the catalytic particle.

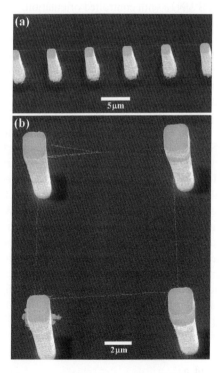

Figure 11. (a) SEM image of a suspended SWNT power line. (b) SEM image of a square of suspended SWNT bridges. Figures reproduced with permission from reference 75.

Figure 12. SEM images of suspended SWNTs grown in various electric fields. The spacing between the edges of the outer poly-Si electrodes is 40 μm. Figures reproduced with permission from reference 31.

6 Summary and Outlook

Due to its equipment accessibility and scalability, CVD synthesis of SWNTs has gained popularity over the past six years. Among the different CVD setups for

SWNT synthesis, floating catalyst configuration is apparently advantageous. There are two problems for fixed-bed configuration: (1) the growth of SWNTs on the surface of the catalyst makes it more and more difficult for feeding gas to reach the bottom part, which reduces the overall efficiency of the catalyst. This limits the amount of the catalyst that can be loaded into the furnace. (2) it is not a continuous operation. Fluidized-bed catalytic reactor appears able to solve these problems.

Many studies have focused on the optimization of the catalyst. However, the lack of a consensus method which can accurately define the amount of SWNTs in the product is the biggest problem for catalyst optimization. A protocol combining TGA measurement and spectroscopy characterization could offer a viable approach.

Overall, although great progresses have been made in the last few years in developing various CVD methods for SWNT synthesis, there still remain several challenging issues. To date the resulting raw material consists of an entangled mat of SWNTs, and the structure of individual tubes varies randomly from zigzag, armchair or chiral form, which coexist in the product. Currently a method to selectively grow metallic or semiconducting SWNTs is still unavailable, although post-growth procedure can partially resolve this problem. Another problem is that the SWNTs produced by CVD method usually contain more defects than that from arc discharge or laser ablation method. These remaining problems are the current research focus in this research field.

Nevertheless, in recent years the yield of CVD growth of SWNTs was greatly enhanced and commercial production of SWNTs has been realized. The availability of large amount of SWNT has benefited its fundamental study and application. Many preliminary application tests have been successful and shown great promise. For instance, SWNT field emitter has been fabricated by CVD of hydrocarbon [81]. It exhibits an ultralow threshold for field-emission. SWNTs have also been grown directly onto AFM tips[82] to offer significant improvements in lateral AFM resolution compared with commercial silicon AFM tips. It is reasonable to expect that in the next few years the quality of the CVD SWNTs will improve and the cost will be reduced significantly, making SWNTs available to more researchers who are developing application using bulk amount of SWNTs.

References:

1. Bethune, D.S., et al., *Cobalt-catalysed growth of carbon nanotubes with single-atomic-layer walls.* Nature, 1993. **363**: p. 605.
2. Iijima, S. and T. Ichihashi, *Single-shell carbon nanotubes of 1-nm diameter.* Nature, 1993. **363**: p. 603.
3. White, C.T., D.H. Robertson, and J.W. Mintmire, *Helical and rotational symmetries of nanoscale graphitic tubules.* Physical Review B, 1993. **47**(9): p. 5485.

4. Saito, R., et al., *Electronic-structure of graphene tubules based on C_{60},* PHYSICAL REVIEW B, 1992. **46**(3): p. 1804.
5. Saito, R., et al., *Electronic structure of chiral graphene tubules,* Applied Physics Letters, 1992. **60**(18): p. 2204.
6. Hamada, N., S. Sawada, and A. Oshiyama, *New one-dimensional conductors-graphitic microtubules,* Physical Review Letters, 1992. **68**(10): p. 1579.
7. Tanaka, K., et al., *Electronic-properties of bucky-tube model,* Chemical Physics Letters, 1992. **191**(5): p. 469.
8. Journet, C., et al., *Large-scale production of single-walled carbon nanotubes by the electric-arc technique.* Nature, 1997. **388**: p. 756.
9. Thess, A., et al., *Crystalline Ropes of Metallic Carbon Nanotubes.* Science, 1996. **273**: p. 483.
10. Ajayan, P.M., *Nanotubes from Carbon.* Chemical Review, 1999. **99**(7): p. 1787-1800.
11. Dai, H.J., *Carbon nanotubes: opportunities and challenges.* Surface Science, 2002. **500**(1-3): p. 218-241.
12. Rao, C.N.R., et al., *Nanotubes.* Chemphyschem, 2001. **2**(2): p. 78-105.
13. Odom, T.W., et al., *Structure and Electronic Properties of Carbon Nanotubes.* Journal of Physical Chemistry B, 2001. **104**(13): p. 2794-2809.
14. Rakov, E.G., *Methods for preparation of carbon nanotubes.* Russian Chemical Reviews, 2000. **69**(1): p. 35-52.
15. Smalley, R.E. and B.I. Yakobson, *The future of the fullerenes.* Solid State Communications, 1998. **107**(11): p. 597-606.
16. Ajayan, P.M. and T.W. Ebbesen, *Nanometre-size tubes of carbon.* Repots on Progress in Physics, 1997. **60**: p. 1025.
17. Applied Physics a-Materials Science & Processing, 1998. **67**(1): p. 1-119.
18. Carbon, 2002. **40**(10): p. 1619-1842.
19. Dresselhaus, M.S., G. Dresselhaus, and P.C. Eklund, *Science of Fullerenes and Carbon Nanotubes.* 1996, San Diego: Academic Press.
20. Dresselhaus, M.S., G. Dresselhaus, and P. Avouris, *Carbon Nanotubes Synthesis, Structure, Properties, and Applications.* Topics in Applied Physics. Vol. 80. 2001: Springer.
21. Harris, P., *Carbon Nanotubes and Related Structures: New Materials for the Twenty-First Century.* 2001: Cambridge University Press.
22. Saito, R., G. Dresselhaus, and M.S. Dresselhaus, *Physical Properties of Carbon Nanotubes.* 1998: Imperial College Press. 258.
23. Pierson, H.O., *Handbook of Chemical Vapor Deposition.* 1992, Park Ridge, New Jersey: Noyes Publication.
24. Hampden-Smith, M.J. and T.T. Kodas, *Chemical Vapor Deposition of Metals.* Chemical Vapor Deposition, 1995. **1**(1): p. 8-23.
25. Ashfold, M.N.R., et al., *Thin film diamond by chemical vapour deposition methods.* Chemical Society Reviews, 1994. **23**(1): p. 21.

26. Snyder, C.E., et al., *Carbon Fibrils*. 1989: US.
27. Tibbetts, G.G., *Carbon fibers produced by pyrolysis of natural gas in stainless steel tubes.* Applied Physics Letter, 1983. **42**(8): p. 666-668.
28. *Carbon-Carbon Materials and Composites*, ed. H.D. Buckley and D.D. Edie. 1993, Park Ridge: Noyes Publications.
29. Tennent, H.G., *Carbon fibrils, method for producing same and compositions containing same.* 1987, Hyperion Catalysis International, Inc.: US Patent #4663230.
30. Dai, H., et al., *Single-wall nanotubes produced by metal-catalyzed desproportionation of carbon monoxide.* Chemical Physics Letters, 1996. **260**: p. 471-475.
31. Zhang, Y.G., et al., *Electric-field-directed growth of aligned single-walled carbon nanotubes.* Applied Physics Letters, 2001. **79**(19): p. 3155-3157.
32. Su, M., et al., *Lattice-oriented growth of single-walled carbon nanotubes.* Journal of Physical Chemistry B, 2000. **104**(28): p. 6505-6508.
33. Franklin, N.R. and H. Dai, *An Enhanced CVD Approach to Extensive Nanotube Networks with Directionality.* Advanced Materials, 2000. **12**(12): p. 890-894.
34. Cheung, C.L., et al., *Diameter-controlled synthesis of carbon nanotubes.* Journal of Physical Chemistry B, 2002. **106**(10): p. 2429-2433.
35. Li, Y.M., et al., *Growth of single-walled carbon nanotubes from discrete catalytic nanoparticles of various sizes.* Journal of Physical Chemistry B, 2001. **105**(46): p. 11424-11431.
36. Tang, Z.K., et al., *Mono-sized single-wall carbon nanotubes formed in channels of AlPO4-5 single crystal.* Applied Physics Letters, 1998. **73**(16): p. 2287.
37. Cassell, A.M., et al., *Large scale CVD synthesis of single-walled carbon nanotubes.* Journal of Physical Chemistry B, 1999. **103**(31): p. 6484-6492.
38. Kong, J., A.M. Cassell, and H.J. Dai, *Chemical vapor deposition of methane for single-walled carbon nanotubes.* Chemical Physics Letters, 1998. **292**(4-6): p. 567-574.
39. Hafner, J.H., et al., *Catalytic growth of single-wall carbon nanotubes from metal particles.* Chemical Physics Letters, 1998. **296**(1-2): p. 195-202.
40. Su, M., B. Zheng, and J. Liu, *A scalable CVD method for the synthesis of single-walled carbon nanotubes with high catalyst productivity.* Chemical Physics Letters, 2000. **322**(5): p. 321-326.
41. Li, Q.W., et al., *A scalable CVD synthesis of high-purity single-walled carbon nanotubes with porous MgO as support material.* Journal of Materials Chemistry, 2002. **12**(4): p. 1179-1183.
42. Tang, S., et al., *Controlled growth of single-walled carbon nanotubes by catalytic decomposition of CH4 over Mo/Co/ MgO catalysts.* Chemical Physics Letters, 2001. **350**(1-2): p. 19-26.

43. Liu, B.C., et al., *Catalytic growth of single-walled carbon nanotubes with a narrow distribution of diameters over Fe nanoparticles prepared in situ by the reduction of LaFeO3.* Chemical Physics Letters, 2002. **357**(3-4): p. 297-300.
44. Flahaut, E., et al., *Synthesis of single-walled carbon nanotubes using binary (Fe, Co, Ni) alloy nanoparticles prepared in situ by the reduction of oxide solid solutions.* Chemical Physics Letters, 1999. **300**(1-2): p. 236-242.
45. Harutyunyan, A.R., et al., *CVD synthesis of single wall carbon nanotubes under "soft" conditions.* Nano Letters, 2002. **2**(5): p. 525-530.
46. Colomer, J.F., et al., *Large-scale synthesis of single-wall carbon nanotubes by catalytic chemical vapor deposition (CCVD) method.* Chemical Physics Letters, 2000. **317**(1-2): p. 83-89.
47. Zheng, B., Y. Li, and J. Liu, *CVD synthesis and purification of single-walled carbon nanotubes on aerogel-supported catalyst.* Applied Physics a-Materials Science & Processing, 2002. **74**(3): p. 345-348.
48. Kitiyanan, B., et al., *Controlled production of single-wall carbon nanotubes by catalytic decomposition of CO on bimetallic Co-Mo catalysts.* Chemical Physics Letters, 2000. **317**(3-5): p. 497-503.
49. Alvarez, W.E., et al., *Synergism of Co and Mo in the catalytic production of single- wall carbon nanotubes by decomposition of CO.* Carbon, 2001. **39**(4): p. 547-558.
50. Bladh, K., L.K.L. Falk, and F. Rohmund, *On the iron-catalysed growth of single-walled carbon nanotubes and encapsulated metal particles in the gas phase.* Applied Physics a-Materials Science & Processing, 2000. **70**(3): p. 317-322.
51. Zheng, B., et al., *Efficient CVD Growth of Single-Walled Carbon Nanotubes on Surface Using Carbon Monoxide Precursor.* Nano Letters, 2002. **2**(8): p. 895-898.
52. Maruyama, S., et al., *Low-temperature synthesis of high-purity single-walled carbon nanotubes from alcohol.* Chemical Physics Letters, 2002. **360**(3-4): p. 229-234.
53. Hornyak, G.L., et al., *A temperature window for chemical vapor decomposition growth of single-wall carbon nanotubes.* Journal of Physical Chemistry B, 2002. **106**(11): p. 2821-2825.
54. Cheng, H.M., et al., *Large-scale and low-cost synthesis of single-walled carbon nanotubes by the catalytic pyrolysis of hydrocarbons.* Applied Physics Letters, 1998. **72**(25): p. 3282-3284.
55. Satishkumar, B.C., et al., *Single-walled nanotubes by the pyrolysis of acetylene-organometallic mixtures.* Chemical Physics Letters, 1998. **293**(1-2): p. 47-52.
56. Ci, L., et al., *Controllable growth of single wall carbon nanotubes by pyrolizing acetylene on the floating iron catalysts.* Chemical Physics Letters, 2001. **349**: p. 191-195.

57. Nikolaev, P., et al., *Gas-phase catalytic growth of single-walled carbon nanotubes from carbon monoxide.* Chemical Physics Letters, 1999. **313**: p. 91-97.
58. Bronikowski, M.J., et al., *Gas-phase production of carbon single-walled nanotubes from carbon monoxide via the HiPco process: A parametric study.* Journal of Vacuum Science & Technology A, 2001. **19**(4): p. 1800.
59. Ago, H., et al., *Gas-phase synthesis of single-wall carbon nanotubes from colloidal solution of metal nanoparticles.* Journal of Physical Chemistry B, 2001. **105**(43): p. 10453-10456.
60. Ci, L., et al., *Double wall carbon nanotubes promoted by sulfur in a floating iron catalyst CVD system.* Chemical Physics Letters, 2002. **359**: p. 63.
61. Zhu, H.W., et al., *Direct Synthesis of Long Single-Walled Carbon Nanotube Strands.* Science, 2002. **296**: p. 884.
62. Kong, J., et al., *Synthesis of individual single-walled carbon nanotubes on patterned silicon wafers.* Nature, 1998. **395**: p. 878-881.
63. Dai, H.J., et al., *Controlled chemical routes to nanotube architectures, physics, and devices.* Journal of Physical Chemistry B, 1999. **103**(51): p. 11246-11255.
64. Kim, W., et al., *Synthesis of Ultralong and High Percentage of Semiconducting Single-walled Carbon Nanotubes.* Nano Letters, 2002. **2**(7): p. 703-708.
65. Kong, J., et al., *Synthesis, integration, and electrical properties of individual single-walled carbon nanotubes.* Applied Physics A-Materials Science & Processing, 1999. **69**(3): p. 305-308.
66. Soh, H.T., et al., *Integrated nanotube circuits: Controlled growth and ohmic contacting of single-walled carbon nanotubes.* Applied Physics Letters, 1999. **75**(5): p. 627-629.
67. Franklin, N.R., et al., *Patterned growth of single-walled carbon nanotubes on full 4- inch wafers.* Applied Physics Letters, 2001. **79**(27): p. 4571-4573.
68. Gu, G., et al., *Growth of single-walled carbon nanotubes from microcontact-printed catalyst patterns on thin Si3N4 membranes.* Advanced Functional Materials, 2001. **11**(4): p. 295-298.
69. Li, Y., et al., *Preparation of monodispersed Fe-Mo nanoparticles as the catalyst for CVD synthesis of carbon nanotubes.* Chemistry of Materials, 2001. **13**(3): p. 1008-1014.
70. Anantram, M., et al., *Nanotubes in nanoelectronics: transport, growth and modeling.* Physica E, 2001. **11**(2-3): p. 118-125.
71. Delzeit, L., et al., *Multilayered metal catalysts for controlling the density of single-walled carbon nanotube growth.* Chemical Physics Letters, 2001. **348**(5-6): p. 368-374.
72. Hongo, H., et al., *Chemical vapor deposition of single-wall carbon nanotubes on iron-film-coated sapphire substrates.* Chemical Physics Letters, 2002. **361**(3-4): p. 349-354.

73. Bandow, S., et al., *Effect of the Growth Temperature on the Diameter Distribution and Chirality of Single-Wall Carbon Nanotubes.* Phys. Rev. Lett., 1998. **80**(17): p. 3779.
74. Peigney, A., C. Laurent, and A. Rousset, *Influence of the composition of a H2-CH4 gas mixture on the catalytic synthesis of carbon nanotubes-Fe/Fe3C-Al2O3 nanocomposite powders.* Journal of Materials Chemistry, 1999. **9**(5): p. 1167-1177.
75. Cassell, A.M., et al., *Directed Growth of Free-Standing Single-Walled Carbon Nanotubes.* Journal of the American Chemical Society, 1999. **121**(34): p. 7975-7976.
76. Delzeit, L., et al., *Growth of carbon nanotubes by thermal and plasma chemical vapour deposition processes and applications in microscopy.* Nanotechnology, 2002. **13**(3): p. 280-284.
77. Walters, D.A., et al., *In-plane-aligned Membranes of Carbon Nanotubes.* Chemical Physics Letters, 2001. **338**(1): p. 14-20.
78. Smith, B.W., et al., *Structural anisotropy of magnetically aligned single wall carbon nanotube films.* Applied Physics Letters, 2000. **77**(5): p. 663-665.
79. Sinnott, S.B., et al., *Model of carbon nanotube growth through chemical vapor deposition.* Chemical Physics Letters, 1999. **315**(1-2): p. 25-30.
80. Dai, H., *Nanotube Growth and Characterization*, in *Topics in Applied Physics*, M.S. Dresselhaus, G. Dresselhaus, and P. Avouris, Editors. 2001, Springer. p. 29-51.
81. Matsumoto, K., et al., *Ultralow biased field emitter using single-wall carbon nanotube directly grown onto silicon tip by thermal chemical vapor deposition.* Applied Physics Letters, 2001. **78**(4): p. 539-540.
82. Cheung, C.L., et al., *Growth and fabrication with single-walled carbon nanotube probe microscopy tips.* Applied Physics Letters, 2000. **76**(21): p. 3136-3138.
83. An, L., et al., *Synthesis of nearly uniform single-walled carbon nanotubes using identical metal-containing molecular nanoclusters as catalyst.* Journal of the American Chemical Society, 2002. **124**(46): p. 13688-13689.

NANOCRYSTALS

M. P. PILENI

Laboratoire des Matériaux Mésoscopiques et Nanomètriques, LM2N, UMR CNRS 7070,
Universite P. et M. Curie
Bât F, BP 52, 4 Place Jussieu, 75005 Paris, France

In the review we describe the various parameters controlling the nanocrystal size and shape. Physical properties of isolated nanocrystals are presented. Nanocrystals are assembled in 2D and 3D superlattices inducing collective properties.

1 Soft chemistry syntheses

1.1 Spherical particles

Spherical nanoparticles are usually produced by coprecipitation or chemical reduction [1]. Several parameters are used to control the nanoparticle size [2]. Let us list them:

i) Change in the solubility product of inorganic materials with temperature [3]. The CdS nanoparticle size is tuned by controlling temperature during the coprecipitation.

ii) Electrostatic interactions. Two approaches have been developed: a) One of the reactants interacts with a stabilizing agent such as a polymer whereas the other evolves freely in the solution. Amorphous metals and semiconductors [4] are produced. The particle size is controlled by the ratio of polymer to reactants. b) Functionalized surfactants (the counter ion of the surfactant is a reactive agent) are used [5]. These surfactants form colloidal assemblies where the chemical reaction takes place [6].

iii) Sterical confinements. Various ways to confine the reactants are used:

  a) The pore size of the zeolites controls that of the nanoparticles and amorphous material is produced [7].

  b) Surfactants are able to self assemble to form water in oil droplets [8] and the reactants are confined inside the droplet. For most materials, the droplet size controls that of the produced nanomaterial [2]. When the reactants are in their ionic form, amorphous nanoparticles are formed. Conversely with functionalized surfactants, nanoparticles characterized by a very high crystallinity are obtained. Under these conditions, both electrostatic interactions and steric confinements act to produce well-defined nanocrystals.

1.2 Control of the particle shape

A colloidal solution is used as template to control the particle shape [9]. The literature contains various examples showing that it is possible to produce nanowires, nanorods, triangles, pyramid, and disks. However, no general way to control the shape of nanocrystals has been published [10]. From a careful study of the various literature data, it seems that the control of shape in the nanoscopic scale follows that of macroscopic crystal growth [11]. Surfactants, salts and pH are the major factors to control the particle shape. To illustrate this, several examples are given below.

Let us consider as template the colloidal solution composed of $Cu(AOT)_2$-H_2O-isooctane. Depending on the water content, various phases are formed [9, 11, 12] (reverse micelles, interconnected cylinders, lamellar phase, "supra aggregates" made of several microphases), (Fig. 1). Note that few impurities markedly change the phase diagram [13].

On replacing water by hydrazine, $Cu(AOT)_2$ is reduced and copper nanocrystals are formed [14] (Fig. 2).

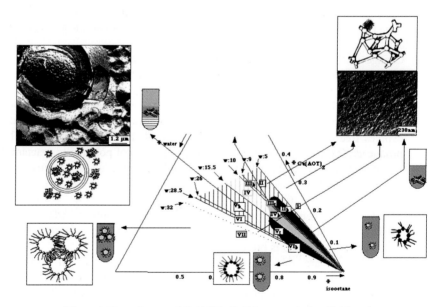

Figure 1. Phase diagram of $Cu(AOT)_2$-H_2O-Isooctane in the oil rich region

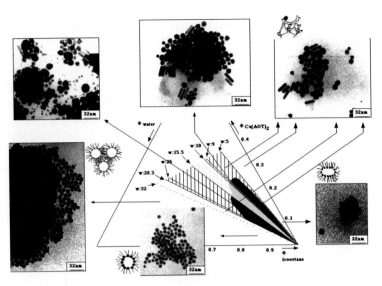

Figure 2. Various shapes of copper nanocrystals made in the various parts of the phase diagram shown in Figure 1

Comparison of Figure 1 and Figure 2 indicates that the colloidal solution used as a template enables a partial control of the nanocrystal shape. This control of the nanocrystal shape by that of the template has been recently confirmed by Simmons et al [15]. In fact, it is well known that addition of phosphatidylcholine to Na(AOT) reverse micelles induces a structural change of the self assembly with a change of the spontaneous curvature and formation of worm-like, cylindrical, reverse micelles that entangle to form gel-like systems [16]. Syntheses of CdS nanoparticles in such colloidal assemblies make it possible to vary the morphology of the nanocrystals from spheres to nanorods with a switch in the crystal structure from cubic to hexagonal. However the role of the template is not as obvious as described above. By adding a slight amount of NaCl to $Cu(AOT)_2$-H_2O-isooctane, instead of obtaining a small amount of cylinders, nanorods are formed [17] (Fig. 3). The aspect ratio is controlled by the chloride concentration [18]. The nanorod structure is a truncated decahedron with 5-fold symmetry [19]. By replacing NaCl by various salts the behavior changes drastically [20, 21]: With NaBr, small nanorods are formed whereas the number of cubic nanorystals increases with increasing Br^- concentration (Fig. 4). With F^- and NO_3^- most of the nanocrystals are spherical (Fig. 4). Two questions arise: i) Does the template change with salt addition? ii) Does salt play the major role in controlling nanocrystal shape? Addition of salt induces a slight change in a phase diagram whereas it remains the same with all the salts. This indicates that salts play the major role in the shape control while the template role is minor. From the literature, various compounds [17, 18, 20-29] such as citrate, sodium hydroxyle, ethylamine, phosphate, chloride and bromide play a key role in the control of the particle anisotropy. Very recently we demonstrate that in presence of surfactant by controlling hydrazine concentration it is possible to make monocrystal nanodisks [30] with a tunable size keeping the same aspect ratio [31]. The faceted particles appear as flat single crystals two (111) faces at the top and the bottom, limited at the edges by three other (111) faces and at the corners by more or less extended (100) faces. Hence, flat pseudo hexagonal nanocrystals are produced (Fig. 5A and 5B). From a larger tilt angle under dark field imaging (Fig. 5C), the flat hexagonal shape of the nanocrystals is more pronounced. From the profile plots made for a rather large number of particles, it is concluded that these flat particles are characterized by an aspect ratio between 4 and 5.5.

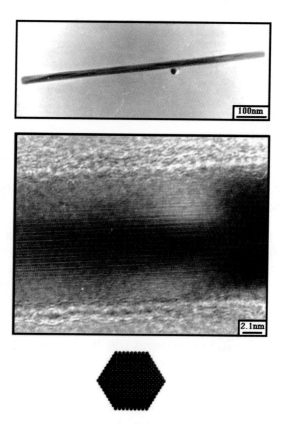

Figure 3. Structure of copper nanorods

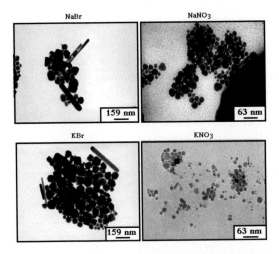

Figure 4. Copper nanocrystals obtained in presence of 10^{-3}M of KBr, NaBr, KNO$_3$ and NaNO$_3$ in Cu(AOT)$_2$-isooctane-water Solution. [Cu(AOT)$_2$] = 5.10^{-2}n w = [H$_2$O]/[AOT] = 1s

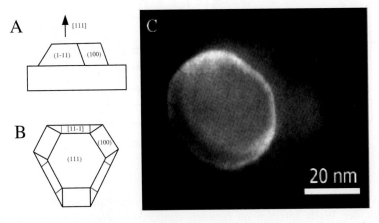

Figure 5. Profile view (A) and top view (B) of a truncated crystal (111) oriented on the substrate. (C) Silver nanodisk observed by TEM under dark field at high tilt angle

2 Fabrication of mesoscopic structures of nanocrystals

Two different approaches can be used to obtain mesoscopic structures of nanocrystals [32]. The first is the self-organization of nanocrystals [33-38] and the second is obtained with the application of external force [39, 40].

2.1 self organization of nanocrystals

Most of the data described involve nanocrystals coated with a surfactant and dispersed in a non-polar solvent. The best 2D and 3D superlattices are obtained for nanocrystals having the lowest size distribution. This is the key parameter in obtaining self-assemblies. In 2D superlattices, the nanocrystals are able to self organize either in compact hexagonal networks [33-35] (Fig. 6A) or form rings (Fig. 6B) [41, 42]. In 3D superlattices either "supra" crystals [33-38] (Fig. 7C) or hexagons [41, 42] (Fig. 6D) are formed.

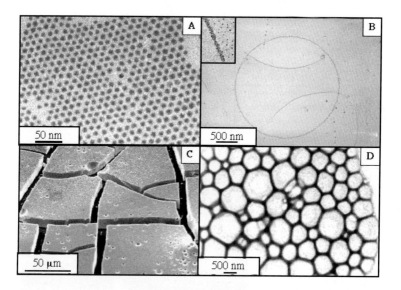

Figure 6. Various self assemblies

To obtain such organizations several factors play an important role:
(i) Van der Walls interactions between nanocrystals control the compacity of the 2D and 3D superlattices: This increases with increasing the particle size [33, 35]. For a given nanocrystal size, the length of the surfactant alkyl chain used to coat the nanocrystals controls the density of the superlattices [43].
(ii) particle-substrate interactions [44]: The best organizations are obtained when particle-substrate interactions are repulsive while the particle-particle ones are attractive. When both particle-substrate and particle-particle interactions are attractive only small domains of nanocrystals organized in a compact hexagonal network are obtained. When the compacity of the monolayer is not very high, the "supra" crystals cannot be formed.

(iii) evaporation rate: When the nanocrystals are dispersed in a solvent having a fast evaporation rate, rings are formed. These rings disappear with increasing the evaporation rate. This was attributed to Marangoni instabilities [41, 42]. The evaporation rate also plays a role in the control of the compacity of the monolayer. It increases with decreasing the solvent evaporation rate [36, 37].
(iv) Substrate temperature. At a low evaporation rate and on increasing substrate temperature the compacity of the monolayer increases. This favors formation of supra crystals [36, 37].
(v) A structural defect in the surfactant chain used to coat the nanocrystals prevents growth of "supra" crystals. This was well demonstrated with $SH-C_{14}H_{29}$ surfactant [43].

2.2 Organization of nanocrystals induced by applying a magnetic field during the deposition process [39].

Whatever the evaporation rate is, when the size distribution of nanocrystals is not low enough, no self-organization is obtained (Fig. 7A).

The observed film is made of aggregates without a defined shape on the substrate. Organization of nanocrystals can be obtained by applying external forces. For magnetic nanocrystals a magnetic field is applied, either perpendicular or parallel to the substrate, during the deposition process. At the end of the evaporation process, mesostructures made of magnetic nanocrystals are obtained. Addition of a solvent drop to the substrate totally destroys these structures and the nanocrystals are dispersed. When the applied magnetic field is perpendicular to the substrate, a large variety of structures depending on the strength of the field is obtained: At a very low applied field, large dots with a rather wide size distribution are formed (Fig. 7B). On increasing the applied magnetic field, well-dispersed dots with a well-defined hexagonal network are observed (Fig. 7C). A further increase in the applied field shows patterns similar to those observed at a lower applied field (Fig. 7C). However the average diameter and inter-dot distance are markedly reduced. The height of the dots decreases with increasing the strength of the applied field. This was recently modeled in our laboratory in terms of magnetic bond number and energy minimization [45, 46]. A further increase in the applied magnetic field induces drastic changes in the mesoscopic structure. Figure 7E shows the presence of dots, worm-like and labyrinth structures. A still further increase in the magnetic field enables formation of a homogeneous labyrinth structure. From these data, it is concluded that, by applying a magnetic field during the cobalt nanocrystal deposition on a substrate, it is possible to form well-defined 3D superlattices. The structure is observed on a very large scale (up to 0.02 mm^2). Similar patterns were previously seen with various systems [47, 48]. The origin of the phenomenon is similar to that obtained with concentrated ferrofluids when they are subjected to an applied field: for magneto-rheological fluids (6% nanocrystal fraction volume),

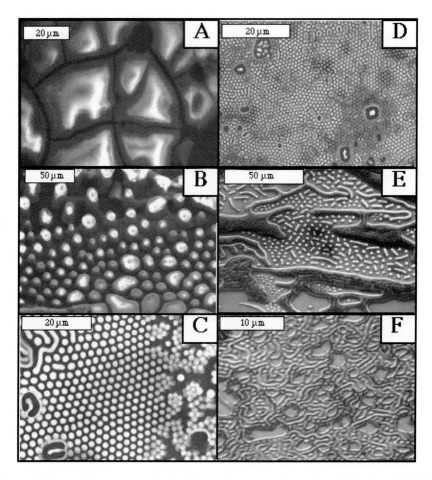

Figure 7. Mesoscopic structures made of cobalt nanocrystals and obtained by applying a magnetic field having various strengths, perpendicular to the substrate, during the evaporation

without an applied field no structure is observed. Columns organized in hexagonal networks or labyrinth patterns are observed by applying a magnetic field perpendicular to the sample [49]. By turning off the magnetic fluids, the mesostructures disappear because the thermal energy is higher than the magnetic energy and the nanocrystals are superparamagnetic. To understand such behaviors the concept of interfacial instability has to be taken into account. In the present study we have to keep in mind that the substrate is immersed in a solution containing coated cobalt nanocrystals dispersed in hexane. Leaving the system under hexane vapor pressure slows the evaporation process down. The interface between the

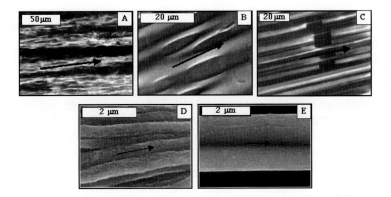

Figure 8. Mesoscopic structures (made) of cobalt nanocrystals (A,B and C) and ferrite nanocrystals (D and E) and obtained by applying a magnetic field having various strengths, parallel to the substrate during the evaporation

hexane vapor and ferrofluid is dynamic: the interface is spatially modulated as a result of the competition between the magnetic interaction and the surface tension.

For coated cobalt nanocrystals dispersed in hexane, when the applied field is parallel to the substrate, long and very compact stripes, with a periodic structure, in the direction of the applied field without any holes and cracks are formed (Figs. 8A, 8B, 8C). A characteristic wavelength, λ_χ, is defined as the average distance between two adjacent stripes [40]. With increasing the strength of the applied field, the characteristic distance decreases to reach a plateau. With ferrite dispersed in aqueous solution, the morphology of the mesoscopic structure markedly changes [50, 51] (Figs. 8D and 8E) compared to that observed with cobalt. The compacity of the ribbons increases with the strength of the applied field.

3 Physical properties of a collection of nanocrystals

Decreasing the nanocrystal size induces changes in the physical properties. This is valid for semiconductors [52] and magnetic or non-magnetic metal nanocrystals [53]. Nanometer size crystals exhibit behavior intermediate between that of the bulk material and molecules. It is impossible here to give a general overview of the various physical properties of nanocrystals. However, we present some examples that illustrate the major behaviors.

3.1 Semiconductors

II-VI quantum dot semiconductors are characterized by specific spectroscopic properties [3]: The absorption and photoemission spectra are red shifted with increasing the particle size. Let us consider CdTe quantum dots characterized by a large Bohr exciton diameter (15 nm). Usually the CdTe quantum dots are made by rf magnetron sputtering in a glass matrix [54, 55]. Recently, we have been able to make CdTe nanocrystals [56] by soft chemistry. As expected, a shift of the absorption and the direct fluorescence with increasing the nanocrystal size is observed.

Diluted magnetic semiconductors (DMS) [57] are semiconductors where host cations (II) are randomly substituted by magnetic ions, Mn^{2+}. The presence of localized magnetic ions in a semiconductor alloy leads to exchange interactions between s-p band electrons and the Mn^{2+} d electrons. This sp-d exchange interaction constitutes a unique interplay between semiconductor physics and magnetism. It plays a double role in determining optical properties:

(i) The band gap of the compound is altered depending upon the concentration of manganese ions.
(ii) The 3d levels of transition metal ions are located in the band gap region and d-d transitions dominate the spectrum. In the nanometer-size crystallites of semimagnetic semiconductors, many of these properties are influenced by the quantum confinement of the electronic states and differ from those of the bulk crystals. For $Cd_{1-y}Mn_yS$, a change in the band gap variation with composition of DMS with no straight line is observed: The minimum is more pronounced when the particle size decreases [58-60] (Fig. 9). It is explained as an increase in the exchange interactions between the d electrons of Mn^{2+} and the band electrons with decreasing the particle size. From magnetization measurements and by EPR it is deduced that the Mn^{2+}-Mn^{2+} interactions also increase with decreasing the particle size. Formation of a $Cd_{1-y}Mn_yS$ nanocrystals core-shell differing by its Mn^{2+} composition enables (a) photoemission (process) due to isolated Mn^{2+} in tetrahedral sites [61, 62].

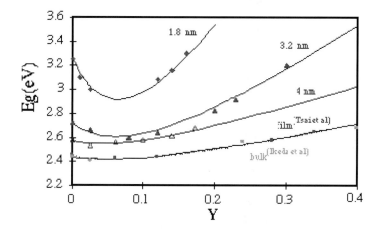

Figure 9. Variation of $Cd_{1-y}Mn_yS$ band gap with y (composition) and at various nanocrystal sizes

3.2 Metal nanocrystals

With respect to metal nanocrystals, the basic theory of the optical properties of metals is discussed elsewhere [53]. In the UV-visible spectral range, the broad absorption bands of metal nanocrystals are due to plasmon resonance excitations or interband transitions. The absorption spectrum of silver nanocrystals dispersed in a dilute solution is characterized by a well-defined plasmon resonance excitation which is well modeled with Mie Theory [63-70]. Hence, both the experimental [69, 71] and simulated [68, 72] absorption spectra show a well defined plasmon resonance peak with a decrease in its band intensity and increase in bandwidth with decreasing particle size. The experimental peaks shapes are nearly Lorentzian, and discrepancies at higher energies are due to inter band transitions in the experimental system (4d-5sp) [73].

4 Collective properties of mesoscopic structures of nanocrystals

Magnetic [50, 51, 74-82], optical [83-86] and transport [87, 88] collective properties are observed. Let us list examples of these collective properties:

4.1 Collective magnetic properties

The collective magnetic properties are mainly based on induced dipolar interactions of nanocrystals deposited on a substrate. The shape of the mesoscopic structure also plays a role. In the following these two approaches are described. With cobalt nanocrystals, dipolar interactions induce changes in the magnetization curves whereas with ferrite formation of tubes of nanocrystals plays one of the major role.

4.1.1 Cobalt nanocrystals

At 3K, the 6-nm and 8-nm coated cobalt nanocrystals are ferromagnetic [74-76]. The magnetization curves are recorded for nanoparticles dispersed in hexane (Figs. 10A and 10B) and deposited on a graphite substrate (Figs. 10C and 10D). At a low volume fraction of nanocrystals dispersed in hexane, the saturation magnetization, M_s, is not reached (Figs. 10A and 10B).

However it is attained when the same nanocrystals are deposited on the substrate. Furthermore, when the applied field is parallel to the substrate, the magnetization curves are squarer with an increase in the reduced remanence compared to that obtained for nanocrystals dispersed in solution. As expected, the reduced remanence increases with the particle size. To explain such differences in the magnetization curves between particles dispersed in solution and deposited on a substrate, we have to make allowance for induced dipolar interactions. Computations are made using a simple model including the elements required to determine the hysteresis loop and by taking into account the dipolar interactions

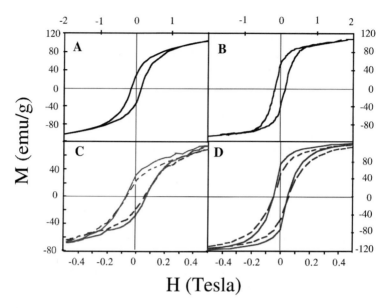

Figure 10. Hysteresis loop of 6-nm and 8-nm nanocrystals: dispersed in a solution (A and B); deposited on a substrate with an applied field either parallel (full curves); perpendicular (dotted curves) to the substrate (C and D)

between nanocrystals [77]. The particles are modeled as Stoner–Wohlfarth particles [89, 90] and an important simplification is made: only the component parallel to the applied field of the total dipolar field is taken into account. This allows determining a coupling constant α_d defined as:

$$\alpha_d = \frac{\pi}{12} \frac{M_s^2}{K} (D/d)^3$$

where d is the nearest neighbor distance. The coupling constant is 0.035 and 0.05 for 6 nm and 8 nm respectively. Simulated magnetization curves change with the coupling constant and with the direction of the applied field. When the applied field is parallel to the substrate, the hysteresis loop is squarer (Fig. 11A) than that obtained when the applied field is perpendicular (Fig. 11B) to the substrate. This is explained in terms of the effective applied field: when the applied magnetic field is parallel to the substrate, the effective field is the sum of the applied field and that induced by dipolar interactions. Conversely, when the applied magnetic field is applied perpendicular to the substrate, the induced dipolar interactions result in a decrease in the effective field compared to the applied field. For a given coupling constant, the ratio of the calculated reduced remanences when the applied field is perpendicular and parallel to the substrate can be deduced. For 6 nm and 8 nm they are 0.65 and 0.66 respectively. Similar behavior is observed experimentally (Fig. 10C and Fig. 10D). These values are in good agreement with those obtained by calculation. This clearly indicates that the fact that the hysteresis loop is squarer for particles deposited on a substrate, compared to those dispersed in solution, is due to dipolar interactions.

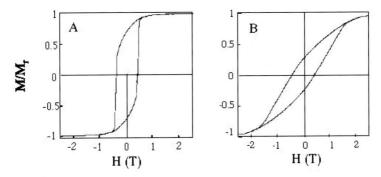

Figure 11. Simulated hysteresis loop obtained with 0.1 as coupling constant and with the applied field parallel (A) and perpendicular (B) to the substrate

4.1.2 Mesostructure shape

As shown in Figure 8, whatever the applied field is during the deposition process, ribbons are formed. The hysteresis loop is recorded at 3K with an applied magnetic field parallel to the substrate. The reduced remanence increases with the strength of the magnetic field applied during the deposition process. This is not due to the orientation of the easy axis as expected [51] but to the organization of ferrite nanocrystals in tubes with a magnetic behavior similar to that observed with nanowires [82].

4.2 Collective transport properties

The silver nanocrystals are deposited on an Au(111) substrate. The particle density on the substrate is controlled by the nanocrystal concentration of the deposited solution. For a single particle, the STS measurement (Fig. 12A) shows, that on increasing the applied voltage, small capacitances of the junctions are charged and the detected current is close to zero. Above a certain threshold voltage, the electrons pass through the system and the current increases with the applied voltage. The plot of dI/dV versus V (Fig. 12B) clearly shows that the differential conductivity minimum reaches zero at zero bias. This non-linear I(V) spectrum and the zero value of dI/dV at zero voltage are characteristic of the well-known Coulomb blockade effect. The I(V) curve shows a gap of 2 V at zero current (Fig. 12A). This indicates that the ligands are sufficiently good electrical insulators to act as tunnel barriers between particles and the underlying substrate. A similar value of the Coulomb gap was obtained previously for isolated silver particles having similar sizes [91]. The particle-substrate distance is fixed by the dodecanethiol coating, thus the two tunnel junctions are characterized by fixed parameters. Similar Coulomb blockade behavior has been observed previously [87, 92].

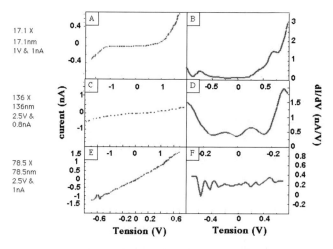

Figure 12. I(V) curves and their derivatives recorded for isolated nanocrystals (A and B), organized in compact hexagonal networks (C and D) and forming 3D superlattices (E and F)

The I(V) curve recorded with a close packed monolayer made of silver nanocrystals organized in a hexagonal network shows that, for large biases, the detected current is reduced by more than one order of magnitude compared to that observed for isolated particles (Fig. 12C). This indicates an increase in the ohmic contribution to the current. The shape of the I(V) curve markedly changes and it decreases non-linearly. In addition, the Coulomb gap is very low (0.45 V) compared to that obtained with isolated particles (2 V). This decrease in the Coulomb blockade gap indicates that electron-tunneling transport occurs between neighboring particles, which form a connected electrical network parallel to the substrate. This makes the effective capacitance between the particles and the substrate larger. The differential conductivity dI/dV has a non zero minimum at zero bias in (Fig. 12D), as observed for Coulomb blockade behavior. From this it can be concluded that when particles are arranged in 2D superlattices, the I(V) characteristic curve exhibits both ohmic and Coulomb blockade contributions. This ohmic contribution can originate from additional tunneling paths (taking place) between the particles and their neighbors. This indicates that lateral tunneling between adjacent nano-crystals contributes significantly to the electron transport. These data are in good agreement with those already published [93, 94].

When "supra" crystals are deposited on an Au(111) substrate, the I(V) curve shows linear ohmic behavior (Fig. 12E). The detected current, above the site point, markedly increases compared to data obtained with a monolayer made of nanocrystals. Of course the dI/dV(V) curve is flat (Fig. 12F). This shows an ohmic character without a Coulomb blockade or staircases indicating an ohmic connection through multi-layers of nanoparticles. From these data it is concluded that organization of nanocrystals enables forming conductive material.

4.3 Collective optical properties

Reflectivity measurements of silver nanocrystals, deposited on various substrates such as HOPG, gold, silicon and $Al_{0.7}Ga_{0.3}As$, vary drastically with the polarization direction and the substrate [84]. Under s-polarized light, the reflectivity spectra are recorded at various angles and are found to be similar for all incidence angles. Under p polarization, the reflectivity spectra markedly change with the incidence angle (Fig. 13). At a low angle (20°), the reflectivity spectra are similar to those obtained under s polarization (dotted line). This is due to the fact that under s and p polarization at low incidence angles, the electric field vector along the particle film is predominant. At high incidence angles, the perpendicular electric field component becomes much larger than the parallel one. Figures 13 show the reflectivity spectra of a monolayer made of silver nanocrystals organized in a hexagonal network and deposited on various substrates. It is seen that there is a great similarity between these reflectivity spectra and those calculated (Fig. 14) for the same substrates as recorded experimentally. The major difference is the appearance of a peak at high energy in the experimental data (Fig. 13) which is not reproduced by calculation. From these data, it is concluded that the film optical properties and dipolar interactions between nanocrystals induce appearance of a plasmon resonance at high energy indicative of collective optical properties. Furthermore, the major change in the responses is due to the reflective index of the substrate. Indeed, the optical response of the nanocrystal film is not sensitive enough and the image forces can be neglected. Thus, the positions of the two plasmon resonances, characteristic of the film optical anisotropy, do not depend on the nature of the substrate. Hence, they can be identified from the reflectivity spectrum, by the calculations. These data were recently confirmed by photoemission process [85]. These results can be generalized to various substrates, especially to absorbing ones, whose optical response is not obvious.

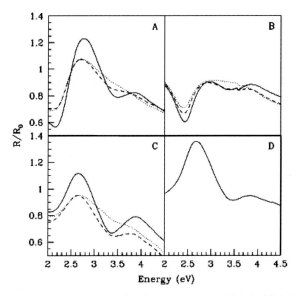

Figure 13. Reflectivity curves recorded at various incidence angles: 20° (...); 45° (---), 60° (___) of a film (made) of silver nanocrystals organized in a compact hexagonal network on various substrates: HOPG (A); Au (B); Si (C) and AlGaS (D)

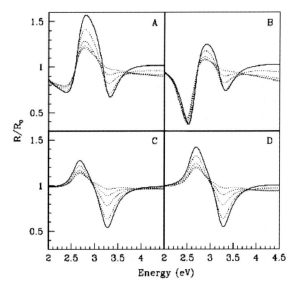

Figure 14. Simulated reflectivity curves recorded at various incidence angles: 20° (...); 45° (---), 60° (___) of a film made of silver nanocrystals organized in a compact hexagonal network on various substrates: HOPG (A); Au (B); Si (C) and AlGaS (D)

5 Conclusion

This paper shows that it is possible to control size and shape of nanocrystals. Their physical properties depend on the particle size and shape. Furthermore they markedly change with their organization in 2D and 3D superlattices

References

1. Klabunde K. J., Nanoscale Materials Chemistry Pub. Wiley-Interscience (ed), New York, Chichester, Weinheim, Brisbane, Singapore, Toronto (2001)
2. Pileni M. P., *J. Phys. Chem.* **97** (1993) pp. 9661
3. Rossetti R., Ellison J. L., Bigson J. M and Brus L. E., *J. Chem. Phys.* **80** (1984) pp. 4464J.
4. Henglein A., *J. Chem. Phys.* **84** (1987) pp. 441
5. Pileni M. P., *Langmuir.* **13** (1997) pp. 3266
6. Pileni M. P., Motte L., Petit C., *Chem. Mat.* **4** (1992) pp. 338
7. Hilinski E. F., Lucas P. A. and Wang Y., *J. Chem. Phys.* **89** (1988) pp. 3435
8. Pileni M. P., Reverse Micelles Pub.Elsevier (ed): Amsterdam New-York, Oxford, Shannon, Tokyo 1989
9. Pileni M. P., *Langmuir* **27** (2001) pp. 7476
10. Pileni M. P., Nature Materials in press
11. Tanori J., Gulik T. and Pileni. M. P., *Langmuir.* **13** (1997) pp. 632
12. Lisiecki I., André P., Filankembo A., Petit C., Tanori J., Gulik-Krzywicki T., Ninham B. W. and Pileni M. P., *J. Phys. Chem.* **103** (1999) pp. 9168 and 9176
13. Filankembo A., André P., Lisiecki I., Petit C., Gulik-Krzywicki, T., Ninham B. W., and Pileni M. P., *Colloids and Surfaces A: Physicochemical and Engineering Aspects.* **174** (2000) pp. 221
14. Tanori J. and Pileni M. P., *Langmuir.* **13** (1997) pp. 639
15. Simmons B. A., Li S., John V. T., McPherson G. L., Bose. A., Zhou W. & He, *Nanoletters.* **2** (2002) pp. 263
16. Scartazzini R. and Luisi P., Organogels from lecithins. *J. Phys. Chem.* **92** (1988) pp. 829
17. Tanori. J. and Pileni M. P., *Adv Mat.* **7** (1995) pp. 862
18. Pileni M. P., Gulik-Krzywicki T., Tanori J., Filankembo A. and Dedieu J. C., *Langmuir.* **14** (1998) pp. 7359
19. Lisiecki I., Filankembo A., Sack-Kongehl H., Weiss K., Pileni M. P. and Urban J., *Phys. Rev. B* **61** (2000) pp. 4968
20. Filankembo A., Pileni, M. P., *J. Phys. Chem.*, **104** (2000) pp. 5867
21. Filemkenbo A., Giorgio S. and Pileni M. P., sent for publication
22. Jana N. R., Gearheart L., and Murphy C. J., *Chem. Comm.*, (2001) pp. 617
23. Chen C. C., Chao C. Y. and Lang Z. H., *Chem. Mat.*, **12** (2000) pp. 1516

24. Henglein A., and Giersig M., *J. Phys. Chem.*, **104** (2000) pp. 6767
25. Li Y., Ding Y., and Wang Z., *Adv. Mat.*, **11** (1999) pp. 847
26. Zhan J., Yang X., Wang D., Li S., Xu Y., Xia Y. and Quian, Y., *Adv. Mat.* **12** (2000) pp. 1348
27. Sugimoto T. and Matjevic E., *J.Colloid Interface .Sci.* **74** (1980) pp. 227
28. Hsu W. P. and Matjevic E., *Langmuir.* **4** (1988) pp. 431
29. Ocana M. and Matjevic E., *J. Mater. Res.* **5** (1990) pp. 1083
30. Maillard M., Giorgio S. and Pileni M. P., *Adv. Mat.* **14** (2002) pp. 1084
31. Maillard M., Giorgio S. and Pileni M.P., *J. Phys. Chem.* (2002) in press
32. Pileni, M. P, *J. Phys. Chem.* **105** (2001) pp. 3358
33. Motte L., Billoudet F. and Pileni M. P., *J. Phys. Chem.* **99** (1995) pp. 16425
34. Motte L., Billoudet F., Lacaze E., and Pileni M. P., Adv. Mater. **8** (1996) pp. 1018
35. Motte L., Billoudet F., Douin J., Lacaze E. and Pileni M. P., *J. Phys. Chem.* **101** (1997) pp. 138
36. Courty A., Fermon C. and Pileni M. P. *Adv. Mater.* **13** (2001) pp. 254
37. Courty A., Azaspin O., Fermon C. and Pileni M. P. *Langmuir* **17** (2001) 1372
38. Lisiecki I., Albony P. A. and Pileni M. P. *Adv. Mat.* **15** (2003) 712
39. Legrand J., Ngo A. T., Petit C. and Pileni M. P., *Adv. Mater.* **13** (2001) pp. 58
40. Petit C., Legrand J., Russier V., and Pileni M. P., *J. Appl. Phys.* **91** (2002) pp. 1502
41. Maillard M., Motte L., Ngo A. T. and Pileni M. P., *J. Phys. Chem.* **104** (2000) pp. 11871
42. Maillard M., Motte L. and Pileni M. P., *Adv. Mater.* **13** (2001) pp. 200
43. Motte L. and Pileni M. P., *J. Phys. Chem.* **102** (1998) pp. 4104
44. Motte L., Lacaze E., Maillard M and Pileni M. P., *Langmuir.* **16** (2000) pp. 3803
45. Richardi J., Ingert D. and Pileni M. P., *J. Phys. Chem.* **106** (2002) pp. 1521
46. Richardi J., Ingert D. and Pileni M. P., *Phys. Rev. E* **66** (2002) 46306
47. Seul M.and Andelman D., *Science.* **267** (1995) pp. 476
48. Hong C. Y., Horng H. E., Jang I. J., Wu J. M., Lee S. L., Yeung W. B. and Yang H. C., *J. Appl. Phys.* **83** (1998) pp. 6771
49. Wu K. T. and Yao Y. D., *J. Mag. Mag. Mat.* **201** (1999) pp. 186
50. Ngo A. T. and Pileni M. P., *Adv. Mat.* **12** (2000) 276
51. Ngo A. T. and Pileni M. P., *J. Phys. Chem. B.* **105** (2001) pp. 53
52. Pileni M. P., *Catalysis Today.* **58** (2000) pp. 151
53. Feldheim D. L. and Foss C. A., Metal nanoparticles: synthesis, characterization and applications Publ. Marcel dekker (ed), New York, Basel 2002
54. Potter B. G. and Simmons H., *Phys. Rev. B.* **43** (1991) pp. 2234
55. Paula A. M., Barbosa L. C., Cruz C. H. B., Alves O. L, Sanjurjo, J. A. and Cesar C. L., *Appl. Phys. Lett.* **69** (1996) pp. 357

56. Ingert D., Feltin N., Levy L., Gouzerh P. and Pileni M. P., *Adv. Mat.* **11** (1999) pp. 220
57. Furdyna J. K. and Kossut J., Semiconductors and semimetals Vol 25, Academic Press 1988.
58. Levy L., Hochepied J. F. and Pileni M. P., *J. Phys. Chem.* **100** (1996) pp. 18322
59. Levy L., Feltin N., Ingert D. and Pileni M. P., *J. Phys. Chem.* **101** (1997) pp. 9153
60. Feltin N., Levy L., Ingert D. and Pileni M. P., *J. Phys. Chem.* **103** (1999) pp. 4
61. Feltin N., Levy L., Ingert D. and Pileni M. P., *Adv. Mater.* **11** (1999) pp. 398
62. Levy L., Feltin, N., Ingert, D. and Pileni M. P., *Langmuir.* **15** (1999) pp. 3386
63. Bohren C. F. and Huffman D. R., Absorption and Scattering of Light by Small Particles. New York:Wiley, 1983.
64. Charle K P., Frank F. and Schulze W., *Ber Bunsenges Phys Chem*, **88** (1984) pp. 354
65. Mie, G Ann Phys **25** (1908) pp. 377
66. Creighton J. A., Eaton. D. G., *J Chem Soc Faraday Trans 2.* **87** (1991) pp. 3881
67. Hovel H., Fritz S., Hilger A., Kreibig U., Vollmer M., *Phys. Rev. B.* **48** (1993) pp. 18178
68. Persson B. N., *J. Surface Science* **281** (1993) pp. 153
69. Ruppin. R., *Surface Science.* **127** (1983) pp. 108
70. Petit C., and Pileni M. P., *J. Phys .Chem.* **97** (1993) pp. 12974
71. Taleb A., Petit C., Pileni M. P., *J. Phys. Chem. B.* **102** (1998) pp. 2214
72. Alvarez M. A., Khoury J. T., Schaaf T. G., Shafigullin M. N., Vezmar, I., and Whetten. R. L., *J. Phys. Chem. B.* **101** (1997) pp. 3706
73. Yamaguchi T., Ogawa M., Takahashi H., Saito N., Anno E., *Surface Science* **129** (1983) pp. 232
74. Petit C., Taleb A. and Pileni M. P., *Adv. Mat.* **10** (1998) 259
75. Petit C., Taleb A. and Pileni M. P., *J. Phys. Chem.* **103** (1999) pp. 1805
76. Legrand J., Petit C., Bazin D. and Pileni M. P., *J. Appl. Surf. Sci.* **164** (2000) pp. 186
77. Russier V., Petit C., Legrand J. and Pileni M. P., *Phys. Rev. B.* **62** (2000) pp. 3910
78. Russier V., Petit C. and Pileni M. P., *J. Appl. Phys.* in press
79. Petit C., Russier V. and Pileni M. P., *J. Phys. Chem.* in press
80. Ngo A. T. and Pileni M. P., *J. Applied Phys.* **92** (2002) 4649
81. Ngo A. T. and Pileni M. P., *New. J. Physics.* **4** (2002) 87
82. Lalaronne Y., Motta L., Russier V., Bonvill P. and Pileni M. P., submitted for publication.
83. Taleb A., Russier V., Courty A. and Pileni M. P., *Phys. Rev. B.* **59** (1999) pp. 13350

84. Pinna N., Maillard M., Courty A., Russier V. and Pileni M. P., *Phys. Rev. B.* **66** (2002) pp. 45415
85. Maillard M., Monchicourt P. and Pileni M. P., to be send for publication.
86. Russier V. and Pileni M. P., *Surg. Sci.* **425** (1999) 313
87. Petit C., Cren T., Roditchev D., Sacks W., Klein J. and Pileni M. P., Adv. Mater. **11** (1999) pp. 1108
88. Taleb A., Silly F., Gusev A. O., Charra F. and Pileni M. P., *Adv. Mater.* **12** (2000) pp. 633
89. Stoner E. C. and Wohlfarth E. P., *Phil. Trans. Roy. Soc.* A2, **40** (1948) pp. 599 reprinted in *IEEE Trans. Magn.* **27** (1991) pp. 3475
90. Pfeiffer H., *Phys. Stat. Sol. (a)*, **122** (1990) pp. 377
91. Park K. H., Shin M., Ha J. S., Yun W. S. and Ko Y., *J. Appl. Phys. Lett.* **75** (1999) pp.139.
92. Simon U., *Adv. Mat.* **10** (1998) pp. 1487
93. Rimbert A. J., Ho T. R. and Clarke J., *Phys. Rev. Lett.* **74** (1995) pp. 4714
94. Ohgi T., Sheng H. Y., Nejoh H., *Appl. Surf. Sci.* **130** (1998) pp. 919

INORGANIC FULLERENE-LIKE STRUCTURES AND INORGANIC NANOTUBES FROM 2-D LAYERED COMPOUNDS

R. TENNE

Department of Materials and Interfaces, Weizmann Institute of Science,
Rehovot 76100, Israel

1 Introduction

Graphite, which has a 2-D layered structure, is the stable form of bulk carbon in ambient conditions. On the other hand fullerenes [1] and nanotubes [2] are known to be the thermodynamically favorable form of carbon if the number of atoms in each particle is not allowed to grow beyond say 0.1 micron. The driving force for the formation of such closed-cage nanostructures can be ascribed to the peripheral atoms in the graphitic lattice, which are only two-fold bonded. The ratio between such peripheral atoms and the three-fold bonded (bulk) carbon atoms increases as $1/d$ when the diameter (d) of the nanoparticle shrinks. To annihilate these edge atoms, and thus reduce the overall energy of the nanoparticle, pentagons are inserted into the otherwise honeycomb (hexagonal) lattice, which leads to the folding of the planar nanostructure. Once twelve such pentagons occur in the nanoparticles, a closed cage (fullerenic) nanostructure is obtained. In order to overcome the bending energy associated with the folding of the planar sheet high temperatures or other sources of energetic excitation are initially required. The strain energy is more than compensated by the annihilation of the dangling bonds, once the nanoparticle is fully closed and is therefore seamless. It was hypothesized [3] that this behavior is not limited to graphite and is likely to be a generic property of highly anisotropic (2-D) layered compounds. The formation of closed polyhedra and nanotubes was initially observed in nanoparticles of WS_2 [3] and MoS_2 [4]. The synthesis of nanotubes from different 2-D compounds, like BN [5], $NiCl_2$ [6], V_2O_5 [7], and numerous others were subsequently reported. A few reviews have been published on this topic in recent years [8]. Fig. 1 shows a transmission electron microscopy (TEM) image of a typical multi-layer WS_2 nanotube obtained by the reaction of WO_3 powder with H_2S gas under a reducing atmosphere. Similarly, nested structures ("onions") of MoS_2 were obtained by reacting MoO_3 vapor with H_2S gas in a reducing atmosphere. These new structures received the generic name - inorganic fullerene-like structures (*IF*). In analogy to carbon, these observations suggest that the binary (multinary) phase diagram of elements, which form layered compounds, like Mo and S, or Ni and Cl, include the new phase of hollow and closed nanostructures. This phase is expected to be situated in proximity to the bulk 2-D phase. If the otherwise plate-like crystallites are not allowed to grow beyond a certain size (less than say 0.2 microns), they spontaneously form the phase of closed cage nanostructures. Globally however, the *IF* phase is less stable than the bulk planar structure. Substantial progress has been achieved over the past few years in the synthesis; elucidation of the nanoparticles structure, and the study of some of their physical properties. This progress is possibly epitomized by the synthesis of 1 kg of a pure phase of WS_2 nanoparticles (0.1-0.2 micron large) with fullerene-like structure in a matter a few weeks period, in the present lab. This chapter is dedicated for a short review of the existing knowledge in this field. The main open questions in this field are briefly discussed in the text, too. Numerous potential applications are offered to these new nanophases in tribology. Possible applications in such diverse fields as battery, photocatalysis, nanoelectronics, etc. are being contemplated.

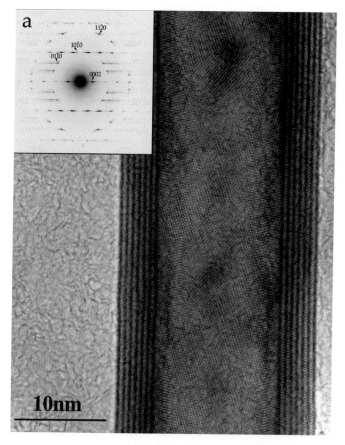

Fig. 1. High-resolution TEM image of a typical multi-wall WS_2 nanotube.

2 Synthesis of inorganic nanotubes and fullerene-like nanoparticles

Strategies for the synthesis of macroscopic amounts of various nanotubes were recently developed. However, precise size and shape-control of the nanoparticles will require further work. Generally, the synthesis of multiwall inorganic nanotubes or fullerene-like particles does not require a catalyst, which saves subsequent cleaning and separation of the nanoparticles phase.

2.1 Metal chalcogenides (M=Mo,W,Nb,Ta,Zr,Ti,Hf,In; X=S,Se)

The synthesis of large amounts of WS_2 [9] and MoS_2 [10] multiwall inorganic nanoparticles with fullerene-like structures were reported. Similarly, strategies for the synthesis of large amounts of nanotubes from the same compounds were also reported [11-15]. Each of these techniques is very different from the others and produces nanotube and fullerene-like material of somewhat different characteristics. This fact by itself indicates that *IF* materials (including nanotubes) of 2-D metal-dichalcogenides are a genuine part of the phase-diagram of the respective constituents. It also suggests that, with minor changes, these techniques can be used for the synthesis of nanotubes from other inorganic layered compounds. However, the synthesis of single wall inorganic nanotubes and polyhedra with precise size and shape control remains to be demonstrated, although major progress has been accomplished in this direction in recent times.

One of the early synthetic methods for MS_2 nanotubes [10] exploited the chemical vapor transport method, which is the standard growth technique for high quality single crystals of layered metal-dichalcogenides (MX_2) compounds. According to this method, a powder of MX_2 (or M and X in 1:2 ratio) is placed on the hot side of an evacuated quartz ampoule, together with a transport agent, like bromine or iodine. A temperature gradient of 20-50 °C is maintained along the ampoule. After a few days, a single crystal of the same compound grows on the colder side of the ampoule. Accidentally, microtubes and nanotubes of MoS_2 were found in the cold end of the ampoule. These preliminary studies were extended to other compounds, like WS_2 and the method was optimized with respect to the production of nanotubes rather than to the bulk crystals. Nanotubes produced by chemical vapor transport were found to consist of the 2H polytype [12c], while microtubes of diameters exceeding 1 mµ and length of a few hundreds micron were found to crystallize in the 3R polytype packing [12d and c] (see Sec. 4). Most remarkably, upon addition of C_{60} as a catalyst, single wall MoS_2 nanotubes could be obtained [16]. These nanotubes are characterized by their remarkable uniformity as shown in Fig. 2 and they are also very long (>100 micron). In fact, each nanotube in this array has the same diameter (0.96 nm) and orientation- (3,3) armchair. It is therefore not surprising that they self-assemble into a highly organized hierarchy of structures from the level of a single nanotube to a macroscopic superstructure. Iodine, which serves as a transport agent in this reaction intercalates in the voids between the nanotubes. The detailed structure of these nanotubes is still under debate, and it is likely to be solved by combining more detailed structural analysis together with *ab-initio* calculations. Optical and electrical measurements indicate that their properties are very different from that of multi-wall nanotubes reported before (see Sec. 4). Single and double wall WS_2 nanotubes of a larger diameter were synthesized as templates on multi-wall carbon nanotubes [17]. In the first stage, a mixture of carbon nanotubes and H_2WO_4 was reacted at 350 ºC. A thin coating (1.2 nm) of WO_3 on top of the carbon nanotubes was obtained. Subsequent annealing of the template in H_2S atmosphere at 900 ºC led to the formation of the single (double) wall WS_2 nanotubes.

An alternative method for the synthesis of MoS_2 nanotubes by a templated growth, was reported [13]. This synthesis is based on a generic deposition strategy, proposed first by Martin [18a] and further perfected by Masuda and co-workers [18b]. Here, perforated alumina membranes are first produced by anodization of aluminum foil

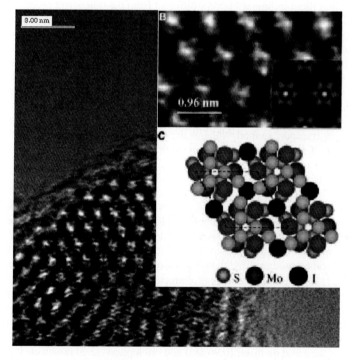

Fig. 2. Head-on TEM image of a single-wall MoS_2 nanotubes 0.96 nm in diameter each (adapted from Ref. [18]).

in an acidic solution. The dense pattern of cylindrical pores self-organize into hexagonal pattern [18b]. These pores serve as a template for the deposition of nanofilaments from a variety of materials. Thermal decomposition of a $(NH_4)_2MoS_4$ precursor, which was deposited from solution, at 450 °C and subsequent dissolution of the alumina membrane in a KOH solution led to the isolation of large amounts of MoS_2 nanotubes [13]. However, due to the limited thermal stability of the alumina membrane, the annealing temperatures of the nanotubes were relatively low and consequently their structure was imperfect. In fact, the MoS_2 nanotubes appeared more like bamboo-shaped hollow fibers.

More recently, Nath and Rao prepared nanotube phases from various layered metal dichalcogenide compounds by first synthesizing the MX_3 or $(NH_4)_2MX_4$ compound, and subsequently annealing the product at temperatures higher than 1000 °C [19]. Fig. 3 shows a scanning electron microscope (SEM) image of typical NbS_2 (a) and TaS_2 (b) nanotubes phases obtained by this method [19]. This synthetic strategy has a number of merits. First, thermal decomposition of the amorphous ammonium salt results in MX_3 nanoparticles. Therefore either precursor leads to the same product. The binary M-X phase diagram shows that MX_3 is situated always to the right (X-rich) side of MX_2 [20]. Therefore, high-temperature annealing of the MX_3 powder leads to the loss of one X atom and formation of MX_2 nanotubes.

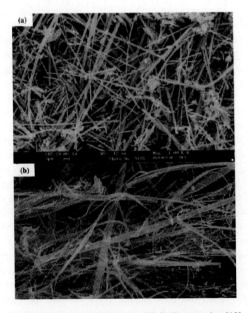

Fig. 3. SEM images of NbS_2 (a) and TaS_2 (b) nanotubes [19].

Further insight into the growth mechanism of the nanotubes is provided by the work of Hsu et al. [15]. Here MoS_2 powder was heated to 1300 °C in the presence of H_2S gas. The reaction was confined by covering the heated powder with a Mo foil. The nanotubes grew on the Mo foil, which was appreciably colder than the heated substrate. By carefully inspecting the product collected on the Mo foil, the authors found that in fact the nanotubes crystallized from an amorphous MoS_3 nanorods. Therefore, they propose that heating of the MoS_2 powder in the presence of H_2S gas leads to the formation of MoS_3, which vaporizes at the high temperature of the substrate, and is collected on the colder Mo foil. The thermal gradient on the Mo foil collector leads to the directional growth of a MoS_3 nanorod, which slowly decomposes and transforms into MoS_2 nanotube. This mechanism is likely to hold also for the formation of other MX_2 nanotubes reported by Nath and Rao [19]. Heating a dispersed powder of amorphous MoS_3 (WS_3) nanoparticles to temperatures below 1000 °C leads to the formation of nested MoS_2 (WS_2) nanoparticles (onions). According to the Mo-S phase diagram MoS_3 is unstable above 850 °C and it crystallizes into MoS_2 nanoparticles. Since the powder is carefully dispersed on the substrate, the coarsening of the nanoparticles is prevented and consequently the intermediate stable phase of MoS_2 (WS_2) nanoparticles with fullerene-like structure is obtained. These observations clearly indicate that the *IF* phase is located between the existence zones of the bulk MS_3 and MS_2 phases, which was discussed before [8a] (see also Sec. 3).

In a related experiment, amorphous MoS_3 nanoparticles were electrodeposited from $(NH_4)_2MoS_4$ solution. Subsequently, short electrical pulses from a tip of a scanning tunneling microscope (STM), which was situated in close proximity to an MoS_3 nanoparticle, led to the abstraction of sulfur from the nanoparticles and crystallization of a few closed MoS_2 layers enfolding the MoS_3 amorphous core of the nanoparticles [21]. Furthermore, leaving a film consisting of WS_3 nanoparticles for three years in the drawer led to a slow loss of sulfur from the film and crystallization of many WS_2 nanoparticles with fullerene-like structure [8a,22]. This diffusion-controlled process can be understood again by referring to the W-S phase diagram, which shows that the zone boundary between these two phases holds down to a room temperature. In another experiment, a dispersed powder of fullerene-like WS_2 nanoparticles were heated for 32 hours in a reducing atmosphere, and were found to be completely stable. This experiment serves to distinguish the *IF*-WS_2 phase from the stable 2H-WS_2 bulk phase. These studies provided instructive information on the growth mechanism and stability zone of the *IF* phases. A novel room-temperature method for producing of nested fullerene-like MoS_2 with an *a*-MoS_3 core using a sonoelectrochemical probe was described [23]. MoS_2 nanotubes also occur occasionally in this product.

The growth mechanism of *IF*-MS_2 (M=Mo,W) nanophase materials by sulfidization of the respective oxide nanoparticles, has been studied in quite a detail [9-11]. Here MO_3 nanoparticles rapidly react with H_2S gas on the surface at temperatures between 800-950 °C. Once the first sulfide layer enfolds the oxide nanoparticle, its surface is completely passivated and hence sintering of the nanoparticles is avoided. In the next step, which may last a few minutes, reduction of the oxide nanoparticle core into

MoO$_{3-x}$ phase by hydrogen takes place. In the third step, which is rather slow and may take a few hours, depending on the size of the nanoparticles and the temperature, a slow diffusion-controlled reaction of the reduced oxide core into sulfide takes place. It is important to note that the sulfidization of the suboxide core occurs along one single growth front in a highly organized fashion. Despite the lack of a complete commensuration between each two adjacent sulfide layers, it is believed that the upper sulfide layer serves as a template to the underlying growing sulfide layer as indicated by TEM analysis [24]. The small difference in densities between the oxide and the denser sulfide phases (7.1 vs. 7.6 g/cm^3, respectively), implies that a small but discernible hollow core is formed in the fully sulfidized *IF* nanoparticle. Further study of the reaction suggested a very unique mechanism for the sulfidization of the topmost oxide surface [9b]. In the first step one oxygen atom is abstracted from the oxide surface by reaction with hydrogen. In the subsequent step, abstraction of another oxygen atom from the same site promotes a shear movement of the metal atoms. Concomitantly, sulfur atom replaces these two oxygen atoms, leading to a reduction of the metal atom and insertion of a sulfur atom to the metal-oxide surface. At intermediate temperatures (700-850 °C), this reaction proceeds in a highly concerted fashion, leading to a perfectly crystalline monolayer of sulfide encapsulating the oxide nanoparticle, which is obtained in a matter of a second or less.

The synthesis of large amounts of WS$_2$ nanotubes, tens to hundreds micron long and with diameters in the range of 10-20 nm, has been recently reported [11c]. Moreover, in a few cases, the nanotubes were deposited on the quartz wall of the reactor self assembling into a foil a few cm long. In fact, the same fluidized-bed reactor and the same precursors are used for the synthesis of the fullerene-like (onion) WS$_2$ nanoparticles and the respective nanotubes. Only, minor changes in the supply of the precursors are necessary in order to switch from one product to the other. The nanophase tungsten oxide precursor is provided from a supplier or prepared in house. Unfortunately, the WS$_2$ nanotubes phase can not be obtained in a pure form, so far. Further work is needed in order to elucidate their detailed growth mechanism and thereupon achieve improved control of the reaction parameters. The length of the nanotubes is determined by the interplay between three reactions: tip growth of the oxide nanowhisker, reduction and sulfidization. Thus, in a strong reducing atmosphere (5% hydrogen in the gas mixture), fullerene-like WS$_2$ nanoparticles or short nanotubes, are obtained. If the reduction of the growing oxide nanowhiskers and its sulfidization rate are slowed-down by using milder reducing conditions (1% H$_2$); trimming the H$_2$S gas flow, and increasing the supply of the WO$_3$ nanoparticles to the reaction zone, long and highly crystalline nanotubes are obtained. It was shown before that annealing WO$_3$ powder in H$_2$ gas atmosphere leads to the formation of WO$_{2.72}$ (W$_{18}$O$_{49}$) needles by a process described as an evaporation-condensation process [23]. The W$_{18}$O$_{49}$ phase consists of pentagonal columns and hexagonal channels, promoting asymmetric growth along the column axis (||b), thus favoring whisker growth. Reduction of the WO$_3$ yields water molecules, which are known to form volatile tungsten oxide clusters and promote the nanowhisker growth [25,26]. The growth morphologies of these whiskers can be related to their crystallographic structures using the *Bravais-Friedel-Donnay-Harker* (BFDH) law as the rule of thumb [27]. According to this law the growth rate of a crystal face is inversely proportional to the interplanar spacing of that face. Indeed, the preferred growth direction of the nanowhisker is the short 0.38 axis of the WO_{3-x} phase, while the larger and more developed faces are those with the largest interplanar spacing. Contrarily, the lattice spacing in the precursor WO_3 does not differ appreciably in the three directions, resulting in no preferred growth direction.

One can therefore visualize the first step in the growth of the WS_2 nanotube as the formation of a sulfide encapsulated oxide nanowhisker. Remarkably, this nanowhisker is formed by the independent yet highly synergistic processes of the tip growth and sulfidization of the outer layer of the nanowhisker. Thus, while reaction with H_2S leads to the growth of a protective WS_2 monolayer on the side wall of the nanowhisker, the nanowhisker tip remains uncovered by this film and continues to grow incessibly like a usual tungsten oxide nanowhisker. This WS_2 monomolecular skin prohibits coalescence of the nanoparticle with neighboring oxide nanoparticles, which therefore drastically slows their coarsening. Simultaneous condensation of $(WO_3)_n$ or $(WO_{3-x} \cdot H_2O)_n$ clusters on the uncovered (sulfur-free) nanowhisker tip, and their immediate reduction by hydrogen gas, lead to a lower volatility of these clusters and therefore to the tip growth. This concerted mechanism leads to a fast growth of the sulfide-coated oxide nanowhisker. After the first layer of sulfide has been formed, in an almost instantaneous process, the conversion of the oxide core into tungsten sulfide is a rather slow diffusion-controlled process. The oxide whisker can be visualized as a template for the sulfide, which grows from outside inwards. The core consisting of the $W_{18}O_{49}$ phase provides a sufficiently open framework for the sulfidization to proceed until the entire oxide core is consumed and is converted into the respective sulfide. Furthermore, the highly ordered nature of the reduced oxide provides a kind of a template for a virtually dislocation-free sulfide layer growth. Further reduction of the oxide core, into e.g. the metal itself, would bring the sulfidization reaction into a halt [3b]. It was indeed shown that in the absence of the sulfide skin, the oxide nanowhisker is reduced rather swiftly to a pure tungsten nanorod in the reactor atmosphere. This phase is opaque to the sulfur atoms and it halts the conversion of the tungsten core into a WS_2 nanotube. Therefore, the encapsulation of the oxide nanowhisker, which tames the reduction of the core, allows for the gradual conversion of this nanoparticle into a hollow WS_2 nanotube. The present model alludes to the highly synergistic nature between the reduction and sulfidization processes during the WS_2 nanotube growth and the conversion of the oxide core into a multiwall tungsten sulfide nanotube. It is important to emphasize, that in this process, the diameter of the nanotube is determined by the precursor diameter. In general two kinds of nanotubes were obtained- those with a closed cap and open ended. The closed cap nanotubes have typically 10 WS_2 layers, while the number of layers is 4-7 for the open ended nanotubes, suggesting that some of the oxide core was volatilized during the slow oxide to sulfide conversion process of the core. Unfortunately, this method does not lend itself to the synthesis of single wall nanotubes, since long oxide nanowhiskers, a few nm thick, would not be mechanically stable.

The multiwall hollow WS_2 nanotubes, which are obtained at the end of the process, are quite perfect in shape, which has a favorable effect on some of their physical and electronic properties. A related method for the synthesis of WS_2 nanotubes, is to first synthesize crystalline and long $W_{18}O_{49}$ nanowhiskers, and subsequently sulfidize these nanowhiskers [11b,14a]. This method yields large amounts of very long (30 microns) WS_2 nanotubes, which are open ended. An alternative growth mechanism for WS_2 nanotubes from WO_3 nanowhiskers was proposed recently [28]. Here all the WS_2 layers terminate at the same position, implying that all the WS_2 layers form simultaneously from the oxide core. This is to be contrasted with the former growth mechanism, where the WS_2 layers were shown to grow from the outer surface of the oxide nanowhisker-inwards, gradually consuming the WO_{3-x} core [11].

In an alternative approach hydrothermal reaction of $(NH_4)_2MoS_4$ dissolved in $(NH_2OH)_2$-H_2SO_4 solution and subsequent annealing of the solid product at 550 °C produced MoS_2 nanotubes [29]. NbS_2 nanoparticles with fullerene-like structure were recently synthesized by reacting $NbCl_5$ vapor with H_2S gas in 550 °C and subsequent annealing at 550 °C in a reducing atmosphere [30]. In another recent study NbS_2 nanotubes were prepared by first synthesizing Ag alloyed $(NbS_4)_xI$ needles and their subsequent decomposition through e-beam or microwave irradiation [31]. The synthesized nanotubes were self-intercalated with Nb and were metallic in character. NbS_2 nanotubes were recently synthesized by first soaking carbon nanotubes in $NbCl_4$ solution; oxidizing the dried solid at 450 °C to form discontinuous coating of NbO_2 on the carbon nanotubes; and finally annealing the product in Ar/H_2S atmosphere at 1050 °C for 30 min [32].

Another strategy for the synthesis of nanomaterials with fullerene-like structures is through microwave irradiation of a suitable precursor [33]. In this process $Mo(CO)_6$ and $W(CO)_6$ powders were vaporized and subsequently mixed with a heated $H_2S(1\%)$/argon atmosphere, under microwave irradiation. When the temperature of the reaction was raised to 580 °C, some, though non-perfect, fullerene-like structures could be observed.

One of the earliest studied nanotubular phases was that of the 2-D compound GaSe [34]. In this compound, each atomic layer consists of a Ga-Ga dimer sandwiched between two outer selenium atoms in a hexagonal arrangement. Using tight binding model, the stable structure of these nanotubes was evaluated and their electronic structure was calculated as a function of the nanotubes diameter. First, it was found that like the bulk material, GaSe nanotubes are semiconductors. Furthermore, the strain energy in the nanotube was shown to increase as $1/d^2$, with d the nanotubes diameter, and the bandgap was found to shrink, as the nanotube diameter becomes smaller.

Although substantial effort was devoted to the synthesis of *IF* nanoparticles from transition metal dichalcogenide compounds, some work has been also dedicated to other metal chalcogenide compounds. Thus SnS_2 nanoparticles with fullerene-like structure were obtained by reacting SnO_2 nanoparticles with H_2S at 650 °C in a reducing atmosphere [35]. More intriguing is the synthesis of gamma-In_2S_3 and VS_2 nanoparticles with fullerene-like structure [36] using the respective oxide nanoparticles as the precursors, and InS nanotubes using solution chemistry methods [37]. In those three cases, the bulk 2-D phase is known to be unstable. Thus the seamless structure of the closed *IF* nanoparticles prevents water and oxygen uptake, and provides kinetic stabilization for the nanoparticles. The kinetic stabilization of such nanophases is of prime importance, since it provides access to phases which could not be studied before.

Bismuth sulfide crystallizes in layered structure, and is a semiconducting material with a direct band gap ranging from 1.3 to 1.7 eV. Applications of this material in photovoltaics and thermoelectrics can be envisaged. Bi_2S_3 powder was synthesized by the direct reaction between $Bi(NO_3)_3$ and $Na_2S.xH_2O$. After drying the power was loaded in a quartz boat together with sulfur and heated to 650 °C [38]. Bi_2S_3 nanotubes were synthesized under argon gas flow and careful purge of the residual oxygen. The reaction is believed to go through chemical vapor transport of the volatile BiS_2. Recently also, Sb_2X_3 (X=S,Se) polygonal microtubes were synthesized by using a modified solvothermal synthesis. Here $SbCl_3$ and X powder were added to a Teflon liner with ethanol as the solvent. The loaded autoclave was heated to 180 °C. The vapor of the reaction vessel was gradually released during the 7-days reaction period through a relief valve [39]. The synthesized microtubules exhibited a rectangular cross-section.

An intriguing case is that of alpha-NiS, which is metallic and forms with the hexagonal nickel arsenide (NiAs) structure at temperatures above 620 K. In this quasi 2-D structure, the nickel and sulfur atoms form separate layers, in which each nickel atom is bound to six sulfur atoms and *vice versa*. Having this fully bonded structure, NiS (NiAs) are not expected to form perfectly crystalline nanotubular structures. Nonetheless, by heating $NiCl_2.6H_2O$ to 60 °C in the presence of ammonia (NH_3) and CS_2, Qian and co-workers successfully synthesized microtubules of the quasi-2-D NiS compound [40]. They speculate that the ammonia molecules intercalate between the NiS layers and binds to the Ni atoms, thus passivating partially the interlayer Ni-S bonds. However, both X-ray diffraction (XRD) and electron diffraction (ED) patterns do not provide sufficient detail to determine the exact structure of this templated microtubule structure. In particular, the exact arrangement of the ammonia molecules within the lattice could not be resolved at this point.

Another strategy for the synthesis of nanotubules was described in Ref. [41]. Here, the quasi one-dimensional (1-D) compound $NaNb_2PS_{10}$ was synthesized first and subsequently dissolved in n-methylformamide (NMF). At low concentrations of the salt birefringence under shear flow was observed, suggesting the formation of a mineral liquid crystalline (MLC) phase. At higher concentrations, a hexagonal mesophase of the anisotropic objects was revealed. By careful elucidation of the structure of such solutions using small angle X-ray scattering, and subsequently freeze fracture electronmicrosopy and TEM, the anisotropic objects were found to be single wall nanotubules, 10 nm in diameter. The walls of the nanotubules are made of chains of the quasi 1-D $Nb_2PS_{10}^-$ anion ligated by the NMF solvent molecules. The use of alkylamines as chelating and structure directing agents for the synthesis of inorganic nanotubes is discussed in greater detail in the next section.

2.2 Metal-oxides and hydroxides

Numerous metal oxide phases are lamellar and could be potentially synthesized in the form of nanotubes. As a few recent studies show, this brand of nanotubes are potentially very useful for cathode materials in intercalation batteries (*vide infra*). Here hydrothermal synthesis; sol-gel processes and other "chemie douce" strategies are found to be particularly useful [7,42]. This observation is not surprising since the formation of various oxide minerals in nature was in fact the result of such reactions on geological time scales. Furthermore, numerous multinary oxide phases are synthesized by such methods for a few decades, now. Nonetheless, the first metal-oxide nanotubes were prepared through conventional high temperature method. In this study, carbon nanotubes templates were coated with an ultra-thin V_2O_5 film by high temperature evaporation. The carbon nanotubes template was then removed by heating the sample in oxygen atmosphere [43]. This strategy was also used for the synthesis of nanotubes from other kind of oxide compounds, such as TiO_2 and SiO_2 [44] (see however the discussion below). Highly crystalline V_2O_5 nanotubes, a few layers thick, were clearly observed. In another study V_2O_5 microtubules were prepared by heating a vanadium target in oxygen atmosphere using CO_2 laser [45]. Unfortunately, this report does not include any detailed TEM analysis, and consequently the structural information which can be drawn from this study is rather limited. More recently V_2O_5 nanotubes of a different sort were prepared by Nesper and co-workers using a sol-gel process followed by a prolonged hydrothermal treatment [7]. In the generic sol-gel process for the synthesis of nanotubes, a metal-organic compound is dissolved in an alcohol together with a template molecule, like alkylamine, which functions as a structure directing moiety. The addition of a small amount of water leads to a slow hydrolysis of the organic-metal compound, i.e. the formation of metal-oxide sol, but the template structure is retained. Acidifying the solution, leads to condensation, i.e. abstraction of OH groups from the sol, and transforms it into a gel, which consists of an -M-O- polymer with longer chains. In the case of the layered compound V_2O_5 [7], a sol was first prepared by mixing vanadium (V) oxide triisoporpoxide with hexadecylamine in ethanol, and aging the solution while stirring, which resulted in the hydrolysis of the vanadium oxide. Subsequent hydrothermal treatment at 180 °C led to condensation and formation of nanotubes with the formal composition $VO_{2.4} \cdot (C_{16}H_{33}NH_2)_{0.34}$. The V_2O_5 nanotubes adopt a spiral growth mode and form a scroll, which in most cases is open ended. They are highly crystalline, as is indicated by their X-ray and electron diffraction patterns. The c-axis separation is much larger than the bulk metal-metal distance in V_2O_5 (0.6 nm), and varies according to the size of the templating agent, i.e. the length of the alkyl tail of the organic amine molecule. Facile and (quite) reversible Li intercalation into such nanotubes has been demonstrated [7c]. After a few charge/discharge cycles however, the loading capacity of the nanotubes decreased by as much as 30%. Bearing on this early work, a few groups have attempted to use vanadium oxide nanotubes as cathode materials, recently. Thus V_2O_5 [46] and $Mn_{0.1}V_2O_5$ [47] nanotubes (nanoscrolles) were prepared according to the synthetic strategy laid down in Ref. [7], and were used to prepare rechargeable lithium cathodes.

The potential of the cathode V_2O_5 [46] against lithium anode varied gradually from 3.5 to 2.0 V upon discharge. The cathodes were cycled 100 times and have suffered modest losses only in their capacity.

Nanotubes-nanoscrolls of $Mg(OH)_2$ (CdI_2 structure) were obtained by hydrothermal treatment of Mg powder in the presence of ethylendiamine at 180 °C [48]. Scroll-like structures were obtained also from the layered compound GaOOH by sonicating an aqueous solution of $GaCl_3$ [49]. Rolled-up scroll-like structures were obtained also from ultrasonically treated $InCl_3$, $TlCl_3$ and $AlCl_3$ solutions, which demonstrate the generality of this process. Hydrolysis of this group of compounds (MCl_3: M=Al,In,Tl) results in the formation of the layered compounds MOOH, which, upon crystallization, prefer the fullerene-like structure. Nevertheless, MOCl compounds with a layered structure are also known to exist, and its formation during the sonication of MCl_3 solutions has not been convincingly excluded. These studies show again the preponderance of various layered compounds which produce rolled-up (nanoscrolls) structures, especially when low temperature growth techniques, like hydrothermal, or ultrasonic syntheses are being employed. More recently, the same group reported the formation of fullerene-like Tl_2O nanoparticles by the sonochemical reaction of $TlCl_3$ in aqueous solution [50]. Some compounds of the formula M_2O, like Ti_2O, possesses the anti-$CdCl_2$ structure, with the anion layer sandwiched between two cation layers. Currently, the yield of the IF-Tl_2O product is not very high (ca. 10%), but purification of this phase by the selective heating of the sample to 300 °C, has been demonstrated. Interestingly, Tl_2O platelets are very unstable against hydrolysis and oxidation in the ambient. The closed nature of the IF-Tl_2O nanoparticles slows down the intercalation of water (oxygen) across the van der Waals gaps and provides kinetic stabilization of the product even at 300 °C.

Various ternary oxides of transition metals, like Nb, Ta, Ti, etc. possess the layered structure and thus are expected to form nanotubular structures. Thus, nanoscale tubes, with scroll-like structure have been obtained from potassium hexaniobate-$K_4Nb_6O_{17}$ by acid exchange and careful exfoliation in basic solution [51,52]. The exfoliation process results in monomolecular sheets, which are unstable against folding, even at room temperature and consequently form the more stable scroll-like structures. Moreover, although the binary oxides of these transition metals do not possess the lamellar structure, and consequently they can not be expected to form crystalline nanotubular structures, the reduced oxides of these phases can. A typical example belonging to this category is the compound $H_2Ti_3O_7$, which possesses a lamellar structure and form perfectly crystalline nanotubes [53] (see Fig. 4). This nanotubular phase was prepared by the usual sol-gel process and subsequent treatment with a concentrated NaOH solution at 130 °C. The use of "chemie douce" processes for the synthesis of such novel nanoscopic phases, which can find applications in battery research and in catalysis is likely to lead to a surge of research in this direction in the future.

V_2O_5 is the "parent compound" of a whole family of vanadates with a layered structure formed by VO_5 square pyramids, like $Na_2V_3O_7$. In this ternary compound, the vanadium atom assumes a valency of IV rather than the valency of V of vanadium atom in the parent compound V_2O_5. Therefore, each vanadium atom possesses an unpaired spin and the material is an antiferromagnet. $Na_2V_3O_7$ nanotubes have recently been synthesized by a solid-state reaction [54]. These tubes are much more complex than the metal dichalcogenides from both their chemical and structural perspectives. The present work [54] provides a detailed account of the electronic, vibronic and to some extent the magnetic properties of such nanotubes (*vide infra*).

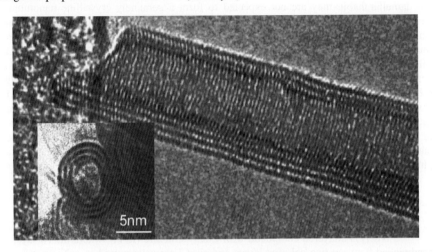

Fig. 4. TEM image of a $H_2Ti_3O_7$ nanotube (adapted from [53]).

The synthesis of a series of layered and hexagonal mesostructures of rare earth (Y, Eu, Tb, Dy, Er, Ho, Tm, Yb, Lu) oxides templated by dodecylsulfate oxides via the homogenous precipitation method using urea was demonstrated [55]. The layered mesophases were shown to have long curved or concentric structures similar to those observed for the layered precursors of vanadium oxide nanotubes [7]. This suggests that rare earth oxides in a nanotubular form could be synthesized, as has been indeed accomplished later-on by this team [56]. Thus, hollow nanotubes with an inner diameter of 3 nm were obtained from erbium, thulium, ytterbium, and lutetium oxide after anionic exchange of the surfactant with acetate ions. In contrast to the V_2O_5 reported earlier, the present nanotubes come as single inorganic layer templated by the surfactant. Ytterbium and lutetium oxide nanotubes were reproducibly obtained, whereas the synthesis of erbium and thulium oxide nanotubes were less reproducible. Yittrium oxide did not produce nanotubes, so far. These results indicate that the nanotubes of ytterbium and lutetium oxides with the smallest ionic radii form a stable phase. The present nanotubes, which are made of an oxide/templating moiety can be considered in fact as the inorganic analog of hexagonal mesophases and vesicles [57]. In particular the chemical bonds holding such structures are relatively weak, and the nanotubes are rather soft.

Numerous reports on micro- and nanotubular structures from various oxides, like SiO_2, TiO_2, ZnO, etc., appeared in recent years (see for example Refs. [44], [58], and [59]). Various "chemie douce" strategies or combination of them, which are very useful for the synthesis of oxide phases in general, have been employed for the synthesis of such micro- and nanostructures. Sol-gel, templated growth and hydrothermal synthesis are abundant among them. However, since the above oxide compounds do not crystallize in any lamellar habit, they are not expected to form a genuinely crystalline nanotubular structure (*vide infra* for further details). For example, in the case of the 3-D compound ZnO, crystalline microtubules were obtained [58], but the reported nanotubes were multicrystalline, see for example Ref. [60]. In qualitative terms, the strain involved in folding nanoparticles of a 3-D compound into a defect-free closed hollow structure, is very large. This elastic energy is much too big to be compensated by the healing of edge effects, which is the driving force for the formation of closed nanotubular structures. In several cases, see for example Ref. [61], nested carbon nanostructures (onions), which are often found on TEM grids, were mistakenly identified as nested structures or nanotubes of oxides or other inorganic compounds. Regardless of their crystallinity, such nanostructures are believed to be of a great value for catalytic, photocatalytic, photovoltaic, and perhaps other applications.

2.3 Metal halides

$NiCl_2$ is a layered compound with $CdCl_2$ structure (R3m), where the Ni layer is sandwiched between two chlorine layers and six Cl atoms surround each Ni atom in an octahedral arrangement. Strong ferromagnetic interactions occur between the Ni atoms, orienting the magnetic dipoles in the *a-b* plane of the layer ($\perp c$). Weak antiferromagnetic coupling between the Ni dipoles of adjacent layers leads to weak antiferromagnetic coupling between the adjacent layers in this material with Néel temperature of 51 K. Spherical and polyhedral nanoparticles of $NiCl_2$ and nanotubes thereof have been first reported in [6]. Here $NiCl_2$ powder was made to sublime at 960 °C under the flow of argon gas. The collected fluffy powder was analyzed by TEM. A few other synthetic techniques were devised, but none of them proved to yield reproducible amounts of the *IF*-$NiCl_2$ nanostructures. In one such method, NiO nanoparticles were prepared by oxidizing a Ni foil. The collected powder was reacted at 600 °C with Cl_2 gas in a reducing atmosphere. These and other methods produced polyhedral nanoparticles or nanotubes of the $NiCl_2$ occasionally, only. The method that proved to be most successful one in this respect was the reactive laser ablation [62], in which $NiCl_2$ powder was ablated in the presence of CCl_4 vapors at elevated temperatures (940 °C). The CCl_4 vapors were introduced to the reactor in order to compensate for the loss of the volatile chlorine in the ablated powder. Spherical nanoparticles found at the nanotube's tip are indicative of a vapor-liquid-solid (VLS) growth mechanism. Indeed the binary Ni-Cl phase diagram shows that at high temperatures a liquid phase coexists with the solid $NiCl_2$ phase. However, the sublimation temperature of the $NiCl_2$ phase is also close to this temperature, which did not allow heating of the ablated powder to higher temperatures. Analysis of the spherical tip did not reveal the exact nature of the molten phase, since it contained substantial amounts of carbon, probably from the CCl_4, and $NiCl_2$ nanoplatelets. Electron diffraction (ED) showed that most of the nanotubes were either of zig-zag type or having a small chiral angle.

IF-$NiCl_2$ nanostructures can not be antiferromagnetic, since there is no parity between the number of Ni atoms in the different layers of the nanotubes or the polyhedral structures. Furthermore, closed polyhedral structures with odd number of layers (1 and 3) have been synthesized, which can not be antiferromagnetic. Unfortunately, the synthesis of large amounts of these nanostructures proved to be rather difficult, mainly due to the hygroscopic nature of the compound. These difficulties curbed the study of the electronic and magnetic properties of these IF nanostructures, so far. Irradiation of particles of the layered compound $CdCl_2.nH_2O$ by the electron beam of a transmission electron microscope (TEM) led to the loss of the water molecules and recrystallization of the water-free $CdCl_2$ into closed cage polyhedral structures [63]. Similarly, nested $CdCl_2$ structures were obtained by vaporization/condensation of $CdCl_2.nH_2O$ powder in the oven at 750 °C under argon gas flow. In general, the synthesis of IF nanostructures from layered metal-halide compounds proved to be a much more intricate task than their chalcogenide analogs. The appreciably larger ionicy of these compounds, make them highly hygroscopic. Furthermore, the polarization of the M-X chemical bond results in a larger bending modulus of the halide compounds, compared with their metal dichalcogenides analogues. Finally, the volatility of the halides during the high temperature synthesis and TEM analysis is a further important obstacle for the synthesis of IF nanoparticles from 2-D halide compounds.

2.4   Boron-nitrides and boron-carbon-nitrides

$B_xC_yN_z$ nanotubes and fullerene-like structures have been synthesized by various laboratories in recent years. One of the most useful synthetic methods is by plasma arc technique. BN nanotubes were first reported by the Zettl group [5]. Since BN is an insulator, a composite anode was prepared from a tungsten rod with an empty bore in the center, which was stuffed with a pressed hexagonal BN powder. For the cathode, water-cooled Cu rod was used. The collected gray soot contained limited amount of multiwall BN nanotubes. It is possible that in this case, the tungsten rod serves also as a catalyst. By perfecting this method, double wall BN nanotubes of a uniform diameter (2 nm) were obtained in large amounts [64]. An alternative route employing HfB_2 electrodes in nitrogen atmosphere was reported [65]. This route led to the synthesis of BN nanotubes with varying number of walls, from a single wall to multiple wall nanotubes. The Hf was not incorporated into the tube and probably played the role of a catalyst. Using Ta instead of W as the metal anode, BN nanotubes with flat heads have been observed, alluding to the existence of three $(BN)_2$ squares in the cap [66]. In another synthetic approach, pyrolisis of CH_3CNBCl_3 complex in the presence of a Co catalyst provided $B_xC_yN_z$

nanotubes and nanofibers [67]. Clear evidence in support of the hypothesis that the cap contains 3 $(BN)_2$ squares was obtained in this case. This is to be contrasted with carbon nanotube caps, which contain 6 pentagons at either edge. Recently, long and quite perfect BN nanotubes were obtained by focusing a continuous CO_2 laser onto hexagonal boron nitride powder in N_2 gas atmosphere [68]. Furthermore, ropes consisting of hexagonal array of BN nanotubes have been observed in this study, which is indicative of the uniformity of the nanotubes' diameter (2 nm). The synthesis of $B_xC_yN_z$ nanotubes and BN cage structure has been intensively pursued by Bando's group (see for example Refs. [69]). Thus, a successful strategy for the synthesis of single and multiple wall $B_xC_yN_z$ nanotubes, through chemical substitution of carbon nanotubes, has been demonstrated [63a]. Here, single wall carbon nanotube bundles were thermally treated with BO_3 at 1523-1623 K under nitrogen gas flow. The resulting nanotubes had diameters of 1.2-1.4 nm, i.e. similar to the precursor nanotubes. Electron beam irradiation of hexagonal boron-nitride resulted in octahedral BN onions [69b]. These structures are characterized by the presence of six $(BN)_2$ squares embedded in the hexagonal BN network.

In another interesting study, multiwall nanotubes were prepared and spontaneous segregation of nanotubes of a different chemical composition was observed. Thus, concentric C and BN nanotubes and even C nanotubes followed by BN and again carbon nanotubes, were identified [70]. Concentric BC_2N and carbon nanotubes have been prepared as well. Using laser ablation method, concentric CN and carbon nanotubes with SiO_2 core have been prepared [71]. Conceivably, semiconducting junctions can be fabricated using the composite BN-C nanotubes. Silver nanoparticles encapsulated within boron nitride nanocages were produced by mixing boric acid, urea and silver nitrate and reduction at 700 °C under hydrogen atmosphere [72]. Generalization of this method to the encapsulation of other metallic nanoparticles was discussed.

2.5 Pure elements

Graphite is the most stable polytype of carbon and its hexagonal honeycomb structure is the archetype of layered (2-D) materials, producing fullerenes and nanotubes. Nonetheless, there are other elements that crystallize in lamellar network, like boron, arsine and phosphorous. These materials could therefore be envisaged as precursors for *IF* nanostructures. Lipscomb and co-workers were among the first to realize the analogy between carbon and non-carbon fullerenes, by studying the analogous structures of C_{60} and $B_{32}H_{32}$. In this case, the extra hydrogen atoms are emerging radially outwards from the boron backbone of the fullerene [73]. More recently, they studied the analogy between carbon and BH, or pure boron nanotubes [74]. Boustani et al., used first principle *ab-initio* calculations to derive the stable structure of boron nanotubes and to calculate their properties, see for example Ref. [75] and references therein. They find that the pure boron nanotubes are stable and exhibit metallic conductivity. However, as of now, no such nanotubes or fullerene-like structures have been synthesized.

The structure and properties of black phosphorous (b-P) and arsenic nanotubes was elucidated using *ab-initio* calculations [76a]. Orthorhombic black phosphorous is the most stable form of this element in ambient conditions. It consists of three-fold coordinated phosphorus atoms arranged in a puckered honeycomb (layered) network. The puckering is the result of the sp^3 bonding of the P atoms, with the lone pairs of each two neighboring P atoms in trans configuration (opposing directions). This leads to a substantial strain in the P nanotubes. Despite the large strain, stable structures for the nanotubes were found in the calculations, both in zig-zag and armchair configurations. In analogy to b-P, the phosphorous nanotubes are found to be semiconductors, irrespective of their chirality. The Young's modulus of the P nanotubes was calculated to be around 0.3 TPa, i.e. roughly ¼ that of carbon nanotubes. Fullerenic structures of b-P were found to be unstable *vis a vis* P_4 units [76b]. These findings are consistent with the observation that folding the lamella along two axes (fullerenes) involves larger strain than along one axis only (nanotubes).

Metallic bismuth (R-Bi) has a pseudolayered structure very similar to that of rhombohedral graphite and black phosphorus. In each layer, one Bi atom is connected with three other Bi atoms according to the 8-*N* rule and thus forms a trigonal pyramid. These pyramids further form a folded bismuth layer by vertex-sharing. The distances between one Bi atom and its three neighbors in the same layer and the neighboring layer are 0.3072 and 0.3529 nm, respectively, and the Bi-Bi-Bi bonding angle is 95.5 °. The analogy between the layered structures of R-Bi and graphite/phosphorus suggests that nanotubes of this element may also exist [77, 78]. Bismuth's small electron effective mass ($0.001m_0$) and large mean free path (0.4 mm at 4 K) make Bi nanotube an interesting system for studying quantum confinement effects. In addition, nanoscaled bismuth materials have recently been suggested to have enhanced thermoelectric properties at room temperature. However, owing to the relatively low melting point of Bi (271.3 °C), the synthesis of Bi nanotubes is much more difficult than that of its carbon or MX_2 analogues. Most of the existing high temperature approaches, such as arc-discharge evaporation, laser ablation, or chemical vapor deposition are inappropriate for synthesis of Bi nanotubes. Given the low temperatures to be used, hydrothermal synthesis was found to be a suitable strategy in this context. Thus bismuth nitrate was mixed with hydrazine and the pH was adjusted to 12 by excess ammonia. The reaction mixture was heated in an autoclave to 120 °C for 12 hr. The black solid was found to consist of crystalline multi wall Bi nanotubes with a diameter of 5 nm and a few microns long [78]. However, the Bi nanotubes were found to be very sensitive to electron beam damage, which is attributed to the fast melting of the nanotubes under the irradiation field.

Silicon with its sp^3 bonds is unable to form isolated polyhedral structures, although a few theoretical works indicated that such structures can not be fully excluded. Since one out of four sp^3 bonds of a given Si atom is pointing outwards of the cage, the most stable fullerene-like structure in this case is a network of interconnected cages. This kind of network is realized in alkali-metal doped silicon clathrates, which were identified to have a connected fullerene-like structure [79]. In these compounds Si polyhedra of twelve 5-fold rings and two, or four more 6-fold rings share faces, and form a network of hollow cage structures, which accommodate endohedral metal atoms, like Na. The fact that endohedral sodium atoms where found in the inflated lattice of the clathrates indicated that their structure is stabilized by the charge transfer from the metal atom guest to the host. However, using a combination of successive vacuum "degassing" of Na, followed by density separation and centrifugation, substantial amounts of the essentially Na-free material were prepared [80]. Atomic absorption analysis showed that the content of residual Na was 600 ppm by weight; less than 1 Na per 1400 Si atoms in the structure. The refined crystal structure of the expanded Si_{136} phase indicates that the degree of distortion from tetrahedral symmetry of the silicon atoms is quite small. Theoretical calculations indicate that the new form of silicon should be a wide bandgap semiconductor. This prediction is borne out by experiment: electrical conductivity and optical absorption measurements yield a band gap of 1.9 eV, approximately twice the value of "normal" semiconducting silicon. Solid state NMR measurements and theoretical calculations confirm that low sodium content silicon clathrates can be considered as polymerized fullerenes, while those with high sodium content are more ionic, like endohedral fullerenes or Zintl phases [81].

3 Thermodynamic, structural and topological considerations

Substantial evidence has been gathered to indicate that *IF* phases are the stable form of nanoparticles of layered compounds. Nonetheless, the thermodynamic stability of fullerene-like materials and nanotubes is a rather intricate issue and only since recently is being considered in a systematic way, mostly through the combination of experimental work and *ab-initio* calculations. *IF* structures are not expected to be globally stable, but they are probably the stable phase of a layered compound, when the nanoparticles are not allowed to grow beyond, say a fraction of a micron (arrested growth). The *IF* phase is postulated to be stable in a narrow domain of conditions, which must be very close to the existence zone of the layered compound itself. A number of observations support this conjecture.

WS_3 is stable below 850 °C under excess of sulfur, and is an amorphous solid. This compound will therefore lose sulfur atoms and crystallize into WS_2, which has a layered (2-D) structure, upon heating or in a sulfur poor environment. Isolated nanoparticles of WS_3 which are allowed to crystallize without coarsening result in fullerene-like WS_2 (MoS_2) nanoparticles even at room temperature [22].

Synthesis of nanotubular phases of numerous layered MS_2 compounds (M=Nb,Ta,Zr,Hf,Ti and X=S,Se) from MS_3 precursor was recently demonstrated [19]. In these works, a powder of the MX_3 precursor was prepared first. Subsequent high temperature (>1000 °C) annealing under hydrogen stream led to sulfur abstraction and formation of uniform and long nanotubes of the respective MX_2 compound. This observation indicates that indeed the *IF*-MX_2 phase lies on the border line between the two macroscopically stable phases MX_3 and MX_2. By careful thermal decomposition of the MX_3 phase in an X-free atmosphere, a phase consisting of MX_2 nanoparticles with fullerene-like structure can be obtained. The *IF* nanoparticles phase becomes the predominant product if the coarsening of the nanoparticles is avoided.

Remskar and co-workers have made an extensive use of the chemical vapor transport (CVT) technique for the growth of multiwall WS_2 (MoS_2) [12], and more recently single-wall MoS_2 nanotubes [16]. The CVT is the preferred synthetic method for crystals of various lamellar compounds, like MoS_2, etc. Not surprising therefore, a small deviation from the window of growth conditions, leads to the growth of MoS_2 platelets rather than to the nanotubes phase. It is also found that upon annealing of a loose powder of fullerene-like MoS_2 (WS_2) nanoparticles, above say 1050 °C, the nanoparticles transform into platelets of the same compound. These experimental facts suggest that the lamellar and nanotubular (*IF*) phases share a joint phase boundary in the binary Mo-S phase diagram.

The energetics of MoS_2 nanotubes and stripes of the same compound consisting of multiple MoS_2 layers was calculated and their relative stability was discussed [82]. Three terms were considered in this calculation: 1. the energy of a MoS_2 atomic unit within the layer (i.e. bulk MoS_2 unit); 2. The same unit when it is at the prismatic (100) edge of the stripe; 3. The van der Waals (vdW) energy of interaction between the layers. The energy of the bulk unit (1) was calculated from *ab-initio* theory. When the tube is formed the Mo-S bonds are distorted and consequently the energy of the MoS_2 lamella increases, like $1/d^2$ [83]. The energy of an edge MoS_2 unit was calculated also from *ab-initio* theory. The vdW energy was estimated from the measured surface energy of MoS_2 platelets. Taking these terms together it was found that multiwall nanotubes are stable compared to the multilayer stripe over a limited range of conditions only, as shown in Fig. 5. For too narrow stripes, the bending energy predominates and the tubular structures are not stable. Nanotubes 4-7 layers thick and 6 nm (inner) diameter are found to be more stable than the respective stripes. Despite the fact that several approximations were used in these calculations, good agreement between theory and experiment was obtained. This work presents further evidence that the WS_2 (MoS_2) nanotubular phase exists in the proximity of the bulk 2-D phase of these compounds.

Continuum models were able to provide important clues as to the structure of closed cage nanoparticles and nanotubes, although they are unable to provide information on the detailed atomic structure of a nanoparticle [84]. In accordance with this model, the cross-section of WS_2 nanotubes synthesized via a solid-gas route, were found to be mostly circular, with no evidence for polyherdal cross-sections [85]. The theory provides also important insight into the structure of a fullerene-like nanostructure with multiple layers ("onions"). Here, curvature occurs along two main axes, which entails considerably higher elastic energy than in the case of a nanotube, where uniaxial bending is imposed. The theory finds that caged nanoparticles of MoS_2 (WS_2) with large radius and a few layers thick fold in a rather regular fashion with no cusps in their structure. A transformation from coherently (evenly) folded into dislocated (polyhedral) structure occurs for nanoparticles with the thickness to radius ratio $h/R > 0.2$. Following the growth of fullerene-like WS_2 (MoS_2) nanoparticles from the respective oxide nanoparticles, revealed that such a transformation indeed occurs, when the number of

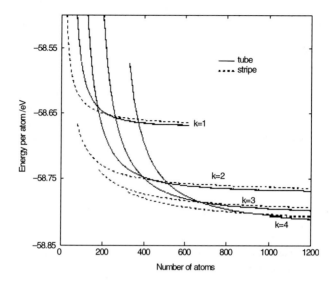

Fig. 5. Calculated energies of MoS$_2$ monolayer (k =1) and multilayer (k >1) stripes in comparison to MoS$_2$ single-wall (k =1) and multiwall (k > 1) nanotubes as a function of size (from [82]).

sulfide layers in the wall is larger than four [9a,10]. The quasi-spherical nanoparticles were shown to abruptly transform into polyherdral structures upon thickening.

This model can be also used to shed some light on the topology of metal halide *IF* nanoparticles. Fullerene-like nanostructures and nanotubes have been recently realized for the layered compounds NiCl$_2$ [6] and CdCl$_2$ [63]. The ionicity of metal halides (ca. 0.75), is appreciably larger than that of MoS$_2$ (0.12). Consequently, the bending and shear moduli of the layered halides is appreciably larger than that of MoS$_2$. The faceting of the fullerene-like nanoparticles of metal-halides [6,62,63] can be ascribed to their large bending modulus. Indeed, *IF*-CdCl$_2$ nanoparticles obtained via e-beam recrystallization show predominantly hexagonal or rectangular cross-sections, depending on the tilt-angle of the nanoparticles *vis a vis* the electron beam [63]. *IF*-NiCl$_2$ nanoparticles obtained via sublimation/condensation process often come as polyhedra with hexagonal shape [6]. On the other hand, NiCl$_2$ nanotubes, for which the bending energy is appreciably smaller, exhibit mostly a circular cross-section [62]. It is to be noted however, that presently the detailed atomic structure of metal halides nanoparticles with *IF* structure is poorly understood.

An observation of great importance is that nanoparticles with *IF* structure of various compounds could be obtained, although the bulk layered compound is either very difficult to synthesize or is virtually nonexistent. This unexpected result can be attributed to the fact that the *IF* structure is seamless while the bulk (platelet form) expose reactive edges and interacts only very weakly with the ambient. For example, coaxial nanotubes of MoS_2 and WS_2 intercalated with Ag and Au atoms were recently reported [86]. The analogous phase in the bulk material is not known. Chalcogenides of the first row transition metals, like $CrSe_2$ and VS_2, are not stable in the layered structure. Notwithstanding, VS_2 nanoparticles with a fullerene-like structure were found to be stable [36]. The unexpected extra stability of this structure emanates from the closed seamless structure of the *IF*, which does not expose the chemically reactive (100) faces to the hostile environment. Similarly, gamma-In_2S_3, which is an unstable polytype with a layered structure, was found to be stable in the *IF* form [36]. More recently, nanotubes of InS were obtained in a low temperature reaction between tri-butyl indium and H_2S in the presence of thiobenzene catalyst [37]. The layered structure of InS was not known before, showing that novel 2-D phases can be stabilized by the nanotubular structures.

Many of the 2-D compounds, especially the more ionic ones, like the metal-dihalides, are not very stable, or completely unstable in the ambient atmosphere. Upon exposure to the ambient atmosphere they intercalate water or oxygen into the van der Waals gap between the layers and subsequently exfoliate or are oxidized. Some, like $CdCl_2$ intercalate water and remain as layered monohydrate. Tl_2O with its anti-$CdCl_2$ structure is also very unstable in the ambient atmosphere. However, the seamless fullerene-like structure of these nanoparticles inhibits water and oxygen intercalation, thus providing substantially improved stability for the nanoparticles in the atmosphere, compared to their bulk analog [6,50,62,63]. This kinetic stabilization of the nanoparticles can be utilized for the study of these compounds, a task hitherto practically impossible. The stability of GaN nanotubes was considered theoretically [87a]. Although the zincblende polymorph is more stable than the lamellar phase of this compound, it was nevertheless observed that the nanotubes form a stable structure. These ideas open new avenues for the synthesis and study of materials with layered structure, which could not be previously obtained, or could not be exposed to the ambient, and therefore could only be studied to a limited extent. On the other hand, this concept provides a vehicle for the study of nanotubular structures with interesting properties, which were not studied heretofore. Recently GaN nanotubes of the zincblende crystalline structure were obtained by using a faceted ZnO nanowires as templates [87b]. Here growth along $(11\bar{2}0)$ easy axes is accomplished.

Layered compounds come, generally, in more than one stacking polytype [88]. Some like CdI_2 are known to have numerous polytypes, where only the stacking of the layers changes from one polytype to the other. For instance, the two most abundant polytypes of MoS_2 are the 2H and 3R. The 2H polytype stands for a hexagonal structure with two S-Mo-S layers (repeat units) in the unit cell (AbA⋯BaB⋯AbA⋯BaB, etc). The 3R polytype has a rhombohedral unit cell of three repeating units (layers), i.e. (AbA⋯BcB⋯CaC⋯AbA⋯BcB⋯CaC, and so forth). The most common polytype of MoS_2 is the 2H, but the 3R polytype was found, for example, in thin MoS_2 films prepared by sputtering [89]. Here, the large energy input of the sputtered ions alter the reaction kinetics, resulting in the synthesis of films consisting of the somewhat less stable polytype. The existence of various polytypes in layered compounds reflects the small differences in energy between the different polytypes, i.e. the smallness of the shearing energy of the layers, compared with the energy of distorting or breaking/making of a chemical bond. The nanotubes grown by the gas phase reaction between MoO_3 (WO_3) and H_2S at 850 °C were found to be exclusively of the 2H polytype [4,11,90]. The appearance of the 3R polytype in such nanotubes can be probably traced to the strain. For example, a "superlattice" of 2H and 3R polytypes was found to exist in MoS_2 nanotubes grown by chemical vapor transport [12e]. The preference of the rhombohedral polytype in both MoS_2 and WS_2 microtubes was attributed to strain effects [12d]. These observations indicate that, in analogy to the thin films, the kinetics of the growth of the nanotubes influences the strain relief mechanism and therefore the nanotubes can accommodate polytypes which are thermodynamically less stable, and are not regularly found in the bulk compounds.

The trigonal prismatic structure of MoS_2 alludes to the possibility of forming stable point defects consisting of a triangle or a rhombus (rectangle) [3b]. The existence of "bucky-tetrahedra" [22] and "bucky-cubes (bucky-octahedra)" [91], which have four triangles and six rhombi in their corners, respectively, was hypothized. However, the most compelling evidence in support of this idea was obtained in nanoparticles collected from the soot of laser ablated MoS_2 pellet [92]. Theoretical calculations indicate that rectangular and even octahedral elements are likely to be stable in the nanotube tip [83]. These calculations suggest that only small distortions of the Mo-S bond and the S-Mo-S dihedral angles are necessary in order to close the cap by a rectangle or octahedron. Point defects of this symmetry were not observed in carbon fullerenes, most likely because such topological elements are not favorized by the sp^2 bonding of carbon atoms. Since the bending energy of a typical inorganic layered compound is almost an order of magnitude larger than that of graphite, it is difficult to envisage the formation of pentagons as folding inducing elements in such compounds. In a recent study various topological defects giving rise to the folding of straight segments of WS_2 nanotubes were considered [93]. However, more detailed *ab-initio* calculations would be necessary to confirm the stability of such defects. These examples and others illustrate the influence of the lattice structure of the layered compound on the detailed topology of the fullerene-like nanoparticle or the nanotube cap obtained for such compounds.

Since boron and nitrogen are situated to the left and right of carbon, respectively, in the periodic table, the B-N pair is isoelectronic with C-C. Naturally therefore, early attempts to synthesize/calculate the analogue of C_{60} were devoted to the $B_{30}N_{30}$ molecule. Hellas, introduction of pentagons into the polyhedron was found to produce induce instability, since pairs of B-B or N-N like atoms are less stable than the B-N pair. Therefore squares are the preferred element of lower symmetry, which induces folding of *IF*-BN. $B_{12}N_{12}$ polyhedron made of eight hexagons and six squares, was calculated to be a most stable closed cage structure [94]. This hypothesis was invariably confirmed through numerous experimental studies [95-97]. In the case of BC_2N nanotubes two different arrangement of the sheet are possible, leading to two isomers with different structure and distinct in their electrical properties [98]. These works underpin the interplay between *ab-initio* calculations and TEM analysis as a most exquisite tool for determining the atomic structure of *IF* nanoparticles.

Synthesis of coaxial sheathed BN/C nanotubes by arc discharge was demonstrated recently. Using a composite anode in this set up, a spontaneous segregation of nanotubes of a different chemical composition was observed. Sandwich-type BN nanotubes sheathed by carbon nanotubes from its two sides was observed [70]. The segregation of the carbon nanotubes into the inner and outer surfaces of the concentric nanotubes structure is attributed to the lower surface energy of graphite as compared with hexagonal BN. Furthermore, binary sheathed C/BN nanotubes, with the carbon nanotube on the outer surface and the BN nanotubes on the inner surface were observed. Concentric BC_2N and carbon nanotubes have been reported as well. Continuum theory of matter was recently used to analyze this situation [99]. First, it was established that ternary (A/B/A) sheathed nanotubes are generically unstable below a minimum thickness (a few atomic layers) of A. In the case of a BN-C pair the theory concludes that the lower surface energy of carbon and its higher stiffness leads to its growth on the outer surface of the nanotube. For thicker carbon layers the ternary system C/BN/C with carbon occupying both the outer and inner surfaces and the BN nanotubes sandwich between the two is the stable configuration, in agreement with the experiment [70]. Given the fact that some carbon nanotubes are semiconducting, while BN nanotubes are high bandgap materials, semiconducting junctions can be conceivably, fabricated using the composite BN-C nanotubes.

X-ray diffraction studies have shown an expansion of 2-4% in the *c*-axis of multi wall *IF* structures (including inorganic nanotubes) [4,9,11]. The shift of the (00*l*) peak to lower scattering angles (larger *c*-axis spacing) for the *IF* phase compared to the bulk material is a clear distinction of this phase and serves as a quality measure for the synthetic process. The average size of the nanoparticles can be calculated from the peak width. Obviously, full commensuration between the layers of a multi-wall fullerene-like nanoparticle or nanotube is impossible. However, the structural relationship between the different layers, which exists in 2H or 3R polytypes, is locally preserved. The lattice expansion along the c-axis serves to relieve the strain of the folded structures, allowing the *IF* nanoparticles to preserve their high degree of crystallinity.

Numerous reports of various oxide nanotubes have appeared in the literature in recent years [8e,44,51,52,58,59]. Here a clear distinction between those materials having anisotropic 2-D layered structure and those having isotropic 3-D lattice is needed. Clearly, nanotubes of compounds with layered structure can maintain their high degree of crystallinity during the folding process. Soft chemistry methods are gaining importance in the synthesis of oxide nanotubes. Here a structure-directing (template) agent like the electron-rich alkylamine molecules, or porous alumina are frequently used. Together with the inorganic compound a supramolecular architecture with an alternating inorganic-organic layers is obtained. On top of its role as being the template for the growth of the ordered array of the inorganic materials, the alkylamine template contributes to the softening of the composite structure during the folding process. This practice is very useful for lamellar compounds with highly ionic character, such as oxides of various sorts. Since such alkylamine template can not sustain high temperatures, "chimie douce" methods are particularly useful for the preparation of oxide nanotubes. However, for other layered compounds with high degree of ionicity, such as the metal dihalides (e.g. $NiCl_2$), these kind of synthetic techniques are of limited use since they suffer fast hydrolysis. The amount of the templating agent between the layers of the nanotube and its exact structure is rather important in determining the structure and lattice spacing in the nanotubes, but so far this point received little attention. Theoretical calculations could play a great role in predicting affordable topologies, but the size of the unit cell of large diameter nanotubes with the templating agent is perhaps beyond the reach of the presently available computer resources.

The crystallinity of nanotubes with isotropic 3-D lattice, like rutile, has to be scrutinized closely. Nanotubes of these compounds are unlikely to form a fully crystalline, evenly folded network. The large strain involved in the folding process of such nanostructures can be relaxed by introducing a variety of defects, grain boundaries, etc. Using structure directing moieties, like alkyamines, nanotubes (nanoscrolls) from isotropic 3-D compounds, like PbS have been reported [8e,100]. Note however, that the spacing between the molecular layers of the nanotubes is much larger than that of the bulk inorganic compound. This observation suggests that a new kind of a quasi-lamellar composite inorganic-organic structure has been formed, affording thereby an easy folding into a nanotubular shape. Quasi-spherical fullerene-like nanoparticles are even more difficult to obtain. In this case folding occurs along two axes, rather than the single axis folding in nanotubes, leading thereby to exceedingly high strains. Irrespective of its crystallinity, oxide nanotubes are likely to play a major role in variety of potential applications, including photocatalysis, photovoltaic energy conversion, environmental issues, sensors, membranes, battery materials, etc.

4 Physical properties

The physical properties of *IF* materials, which have been discussed in some length in previous reports [8b] will be mentioned here only briefly. Additional information gathered over the last two years will nevertheless be described and discussed in the present report in greater detail. Band structure calculations of various inorganic nanotubes, like BN [101], MoS_2 (WS_2) [83], b-P [76] and other inorganic compounds with layered structure and semiconducting character, led to a few generic observations: Firstly, in contrast to carbon nanotubes, which can be metallic or semiconducting depending on their diameter and chirality, all inorganic nanotubes derived from bulk semiconductor compounds were found to be semiconductors, irrespective of their detailed structure. Bulk BN (MoS_2, WS_2, b-P) materials are known to have an indirect band gap. On the other hand the smallest forbidden gap of zigzag (n,0) nanotubes from the above materials were found to be a direct (Γ-Γ) one. An indirect bandgap (Δ-Γ) was calculated for the armchair (n,n) nanotubes. The third point to be noted is that the strain in the nanotubes was found to scale like $1/d^2$, where d is the nanotube diameter. The fourth point is that overwhelmingly, the band gap of the inorganic nanotubes was found to diminish with decreasing diameter of the (inorganic) nanotubes. In contrast to that, the band gap of semiconducting carbon nanotubes increases with a shrinking diameter of the cage. It should be furthermore emphasized that generically, the bandgap of semiconducting nanoparticles increases with a decrease of its diameter, which is attributed to the quantum size confinement of the electron wavefunction [102]. Optical measurements of MoS_2 (WS_2) nanoparticles with *IF* structure confirmed some of these conclusions [103]. In particular, the excitonic bandgap of *IF*-WS_2 (*IF*-MoS_2) was found to shift to lower energies with decreasing diameter of the nanoparticles. In accordance with the properties of bulk NbS_2 nanotubes of this material were calculated to be metallic irrespective of their diameter and chirality [104]. Preliminary transport measurements tend to confirm the theoretical calculations [19b]. However, they also indicate, that in contrast to the bulk material, the nanotubes do not exhibit superconductivity down to 4.2 K.

In a recent study [105], individual WS_2 nanotubes were studied by a high resolution scanning tunneling microscopy (STM) and their bandgap was determined using scanning tunneling spectroscopy (STS). Earlier studies failed to provide such information, but the improved synthesis provided longer and more uniform nanotubes, which adhered sufficiently tight to the underlying graphite substrate and could be examined by the STM probe without being swept by the tip. The study was undertaken in ambient atmosphere, and it suffered from substantial scattering in the data, which could be attributed to various factors. The experimental work was complemented with band structure calculations of nanotubes with diameter as large as 4 nm and extrapolation to nanotubes of a larger diameter. Notwithstanding the experimental and computational difficulties, satisfactory agreement was obtained between theory and experiment. This study shows (see Fig. 6), that indeed the bandgap decreases with shrinking diameter of the WS_2 nanotubes.

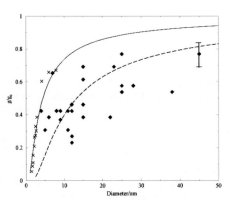

Fig. 6. Comparison of experimental data (diamonds, error bar of +/-100 meV is drawn only for the largest tube for illustration) and calculated values (x) for g/g_0 vs. diameter. The x signs represent the calculated values for nanotubes of diameter <8 nm. Solid curve follows the extrapolation formula, and dashed curve is corresponding fit to experimental points [107].

Generically, iodine doped bulk 2H-MoS$_2$ is known to exhibit very high n-type conductivity. In a recent study, using a room temperature ultra high-vacuum STM/STS instrument, the electronic properties of iodine-doped MoS$_2$ nanotubes were studied and were found to depend on the density of the free electrons of the underlying substrate [106]. Regions of the nanotubes which were suspended on a vacuum gap, between two solid contacts were found to exhibit semiconducting behavior in close agreement with previous data. However, regions of the nanotubes which were in intimate contact with the underlying substrate exhibited a metallic behavior, which indicates that charge transfer from the underlying graphite substrate to the nanotubes took place in this case. More careful measurements are nevertheless necessary, with for example substrates of varying conductivity, to confirm these preliminary findings.

Intercalation of fullerene-like MoS$_2$ and WS$_2$ nanoparticles with alkali metal atoms from the vapor phase has been recently accomplished [107]. For the sake of comparison, bulk platelets of the 2H polytype were also intercalated. Extensive analysis of the intercalated samples was carried-out in a strictly water and air-free environment. Unfortunately, the nanoparticles could not be uniformly intercalated, but they nevertheless revealed substantial changes in their structural, chemical, and physical behavior. For most of the analyzed fullerene-like nanoparticles the intercalation took place only in the 5-10 outermost layers, whereas the inner layers remained in their pristine form. Also some of the nanoparticles were not intercalated, even after very long exposure to the alkali metal vapor. The structure of the intercalated *IF* nanoparticles did not change appreciably, but some disorder was nonetheless observed. Exposure of the intercalated nanoparticles to the ambient atmosphere led to quite remarkable changes. After about one day exposure, already one hydration layer surrounded the alkali atoms, which led to almost 0.3 nm expansion of the interlayer distance (0.6 nm). Longer exposure led to absorption of more water molecules into the lattice, in particular for the sodium intercalated samples. Fig. 7 shows TEM images of a pristine MoS$_2$ fullerene-like nanoparticle and a similar nanoparticle after intercalation and exposure to the ambient atmosphere for a few days. The lattice expansion of the interlayer (c-axis) distance in the ambient-exposed nanoparticle is clearly delineated in this image. Longer exposure to the atmosphere led to a slow deintercalation of the hydrated ions from the *IF* nanoparticles. After a 4-8 weeks exposure period the nanoparticles were deintercalated completely. Structural analysis by TEM showed that the nanoparticles returned essentially to their original structure, but the continuity of the molecular sheets was disrupted from place to place.

While the electrical conductivity of the *IF*-phase increased by 4-8 orders of magnitudes immediately after the intercalation, it acquired a semi-insulating behavior with somewhat increased resistivity values following the lengthy exposure to the ambient atmosphere. Magnetic measurements showed that the samples changed from diamagnetic for the pristine phase to (Heisenberg- i.e. temperature independent) paramagnetic behavior immediately after the intercalation. After long exposure to the atmosphere the samples became diamagnetic again with somewhat increased values. These measurements clearly indicate that electron transfer from the guest to the host lattice occurred in the freshly intercalated phases. Also, the deintercalation of the samples (after long exposure to the atmosphere) seem to have led to a kind of cleaning effect ("outgassing") of the samples from unknown (possibly metallic) impurities. Though not tried, the process could have been repeated a few times, and would possibly lead to higher uploads of the intercalated alkali atoms in the *IF* phase. Interestingly enough, none of the intercalated samples, including the bulk 2H-MoS_2, revealed a transition to a superconductor at low temperatures even for the highest loading of the alkali intercalant.

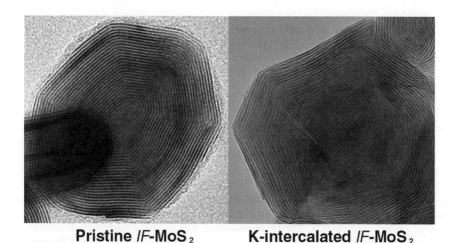

Pristine *IF*-MoS_2   **K-intercalated *IF*-MoS_2**

Fig. 7. TEM images of a fullerene-like MoS_2 nanoparticle before and after Ka intercalation and exposure to the ambient atmosphere [109]. Note the expansion of the interlayer distance from the original (0.61 nm) spacing after exposure of the sample to the ambient atmosphere.

5 Applications

Some potential applications of *IF* materials have emerged already, others like in nanotechnology will probably come in due time. The most mature field of applications of these nanomaterials is in tribology, but few other potential applications are quite promising as well.

5.1 Tribological applications

Many potential applications of the *IF* nanomaterials have been described in great detail in various reviews [8b,108]. These studies will therefore only briefly resuscitated in the present report. The intriguing tribological properties of *IF*-WS$_2$ in various lubricating fluids were first reported by Rapoport et al [109]. More recently, a number of groups have widened the scope of these studies [110-114], which suggests numerous potential applications for these nanomaterials. Thus in [110] thin coatings consisting of *IF*-MoS$_2$ nanoparticles were prepared by a modified arc-discharge method and its tribological properties were evaluated in both inert gas atmosphere and in humid air. Comparison with a sputtered MoS$_2$ film, which consisted of platelets of this material, was also included in this study. The *IF* film exhibited friction coefficients below 0.01 and longevity, even in 45% humidity, while the sputtered film deteriorated rather rapidly. The low friction coefficients of the *IF* film was attributed to rolling friction.

In another study [111], self lubricating porous matrices of brass-graphite were prepared by first densifying the metal powder and sintering. Subsequently, *IF*-WS$_2$ powder (5%) was impregnated into the matrix. For the sake of comparison reference samples of the pristine densified porous matrix and the same matrix impregnated with 2H-WS$_2$ were tested as well. The sample without any solid lubricant exhibited relatively high friction coefficient under low loads (ca. 0.15) and was seized already under 45 kg/cm^2. Both 2H and *IF* impregnated samples exhibited relatively low friction coefficients (0.03) under low loads. The 2H impregnated sample was seized under loads exceeding 60 kg/cm^2. The *IF* impregnated sample was not seized until 130 kg/cm^2 was applied on the contact area. The *IF* impregnated sample distinguished itself also in its prolonged lifetimes, which exceeded by more than one order of magnitude the 2H impregnated sample.

The mechanism of action of the *IF* nanoparticles in the contact surfaces is under detailed investigation. The porous matrix can be described by a multi-scale network of interconnected pores ranging from a few tens of microns to a few nm in size. It is hypothesized from this [111] and the following studies that the nanoparticles are stored in the porous matrix and are gradually released to the contact surface during the long measurements, which sometime last two weeks. Once at the surface, these nanoparticles act as small nano-ballbearings preventing the contact between the asperities of the mating metal surfaces. The low friction coefficients can be attributed to the low shear stress necessary for rolling/sliding of the nanoparticles. The rolling/sliding of the nanoparticles is particularly effective when the nanoparticles are mixed with a lubricating fluid. Submicron cavities at the surface provide a temporal locking of the nanoparticles, which would otherwise disappear from the contact surface, swiftly. While the surfaces of the reference matrices are rapidly clogged by wear debris and become rather smooth, the *IF* impregnated sample preserves substantial part of its porosity during the much longer tribological experiments, suggesting a synergistic role of the matrix and the nanoparticles in lowering the wear and extending the lifetime of these samples.

Prolonged tribological experiments reveal slow exfoliation of the *IF* nanoparticles. A thin superficial film (transfer layer) of mixed carbonaceous material-WS_2 islands and WO_3 is formed on the contact surfaces and provides low friction and wear rates. Addition of lubricating fluids to the contact surface is found to enhance the tribological behavior of the porous samples, with the *IF* impregnated sample benefiting the most. The transfer layer (third body) was first investigated using the surface force apparatus [112]. Further experimental and theoretical studies, which are underway, will be useful in unraveling the detailed mechanism of action of this intricate system.

5.2 Electrode materials

Inorganic nanotubes offer a unique opportunity for intercalation and insertion batteries, which is an important field of endeavor. The reason for this is that nanotubes offer a very large space for intercalation of Li ions or protons in between the layers; in the empty core of the nanotubes, or through adsorption to the nanotube's outer surface. If the nanotubes are prepared open ended and relatively short rapid diffusion in and out of the matrix can be ascertained. Furthermore, being the genuinely stable nanophase of the respective 2-D material, they are likely to undergo many charge/discharge cycles without losing their crystallinity, or being altered in due course. Some early studies of V_2O_5 nanotubes as cathode material in Li intercalation batteries were reported [7c]. Further studies of this nanotubular phase [46] has shown that a respectable performance can be accomplished with these electrodes (150-200 mA/h) even after 150 charge/discharge cycles. The capacity varied rather slowly with the number of cycles, initially increasing and then slowly deteriorating. On per formula basis, this capacity indicates that each vanadium atom in the nanotubes is involved in the oxidation/reduction reactions. A similar study by yet another group was undertaken on nanotubes of the related layered material $Mn_{0.1}V_2O_5$ [47]. In another study, MoS_2 nanotubes were studied as cathode material in Li interaction batteries and exhibited also fast kinetics and respectable voltage (3V) and capacity (250 mAh/g) [115]. The reversible cyclic voltammogram attests to the fast electrokinetics obtained in this system. In another recent study, MoS_2 nanotubes were investigated as cathode material for hydride batteries [116]. MoS_2 nanotubes were synthesized using a procedure similar to the one described in Ref. [19b], i.e. by annealing $(NH_4)_2MoS_4$ at relatively low temperatures (400 °C) under hydrogen/thiophene atmosphere. The MoS_2 nanotubes were impregnated into a Ni foam and studied as electrode material in a 5 M KOH solution. NiOH/NiOOH counter electrode and Hg/HgO reference electrodes were used in the experiment. For the sake of comparison, similar electrode was prepared using a powder of bulk $2H$-MoS_2. While the nanotube electrode exhibited a capacity of 260 mAh/g at 50 mA/g discharge current, the bulk MoS_2 electrode produced a meager 62 mAh/g capacity. After 30 charge/discharge cycles the capacity of the *IF*-MoS_2 electrode decreased by 2%, only.

5.3 Catalysis

MoS_2 particles are used as heterogeneous catalysts in the petrochemical industry for hydrodesulforization of natural oil. MoS_2 particles were also used as catalysts for the reaction: $CO + 3H_2 \rightarrow CH_4 + H_2O$, but revealed poor reactivity and selectivity even at 350 °C. In a recent study, MoS_2 nanotubes were tested as methanation catalyst and exhibited very good reactivity and selectivity at temperatures as low as 200 °C [117]. This important study paves the way for numerous future studies attempting to use these novel nanomaterials as catalysts. Again, the relatively high surface area of this nanophase and the thermodynamic stability of such nanoparticles could lead to fascinating new applications for such nanomaterials.

In a preliminary study it was demonstrated that WS_2 nanotubes can serve as tips for scanning probe microscopy [118]. However, this work has not gone anywhere further for a long time, mostly since a more precise control on the synthesis of the nanotubes was needed, which has been recently demonstrated [11c]. Another difficulty lies in the fact that the nanotubes come in bundles, which are not easy to separate and assemble into a device. Future progress in the applications of inorganic nanotubes for various nanotechnologies will depend to a large extent on the ability to synthesize them in a controlled manner and develop strategies to manipulate them.

6 Conclusions

Inorganic fullerene-like structures and inorganic nanotubes, in particular, are believed to be generically stable structure of nanoparticles of inorganic layered (2-D) compounds. Various synthetic approaches to produce these nanostructures are presented. In some cases, like WS_2, MoS_2, BN and V_2O_5 both fullerene-like nanoparticles and nanotubes are produced in gross amounts. However, size and shape control is still at its infancy. Study of these novel nanostructures has led to the observation of a few interesting properties and some potential applications in tribology, high energy density batteries, and nanoelectronics.

Acknowledgement

We are indebted to Alex Margolin and to Dr. Rita Rosentsveig for the synthesis of the *IF*-WS_2 nanoparticles, and to Yaron Rosenfeld Hacohen for the growth of the $NiCl_2$ fullerene-like nanoparticles and nanotubes. This work was supported in part by the following agencies: Israeli Ministry of Science (Tashtiot program); US-Israel Binational Science Foundation; Israel science Foundation; Israeli Academy of Sciences (First program).

References

1. H.W. Kroto, J.R. Heath, S.C. O'Brein, R.F. Curl, and R.E. Smalley, *Nature* **1985**, *318*, 162.
2. S. Iijima, *Nature* **1991**, *354*, 56.
3. a. R. Tenne, L. Margulis, M. Genut, and G. Hodes, *Nature* **1992**, *360*, 444; b. , L. Margulis, G. Salitra, R. Tenne, and M. Talianker, *Nature* **1993**, *365*, 113.
4. Y. Feldman, E. Wasserman, D.J. Srolovitz, and R. Tenne, *Science* **1995**, *267*, 222.
5. a. N.G. Chopra, J. Luyken, K. Cherry, V.H. Crespi, M.L. Cohen, S.G. Louie, and A. Zettl, *Science* **1995**, *269*, 966; b. Z. W. Sieh, K. Cherrey, N.G. Chopra, X. Blasé, Y. Miyamoto, A. Rubio, M.L. Cohen, S.G. Louie, A. Zettl, and R. Gronsky, *Phys. Rev. B* **1995**, *51*, 11229.
6. Y. Rosenfeld Hacohen, E. Grunbaum, R. Tenne, J. Sloan, and J.L. Hutchison, *Nature* **1998**, *395*, 336.
7. a. M.E. Spahr, P. Bitterli, R. Nesper, M. Müller, F. Krumeich, and H.U. Nissen, *Angew. Chem. Int. Ed.* **1998**, *37*, 1263; b. F. Krumeich, H.-J. Muhr, M. Niederberger, F. Bieri, B. Schneyder, and R. Nesper, *J. Am. Chem. Soc.*, **1999**, *121*, 8324; c. M.E. Spahr, P. Stoschitzki-Bitterli, R. Nesper, O. Haas, and P. Novak, *J. Electrochem. Soc.* **1999**, *146*, 2780.
8. a. R. Tenne, M. Homyonfer, and Y. Fedman, *Chem. Mater.* **1998**, *10*, 3225; b. R. Tenne, *Progress in Inorganic Chemistry*, Ed. Kenneth D. Karlin, John Wiley&Sons **2001**, *50*, 269; c. V.V. Pokropivny, *Powder Metal. Metal Ceramics* **2001**, *40*, 485; d. A.L. Ivanovskii, *Uspekhi Khimii* **2002**, *71*, 203; e. G.R. Patzke, F. Krumeich, and R. Nesper, *Angew. Chem. Intl. Ed.* **2002**, *41*, 2446.
9. a. Y. Feldman, G.L. Frey, M. Homyonfer, V. Lyakhovitskaya, L. Margulis, H. Cohen, G. Hodes, J.L. Hutchison, and R. Tenne, *J. Am. Chem. Soc.* **1996**, *118*, 5362; b. Y. Feldman, V. Lyakhovitskaya, and R. Tenne, *J. Am. Chem. Soc.* **1998**, *120*, 4176 (1998); c. Y. Feldman, A. Zak, R. Popovitz-Biro, and R. Tenne, *Solid State Sci.* **2000**, *2*, 663.
10. A. Zak, Y. Feldman, V. Alperovich, R. Rosentsveig, and R. Tenne, *J. Am. Chem. Soc.* **2000**, *122*, 11108.
11. a. A. Rothschild, G.L. Frey, M. Homyonfer, M. Rappaport, and R. Tenne, *Mater. Res. Innov.* **1999**, *3*, 145; b. A. Rothschild, J. Sloan, and R. Tenne, *J. Am. Chem. Soc.* **2000**, *122*, 5169; c. R. Rosentsveig, A. Margolin, Y. Feldman, R. Popovitz-Biro, and R. Tenne, *Chem. Mater.* **2002**, *14*, 471.
12. a. M. Remskar, Z. Skraba, F. Cléton, R. Sanjinés, and F. Lévy, *Appl. Phys. Lett.* **1996**, *69*, 351; b. M. Remskar, Z. Skraba, F. Cléton, R. Sanjinés, and F. Lévy, *Surf. Rev. Lett.*, **1998**, *5*, 423; c. M. Remskar, Z. Skraba, M. Regula, C. Ballif, R. Sanjinés, and F. Lévy, *Adv. Mater.* **1998**, *10*, 246; d. M. Remskar, Z. Skraba, C. Ballif, R. Sanjinés, and F. Lévy, *Surface Science* **1999**, *435*, 637; e. M. Remskar, Z. Skraba, R. Sanjinés, and F. Lévy, *Appl. Phys. Lett.* **1999**, *74*, 3633.
13. C.M. Zelenski and P.K. Dorhout, *J. Am. Chem. Soc.* **1998**, *120*, 734.

14. a. Y.Q. Zhu, W.K. Hsu, N. Grobert, B.H. Chang, M. Terrones, H. Terrones, H.W. Kroto, and D.R.M. Walton, *Chem. Mater.* **2000**, *12*, 1190; b. Y.Q. Zhu, W.K. Hsu, H. Terrones, N. Grobert, B.H. Chang, M. Terrones, B.Q. Wei, H.W. Kroto, D.R.M. Walton, C.B. Boothroyd, I. Kinloch, G.Z. Chen. A.H. Windle, and D.J. Fray, *J. Mater. Chem.* **2000**, *10*, 2570.
15. W.K. Hsu, B.H. Chang, Y.Q. Zhu, W.Q. Han, H. Terrones, M. Terrones, N. Grobert, A.K. Cheetham, H.W. Kroto, and D.R.M. Walton, *J. Am. Chem. Soc.* **2000**, *122*, 10155.
16. M. Remskar, A. Mrzel, Z. Skraba, A. Jesih, M. Ceh, J. Demsar, P. Stadelmann, F. Levy, and D. Mihailovic, *Science* **2001**, *292*, 479.
17. R. L. D. Whitby, W. K. Hsu, C. B. Boothroyd, P. K. Fearon, H.W. Kroto, and D. R. M. Walton, *Chem. Phys. Chem.* **2001**, *10*, 620.
18. a. C.R. Martin, *Acc. Chem. Res.* **1995**, *28*, 61; b. H. Masuda and K. Fukuda, *Science* **1995**, *268*, 1466.
19. a. M. Nath, A. Govindaraj, and C.N.R. Rao, *Adv. Mater.* **2001**, *13*, 283; b. M. Nath and C.N.R. Rao, *J. Am. Chem. Soc.* **2001**, *123*, 4841; c. M. Nath and C.N.R. Rao, *Chem. Comm.* **2001**, 2236.
20. "Constitution of Binary Alloys, First Supplement", R.P. Elliot, McGraw-Hill Book Comp., New York, 1965, p. 632.
21. M. Homyonfer, Y. Mastai, M. Hershfinkel, V. Volterra, J.L. Hutchison, and R. Tenne, *J. Am. Chem. Soc.* **1996**, *118*, 7804.
22. L. Margulis, S. Iijima, and R. Tenne, *Microscopy, Microanalysis, Microstructures*, **1996**, *7*, 87.
23. Y. Mastai, M. Homyonfer, A. Gedanken, and Hodes G, *Adv. Mater.* **1999**, *11*, 1010.
24. J. Sloan, J.L. Hutchison, R. Tenne, Y. Feldman, M. Homyonfer, and T. Tsirlina, *J. Solid State Chem.* **1999**, *144*, 100.
25. V. K. Sarin, *J. Mater. Sci.* **1975**, *10*, 593.
26. T. Millner and J. Neugebauer, *Nature* **1949**, *163*, 602.
27. a. J.D. Donnay and D. Harker, *Amer. Miner.* **1937**, *22*, 446; b A.S. Myerson in "Molecular Modeling Applications in Crystallization" ed. A.S. Myerson, Cambridge University Press, Cambridge, 1999.
28. R. L. D. Whitby, W. K. Hsu, H. W. Kroto, and D. R. M. Walton, *Phys. Chem. Chem. Phys.* **2002**, *4*, 3938.
29. P. Afanasiev, C. Geantel, C. Thomazeau, and B. Jouget, *Chem. Comm.* **2000**, 1001.
30. C. Schoffenhauer, R. Popovitz-Biro, and R. Tenne, *J. Mater. Chem.* **2002**, *12*, 1587.
31. M. Remskar, A. Marzel, A. Jesih, and F. Lévy, *Adv. Mater.* **2002**, *14*, 680.
32. Y.Q. Zhu, W.K. Hsu, H.W. Kroto, and D.R.M. Walton, *J. Phys. Chem. B* **2002**, *106*, 7623.
33. a. D. Vollath and D.V. Szabo, *Mater. Lett.* **1998**, *35*, 236; b. D. Vollath and D.V. Szabó, *Acta Materialia* **2000**, *48*, 953.
34. M. Cote, M.L. Cohen, and D.J. Chadi, *Phys. Rev. B* **1998**, *58*, R4277.

35. B. Alperson, M. Homyonfer, and R. Tenne, *J. Electroanal. Chem.* **1999**, *473*, 186.
36. M. Homyonfer, B. Alperson, Yu. Rosenberg, L. Sapir, S.R. Cohen, G. Hodes, and R. Tenne, *J. Am. Chem. Soc.* **1997**, *119*, 2693.
37. J.A. Hollingsworth, D.M. Poojary, A. Clearfield, and W.E. Buhro, *J. Am. Chem. Soc.* **2000**, *122*, 3562.
38. C. Ye, G. Meng, Z. Jiang, Y. Wang, G. Wang, and L. Zhang, *J. Am. Chem. Soc.* **2002**, *124*, 15180.
39. X. Zheng, Y. Xie, L. Zhu, X. Jiang, Y. Jia, W. Song, and Y. Sun, *Inorg. Chem.* **2002**, *41*, 455.
40. X. Jiang, Y. Xie, J. Lu, L. Zhu, W. He, and Y. Qian, *Adv. Mater.* **2001**, *12*, 1278.
41. F. Camerel, J.-C. P. Gabriel, P. Batail, P. Davidson, B. Lemaire, M. Schmutz, T. G. Krzywicki, and C. Bourgaux, *Nano Lett.* **2002**, *2*, 403.
42. a. H. Nakamura and Y. Matsui, *J. Am. Chem. Soc.* **1995**, *117*, 2651; b. H.P. Lin, C.Y Mou, and S.B Liu, *Adv. Mater.* **2000**, *12*, 103.
43. P.M. Ajayan, O. Stephan, P. Redlich, and C. Colliex, *Nature* **1995**, *375*, 564.
44. a. B.C. Satishkumar, A. Govindaraj, E.M. Vogel, L. Basumallick, and C.N.R. Rao, *J. Mater. Res.* **1997**, *12*, 604; b. M. Adachi, T. Harada, and M. Harada, *Langmuir* **1999**, *15*, 7097; c. M. Adachi, Y. Murata, M. Harada, and S. Yoshikawa, *Chem. Lett.* **2000**, 942.
45. L. Nanai and T.F. George, *J. Mater. Res.* **1997**, *12*, 283.
46. S. Nordlinder, K. Edström, and T. Gustafsson, *Electrochem. Solid State Lett.* **2001**, *4*, A129.
47. A. Dobley, K. Ngala, T. Shoufeng, P.Y. Zavalij, and M.S. Whittingham, *Chem. Mater.* **2001**, *13*, 4382.
48. a. Y. Li, M. Sui, Y. Ding, G. Zhang, J. Zhuang, and C. Wang, *Adv. Mater.* **2000**, *12*, 818; b. Y. Ding, G. Zhang, H. Wu, B. Hai, L. Wang, and Y. Qian, *Chem. Mater.* **2001**, *13*, 435.
49. S. Avivi, Y. Mastai, G. Hodes, and A. Gedanken, *J. Am. Chem. Soc.* **1999**, *121*, 4196.
50. S. Avivi, Y. Mastai, and A. Gedanken, *J. Am. Chem. Soc.* **2000**, *122*, 4331.
51. G.B. Saupe, C.C. Waraksa, H-N Kim, Y.J. Han, D.M. Kaschak, D.M. Skinner, and T.E. Mallouk, *Chem. Mater.* **2000**, *12*, 1556.
52. R. Abe, K. Shinohara, A. Tanaka, M. Hara, J.N. Kondo, and K. Domen, *Chem. Mater.* **1997**, *9*, 2179.
53. a. G.H. Du, Q. Chen, R.C. Che, Z.Y. Yuan, and L.-M. Peng, *Appl. Phys. Lett.* **2001**, *79*, 3702; b. Q. Chen, G.H. Du, S. Zhang, and L.-M. Peng, *Acta Cryst. B* **2002**, *58*, 587.
54. a. P. Millet, J.Y. Mila, and J. Galy, *J. Solid State Chem.* **1999**, *147*, 676; b. J. Choi, J.L. Musfeldt, Y.J. Wang, H.-J. Koo, M.-H. Whangbo, J. Galy, and P. Millet, *Chem. Mater.* **2002**, *14*, 924.
55. M. Yada, H. Kitamura, M. Machida, and T. Kijima, *Inorg. Chem.* **1998**, *37*, 6470.
56. M. Yada, M. Mihara, S. Mouri, M. Kuroki, and T. Kijima, *Adv. Mater.* **2002**, *14*, 309.

57. *Micelles, Membranes, Microemulsions and Monolayers*, Eds. W.M. Gelbart, A. Ben-Shaul, D. Roux, Springer, New York, 1994.
58. M. Zhang, Y. Bando, and K. Wada, *J. Mater. Res.* **2001**, *16*, 1408.
59. a. L. Vayssieres, K. Keis, A. Hagfeldt, and S.-E. Lindquist, *Chem. Mater.* **2001**, *13*, 4395.
60. Z. Wang and H.L.Li, *Appl. Phys. A* **2002**, *74*, 201.
61. S.M. Liu, L.M. Gan, L.H. Liu, W.D. Zhang, and H.C. Zeng, *Chem. Mater.* **2002**, *19*, 1391 and **2002**, *14*, 2427.
62. Y. Rosenfeld Hacohen, R. Popovitz-Biro, E. Grunbaum, Yehiam Prior, and R. Tenne, *Adv. Mater.* **2002**, *14*, 1075.
63. R. Popovitz-Biro, A. Twersky, Y. Rosenfeld Hachoen, and R. Tenne, *Isr. J. Chem.* **2001**, *41*, 7.
64. J. Cumings and A. Zettl, *Chem. Phys. Lett.* **2000**, *316*, 211, and *Chem. Phys. Lett.* **2000**, *318*, 497.
65. A Loiseau, F. Willaime, N. Demoncy, G. Hug, and H. Pascard, *Phys. Rev. Lett.* **1996**, *76*, 4737.
66. M. Terrones, W.K. Hsu, H. Terrones, J.P. Zhang, S. Ramos, J.P. Hare, R. Castillo, K. Prassides, A.K. Cheetham, H.W. Kroto, and D.R.M. Walton., *Chem. Phys. Lett.* **1996**, *259*, 568.
67. M. Terrones, A.M. Benito, C. Manteca-Diego, W.K. Hsu, O.I. Osman, J.P. Hare, D.G. Reid, H. Terrones, A.K. Cheetham, K. Prassides, H.W. Kroto, and D.R.M. Walton., *Chem. Phys. Lett.* **1996**, *257*, 576.
68. T. Laude, A. Marraud, Y. Matsui, and B. Jouffrey, *Appl. Phys. Lett.* **2000**, *76*, 3239.
69. a. D. Golberg, Y. Bando, W. Han, K. Kurashima, and T. Sato, *Chem. Phys. Lett.* **1999**, *308*, 337; b. D. Golberg, Y. Bando, O. Stéphan, and K. Kurashima, *Appl. Phys. Lett.* **1998**, *73*, 2441.
70. K. Suenaga, C. Colliex, N. Demoncy, A. Loiseau, H. Pascard, and F. Willaime, *Science* **1997**, *278*, 653.
71. Y. Zhang, K. Suenaga, and S. Iijima, *Science* **1998**, *281*, 973.
72. T. Oku, T. Kusunose, K. Niihara, and K. Suganuma, *J. Mater. Chem.* **2000**, *10*, 255.
73. W.N. Lipscomb and L. Massa, *Inorg. Chem.* **1992**, *31*, 2297.
74. A. Gindulyte, W. N. Lipscomb, and L. Massa, *Inorg. Chem.* **1998**, *37*, 6544.
75. I. Boustani, A. Quandt, E. Hernandez, and A. Rubio, *J. Chem. Phys.* **1999**, *110*, 3176.
76. a. G. Seifert and E. Hernandez, *Chem. Phys. Lett.* **2000**, *318*, 355; b. G. Seifert, T. Heine, and P.W. Fowller, *Eur. Phys. J. D* **2001**, *16*, 341.
77. Y. Li, J. Wang, Z. Deng, Y. Wu, X. Sun, D. Yu, and P. Yang, *J. Am. Chem. Soc.* **2001**, *123*, 9904.
78. C. Su, H.-T. Liu, and J.-M. Li, *Nanotech.* **2002**, *13*, 746.
79. a. J.S. Kasper, P. Hagenmuller, M. Pouchard, and C. Cros, *Science* **1965**, *150*, 1713; b. C. Cros, M. Pouchard, and E.P. Hagenmuller, *J. Solid State Chem.* **1970**, *2*, 570.

80. J. Gryko, P. F. McMillan, R. F. Marzke, G. K. Ramachandran, D. Patton, and S. K. Deb, *Phys. Rev. B* **2000**, *62*, R7707.
81. M. Pouchard, C. Cros, P. Hagenmuller, E. Reny, A. Ammar, M. Menetrier, and J.-M. Bassat, *Solid State Sci.* **2002**, *4*, 723.
82. G. Seifert, Th. Köhller, and R. Tenne, *J. Phys. Chem. B* **2002**, *106*, 2497.
83. G. Seifert, H. Terrones, M. Terrones, G. Jungnickel, and T. Frauenheim, *Phys. Rev. Lett.* **2000**, *85*, 146.
84. D.J. Srolovitz, S.A. Safran, M. Homyonfer, and R. Tenne, *Phys. Rev. Lett.* **1995**, *74*, 1779.
85. R. Rosentsveig, A. Margolin, Y. Feldman, R. Popovitz-Biro, and R. Tenne, *Chem Mater.* **2002**, *14*, 471.
86. a. M. Remskar, Z. Skraba, R. Sanjines, and F. Lévy, *Surf. Rev Lett.* **1999**, *6*, 1283; b. M. Remskar, Z. Skraba, P. Stadelmann, and F. Lévy, *Adv. Mater.* **2000**, *12*, 814.
87. S.M. Lee, Y.H. Lee, Y.G. Hwang, J. Elsner, D. Porezag, and T. Frauenheim, *Phys. Rev. B* **1999**, *60*, 7788.
88. J.A. Wilson and A.D. Yoffe, *Adv. Phys.* **1969**, *18*, 193.
89. J. Moser, F. Lévy, and F. Bussy, *J. Vac. Sci. Technol. A* **1994**, *12*, 494.
90. L. Margulis, P. Dluzewski, Y. Feldman, and R. Tenne, *J. Microscopy* **1996**, *181*, 68.
91. R. Tenne, *Adv. Mater.* **1995**, *7*, 965.
92. P.A. Parilla, A.C. Dillon, K.M. Jones, G. Riker, D.L. Schulz, D.S. Ginley, and M.J. Heben, *Nature* **1999**, *397*, 114.
93. R.L.D. Whitby, W.K. Hsu, T.H. Lee, C.B. Boothroyd, H.W. Kroto, and D.R.M. Walton, *Chem. Phys. Lett.* **2002**, *359*, 68.
94. F. Jensen and H. Toftlund, *Chem. Phys. Lett.* **1993**, *201*, 89.
95. M. Terrones, W.K. Hsu, H. Terrones, J.P. Zhang, S. Ramos, J.P. Hare, R. Castillo, K. Prassides, A.K. Cheetham, H.W. Kroto, and D.R.M. Walton., *Chem. Phys. Lett.* **1996**, *259*, 568.
96. D. Golberg, Y. Bando, O. Stéphan, and K. Kurashima, *Appl. Phys. Lett.* **1998**, *73*, 2441.
97. O. Stéphan, Y. Bando, A. Loiseau, F. Willaime, N. Shramchenko, T. Tamiya, and T. Sato, *Appl. Phys. A* **1998**, *67*, 107.
98. Y. Miyamoto, A. Rubio, S.G. Louie, and M.L. Cohen, *Phys. Rev. B* **1994**, *50*, 18360.
99. M. I. Mendelev, D. J. Srolovitz. S.A. Safran, and R. Tenne, *Phys. Rev. B* **2002**, *65*, 075402.
100. S.W. Guo, L. Konopny, R. Popovitz-Biro, H. Cohen, M. Sirota, E. Lifshitz, and M. Lahav, *Adv. Mater.* **2000**, *12*, 302.
101. A. Rubio, J.L. Corkill, and M.L. Cohen, *Phys. Rev. B* **1994**, *49*, 5081.
102. a. A.D. Yoffe, K.J. Howlett, and P.M. Williams, 'Cathodoluminescence studies in the SEM', *Scannjing Electron Microscopy/1973 (Part II)*, 301-308 (1973); b. D.A.B. Miller, D.S. Chemla, and S. Schmitt-Rink, in *Optical Nonlinearities and*

Instabilities in Semiconductors,Haug H, Ed., Academic Press, Orlando, FL, 1988, p. 325; c. M.L. Steigerwald and L.E. Brus, *Annu. Rev. Mater. Sci.* **1989**, *19*, 471.
103. G.L. Frey, S. Elani, M. Homyonfer, Y. Feldman, and R. Tenne, *Phys. Rev. B* **1998**, *57*, 6666.
104. G. Seifert, H. Terrones, M. Terrones, and T. Frauenheim, *Solid State Comm.* **2000**, *115*, 635.
105. L. Scheffer, R. Rosentsveig, A. Margolin, R. Popovitz-Biro, G. Seifert, S.R. Cohen, and R. Tenne, *Chem. Phys. Phys. Chem.* **2002**, *4*, 2095.
106. O. Tal, M. Remskar, R. Tenne, and G. Haase, *Chem. Phys. Lett.* **2001**, *344*, 434.
107. A. Zak, Y. Feldman, H. Cohen, V. Lyakhovitskaya, G. Leitus, R. Popovitz-Biro, S. Reich, and R. Tenne, *J. Am. Chem. Soc.* **2002**, *124*, 4747.
108. R. Tenne, *Encyclopedia of Electrochemistry, Vol6: Semiconductor Electrodes and Photoelectrochemistry*, Eds. Bard and Stratmann; Vol. Ed. S. Licht, Wiley-VCH (2001).
109. L. Rapoport, Yu. Bilik, Y. Feldman, M. Homyonfer, S.R. Cohen, and R. Tenne, *Nature* **1997**, *387*, 791.
110. M. Chhowalla and G.A.J. Amaratunga, *Nature* **2000**, *407*, 164.
111. L. Rapoport, M. Lvovsky, I. Lapsker, V. Leshchinsky, Yu Volovik, Y. Feldman, A. Zak, and R. Tenne, *Adv. Eng. Mater.* **2001**, *3*, 71.
112. Y. Golan, C. Drummond, J. Israelachvili, and R. Tenne, *Wear* **2000**, *245*, 190.
113. L. Cizaire, B. Vacher, T. Le-Mogne, J.M. Martin, L. Rapoport, A. Margolin, and R. Tenne, *Surf. Coating Tech.* **2002**, *160*, 282.
114. W. X. Chen, Z. D. Xu, R. Tenne, R. Rosenstveig, W. L. Chen, H. Y. Gan, and J. P. Tu, *Adv. Eng. Mater.* **2002**, *4*, 686.
115. J. Chen *et al.*, to be published
116. J. Chen, N. Kuriyama, H. Yuan, H.T. Takeshita, and T. Sakai, *J. Am. Chem. Soc.* **2001**, *123*, 11813.
117. J. Chen, S.-L. Li, Q. Xu, and K. Tanaka, *Chem. Comm.* **2002**, 1722.
118. A. Rothschild, S.R. Cohen, and R. Tenne, *Appl. Phys. Lett.* **1999**, *75*, 4025.

SEMICONDUCTOR NANOWIRES: FUNCTIONAL BUILDING BLOCKS FOR NANOTECHNOLOGY

HAOQUAN YAN, PEIDONG YANG

Department of Chemistry, University of California, Berkeley, California 94720

One-dimensional (1D) nanostructures are ideal systems for investigating the dependence of electrical transport, optical properties and mechanical properties on size and dimensionality. They are expected to play an important role as both interconnects and functional components in the fabrication of nanoscale electronic and optoelectronic devices. This article presents an overview of current research activities that center on nanowires whose lateral dimensions fall anywhere in the range of 1 - 200 nm. It is organized into three parts: The first part discusses various methods that have been developed for generating nanowires with tightly controlled dimensions, orientations, interfaces and well-defined properties. The second part highlights a number of strategies that are being developed for the hierarchical assembly of nanowire building blocks. The third part surveys some of the novel physical properties (e.g., optical, electrical, and mechanical) of these nanostructures. Finally, we conclude with some personal perspectives on the future research directions in this field.

1 Introduction

Nanostructures (that is, structures with at least one dimension between 1 and 100 nm) have attracted steadily growing interest due to their fascinating properties, as well as their unique applications relative to their bulk counterparts [1-4]. The ability to generate such structures is now central to the advance of many areas in modern science and technology. There are a large number of new opportunities that could be realized by down-sizing currently existing structures into the regime of <100 nm, or by making new types of nanostructures [1-4]. The most successful examples are in microelectronics, where "smaller" has always meant greater performance ever since the invention of transistors: higher density of integration, faster response, lower cost, and less power consumption [1-4]. It has been recognized that a broad range of interesting and new phenomena are associated with nanometer-sized dimensions; well-established examples include size-dependent excitation [5], quantized (or ballistic) conductance [6], Coulomb blockade or single electron tunneling [7], and metal-insulator transition [8]. It is generally accepted that the quantum confinement of electrons by the potential well of a nanometer-sized structure provides one of the most versatile and powerful means to control the electrical, optical, magnetic, and thermoelectric properties of a solid material [9-11].

In the past decades, significant progress has already been made in the field of zero- (0D) and two-dimensional (2D) nanostructures (i.e., quantum dots and quantum wells, respectively). For example, a rich variety of methodologies have been developed for synthesizing quantum dots or fabricating quantum wells from a broad range of materials with well-controlled dimensions [1]. Using quantum dots as the model system, a wealth of interesting chemistry and physics has been learned by studying the evolution of their fundamental properties with size [1-11]. The past decade has also witnessed the great research advancement in the area of carbon nanotube, a unique type of one-dimensional nanostructures [10,12]. With these nanostructures as the functional components, a variety of nanoscale devices have been fabricated as prototypes by many research groups around the world, with notable examples including quantum dot lasers [9], single electron transistors [6,7,11], logic and memory units [10], as well as light-emitting diodes (LEDs).

Nanoscale one-dimensional (1D) materials have stimulated great interest due to their importance in basic scientific research and potential technology applications [12,13]. Other than carbon nanotubes, 1D nanostructures (nanowires or quantum wires) are ideal systems for investigating the dependence of electrical transport and mechanical properties on size and dimensionality. They are expected to play an important role as both interconnects and functional components in the fabrication of nanoscale electronic and optoelectronic devices. Many unique and fascinating properties have already been proposed or demonstrated for this class of materials, such as superior mechanic toughness [14], higher luminescence efficiency [15], enhancement of thermoelectric figure of merit [16] and lowered lasing threshold [17].

Nanowires (or quantum wires), by definition, are anisotropic nanocrystals with large aspect ratios (length/diameter). Generally, they would have diameters of 1-200 nm and length up to several tens of micrometers. Nanowires differ significantly from spherical nanocrystals by their morphology as well as physical properties. An important issue in the study and application of these 1D materials is how to assemble individual atoms into 1D nanostructures in an effective and controllable way. Although 1D nanostructures could be fabricated using a number of advanced nanolithographic techniques, such as e-beam writing, proximal-probe patterning, and x-ray lithography [18], these processes, however, generally are slow and the cost is high, the development of these techniques into practical routes for fabricating large numbers of 1D nanostructures rapidly and at low-cost still requires great ingenuity. On the other hand, chemical synthesis represents another important approach to 1D structures [19-25]. It's much more promising in terms of cost and potential for high volume production as well as tight dimension control. A significant challenge of the chemical synthesis is how to rationally control the nanostructure assembly so that their size, dimensionality, interfaces, and ultimately their 2-dimensional and 3-dimensional superstructures can be tailor-made towards desired functionality.

This article presents an overview of current research activities that center on nanowires whose lateral dimensions fall anywhere in the range of 1 - 200 nm. As can be readily seen in the literature, there are already fair amount of work being done on metal nanowires and inorganic nanotubes [26,27]. These studies, however, will not be covered in this article due to space limitation. The current review is organized into three parts: the first part discusses various chemical methodologies that have been developed for synthesizing nanowires with tightly controlled dimensions, orientations, interfaces and well-defined properties. The second part highlights a number of strategies that are being developed for the hierarchical assembly of nanowire building blocks. The third part surveys some of the novel physical properties (e.g., optical, electrical, and mechanical) of these nanostructures. Finally, we conclude with some personal perspectives on the future research directions in this field.

2 Nanowire Synthesis

An important issue in the study and application of nanowires is how to assemble individual atoms into such a unique 1D nanostructure in an effective and controllable way. A general requirement for any successful preparative methodology is to be able to achieve nanometer scale control in diameter during anisotropic crystal growth while maintaining a good overall crystallinity. During the past decade, many methodologies have been developed to synthesize 1-dimensional nanostructures [19-25]. Overall, they can be categorized into two major approaches based on the reaction media that were used during the preparation: solution and gas phase based processes.

2.1 Solution-Based Approaches to Nanowires

2.1.1 Template-Directed Synthesis

Template-directed synthesis represents a convenient and versatile method for generating 1D nanostructures. In this technique, the template simply serves as a scaffold against which other kinds of materials with similar morphologies are synthesized. These templates could be nanoscale channels within mesoporous materials or porous alumina and polycarbonate membranes. The nanoscale channels can be filled using solution, sol-gel or electrochemistry to generate 1D nanoscale objects. The produced nanowires can then be released from the templates by selectively removing the host matrix. The template method has a number of interesting and useful features. First, it seems to be very general. The methodology was pioneered by Moskovits and Martin [22]. Researchers have now used this method to prepare nanoscale wires, and tubules made of electronically conductive polymers, metals, semiconductors, oxides, carbon, and many other materials. Furthermore, nanostructures with extraordinarily small diameters can be prepared using this method. For example, Wu and Bein [28] have recently used this method to prepare conductive polymer nanofibers with diameters of 3 nm in MCM-41 mesoporous silica. It is quite difficult to make nanowires with diameters this small using lithographic methods. In addition, because the pores in the porous membranes usually have monodispersed diameters, analogous monodispersed nanostructures can be in principle prepared. Finally, the 1D nanostructures synthesized within the pores can be freed from the template membrane and collected. Alternatively, an ensemble of 1D nanostructures can be obtained.

A number of hosts have been extensively used in this type of templating process. One of them is porous polymeric filtration membrane [22] that has been prepared via the track-etch method. Membranes with a wide range of pore diameters (down to 10 nm) and pore densities approaching 10^9 pores/cm^2 are available commercially. The most commonly used material to prepare porous membranes is polycarbonate; however, a number of other materials are amenable to the track-etch process. Porous alumina membranes are another excellent host materials. They are prepared electrochemically from aluminum metal. The pores in these membranes are arranged in a regular hexagonal lattice. Pore densities as high as 10^{11} pores/cm^2 can be achieved. Many materials have been fabricated into nanowires using this templating process, including various inorganic materials Au, Ag, Pt, TiO_2, MnO_2, ZnO, SnO_2, Bi_2Te_3, conductive polymers polypyrrole, poly(3-methylthiophene), and polyaniline, and carbon nanotubules [29,30]. This templating methodology has been quite successful in term of controlling the material morphology into nanowires. It, however, has difficulty to obtain materials with single crystallinity. For example, although single crystalline nanowires of Bi have been obtained by templating Bi melt against alumina membranes [16,31], most of the nanowires prepared using the templating process were polycrystalline which could limit their potential use in studies or applications related to electronic transport.

Besides porous alumina and polymer membranes, with their high surface areas and uniform pore sizes, mesoporous silica materials have also been widely used as host materials for loading catalysts, polymers, metal and semiconductor nanoparticles that have potential catalytic, environmental, and optoelectrical applications. Both polymer and inorganic nanowires have been successfully synthesized using either MCM-41 or SBA-15 as the templates [28, 32-35]. A simple chemical methodology for the formation of uniform Ag nanowires within mesoporous silica SBA-15 has been developed recently [32]. This process involves $AgNO_3$ solution impregnation followed by thermal decomposition. Transmission electron microscopy studies on these samples show that these continuous Ag nanowires are made of long polycrystalline domains. They have uniform diameters of 5 - 6 nm, and large aspect ratios between 100 and 1000. This process represents a viable approach for synthesizing uniform metallic nanowires (Au, Pt) and may be applicable for making other inorganic nanowires. For example, Ge nanowires has been successfully synthesized within the meso-channels of MCM-41 [35].

It is also recognized that when nanowires/nanotubes are allowed to react with proper chemicals under carefully controlled conditions, they can be transformed into another substance without changing their wire-like morphologies. This template effect provides an effective route to 1D nanostructures that are difficult (or impossible) to directly synthesize or fabricate. The Lieber group was among the first of several laboratories that explored the potential of this method for generating SiC nanowires from carbon nanotubes [36]. One of the major problems associated with this approach is the difficulty to control the composition and crystallinity of the final products. The nanowires synthesized using this approach are often polycrystalline.

An interesting example of single crystalline nanowire transformation is recently reported in the Se-to-Ag_2Se system. Single crystalline Ag_2Se nanowires have been successfully synthesized through a novel topotactic reaction process that used single crystalline nanowires of trigonal selenium as solid templates. During the templating process, the t-Se nanowires react with aqueous $AgNO_3$ solutions at room temperature [37]. An interesting diameter-dependent phase transformation was also discovered in this nanowire system. The Ag_2Se nanowires adopt a tetragonal crystal structure when their diameters were less than ~ 40 nm. With an increase in their diameters, an orthorhombic phase was found to be more favorable. This work represents the first demonstration of a template-directed process for the formation of single crystalline nanostructures in the solution phase and at room temperature.

Overall, the templating approach in solution is a fairly powerful technique to prepare nanowires of various compositions. This technique, however, intrinsically yields products of polycrystalline nature.

2.1.2 Solution-Liquid-Solid Method

To obtain highly crystalline semiconductor nanowires at low temperatures, a solution-liquid-solid (SLS) mechanism (Figure 1) for the growth of InP, InAs, and GaAs nanowhiskers was introduced by Buhro's group at Washington University [24]. This approach uses simple, low-temperature, solution-phase reactions. The materials are produced as near-single-crystal whiskers having widths of 10 to 150 nanometers and lengths of up to several micrometers. This growth mechanism shows that processes analogous to vapor-liquid-solid (VLS) growth (see section 2.2) can operate at low temperatures. Similar synthesis routes for other covalent solids are possible. For example, low-temperature growth of indium nitride nanowires from azido-indium precursors has been developed by the same group [38] at temperature as low as 111-203 °C. It is claimed that the chemical synthetic pathway consists of a molecular component, in which precursor substituents are eliminated, and a non-molecular component, in which the InP crystal lattices are assembled. The two components working in concert comprise the so-called SLS mechanism [39].

More recently, bulk quantities of defect-free silicon (Si) nanowires with nearly uniform diameters ranging from 4 to 5 nm were grown to a length of several micrometers using a supercritical fluid solution-phase approach. In their process, Korgel et al. used solvent-dispersed, monodispersed, alkanethiol-capped gold nanocrystals to direct Si nanowire growth with narrow wire diameter distributions [40]. At temperatures of 500°C and 270 bar, the silicon precursor diphenylsilane decomposes to Si atoms. The Si atoms dissolve into the Au nanocrystals until reaching supersaturation, at which point they are expelled from the particle as a thin nanometer-scale wire. The supercritical fluid medium provides the high temperatures necessary to promote Si crystallization. The growth mechanism shares some similarity with the solution-liquid-solid mechanism proposed by Buhro as well as the well-known vapor-liquid-solid nanowire growth mechanism (see section 2.2). In their supercritical fluid environment, relatively monodispersed Au nanocrystals can be maintained to seed uniform nanowire growth.

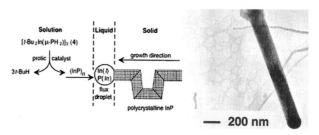

Figure 1. Solution-liquid-solid (SLS) crystal-growth mechanism and representative $Al_xGa_{1-x}As$ nanowhiskers grown by the SLS mechanism. [Reprinted with permission from Ref. 39, Copyright American Chemical Society, 1997].

In addition to these solution routes to elemental and III-V semiconductor nanowires, it has recently been reported that by cleverly exploring the selective capping capabilities of mixed surfactants, it is possible to extend the well-established II-IV semiconductor nanocrystal synthesis to the synthesis of semiconductor nanorods[41,42], a unique version of nanowires with relatively shorter aspect ratios.

2.1.3 Solvothermal Chemical Synthesis

Lately, solvothermal methodology has been extensively examined as one possible route to produce semiconductor nanowires and nanorods. In these processes, a solvent was mixed with certain metal precursors and possibly a crystal growth regulating or templating agent such as amine. This solution mixture was then placed in an autoclave kept at relatively high temperature and pressure to carry out the crystal growth and assembly process. This methodology seems to be quite versatile and has been demonstrated to be able to produce many different crystalline semiconductor nanorods and nanowires as exemplified by extensive work done by Qian's group [43,44]. The solvothermal process itself, however, is inherently complex. The products are usually not pure and the monodispersity of the sample is also far from ideal. A detailed understanding of the reaction and crystal growth mechanism under solvothermal conditions is needed in this regard.

Xia et al. recently demonstrated a related solution-phase approach to the synthesis of uniform nanowires of selenium with lateral dimensions controllable in the range of 10-30 nm, and lengths of up to hundreds of micrometers [45]. The first step of this approach involved the formation of solid selenium in an aqueous solution through the reduction of selenious acid with excess hydrazine by refluxing this reaction mixture at an elevated temperature:

$$H_2SeO_3 + N_2H_4 \rightarrow Se (\downarrow) + N_2 (\uparrow) + 3H_2O$$

The initial product was brick-red-colored spherical colloids of amorphous (a-) selenium with sizes around ~300 nm. When this solution was cooled down to room temperature, the small amount of selenium dissolved in solution precipitated out as nanocrystallites of trigonal (t-) selenium. During the aging of this dispersion that contained a mixture of a-Se colloids and t-Se nanocrystallites, the a-Se colloids slowly dissolved into the solution due to a higher free energy as compared to the t-Se phase. This dissolved selenium subsequently grew as crystalline (t-phase) nanowires on the seeds. For this solid-solution-solid transformation, the 1D morphology of the final products was largely determined by the linear characteristics of the building blocks – that is, the extended, helical chains of selenium in the crystalline t-phase. Each nanowire has a uniform diameter along its longitudinal axis, which was defined by the lateral dimensions (perpendicular to the c-axis) of the seed.

Overall, nanowire synthesis in solution represents a very interesting and rich area for further investigation. While nanowires of many different compositions have been made through these simple solution routes, the monodispersity and crystallinity of the products have been problematic. It is expected, however, with more fundamental understanding of the crystal nucleation and growth in solution, one should be able to prepare highly crystalline nanowires with well-defined, or even monodispersed sizes and aspect ratios.

2.2   Nanowire Growth in Gas Phase

2.2.1   Vapor-Liquid-Solid Nanowire Growth Mechanism

A well-accepted mechanism of nanowire growth via gas phase reaction is the so-called Vapor-Liquid-Solid (VLS) process proposed by Wagner in 1960s during his studies of large single-crystalline whisker growth [46]. According to this mechanism, the anisotropic crystal growth is promoted by the presence of liquid alloy/solid interface. This process is illustrated in figure 2 for the growth of Ge nanowire using Au clusters as solvent at high temperature. Based on Ge-Au binary phase diagram, Ge (from the decomposition of GeH$_4$, for example) and Au will form liquid alloy when the temperature is higher than the eutectic point (363 °C, Figure 2-(I)). The liquid surface has a large accommodation coefficient and is therefore a preferred deposition site for incoming Ge vapor. After the liquid alloy becomes supersaturated with Ge, Ge nanowire growth occurs by precipitation at the solid-liquid interface (Figure 2-(II-III)).

Recently, real-time observation of Ge nanowire growth was conducted in an in-situ high temperature transmission electron microscope (TEM) [47]. The experiment result clearly shows three growth stages: formation of Au-Ge alloy, nucleation of Ge nanocrystal and elongation of Ge nanowire. Figures 3a-f show a sequence of TEM images during the growth of a Ge nanowire in-situ. This real-time observation of the nanowire growth directly mirrors the proposed VLS mechanism in Figure 2.

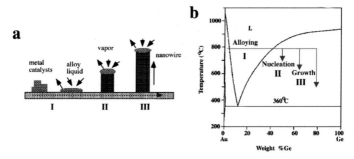

Figure 2. (a) Schematic illustration of vapor-liquid-solid nanowire growth mechanism including three stages (I) alloying, (II) nucleation, and (III) axial growth. The three stages are projected onto the conventional Au-Ge binary phase diagram (b) to show the compositional and phase evolution during the nanowire growth process.

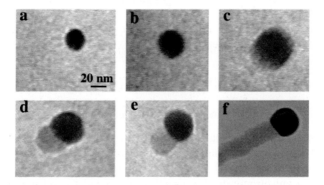

Figure 3. In-situ TEM images recorded during the process of nanowire growth. (a) Au nanoclusters in solid state at 500 °C; (b) alloying initiates at 800 °C, at this stage Au exists mostly in solid state; (c) liquid Au/Ge alloy; (d) the nucleation of Ge nanocrystal on the alloy surface; (e) Ge nanocrystal elongates with further Ge condensation and eventually forms a wire (f). [Reprinted with permission from Ref. 47, Copyright American Chemical Society, 2001].

(I): Alloying process (Fig. 3a-c). Au clusters remain at solid state up to the maximum experimental temperature 900 °C if there is no Ge vapor condensation. With increasing amount of Ge vapor condensation and dissolution, Ge and Au form alloy and liquify. The volume of the alloy droplets increases and the elemental contrast decreases while the alloy composition crosses sequentially, from left to right, a biphasic region (solid Au and Au/Ge liquid alloy) and a single phase region (liquid). This alloying process can be depicted as an isothermal line in the Au-Ge phase diagram (Fig. 2b).

(II): Nucleation (Fig. 3d-e). Once the composition of the alloy crosses the second liquidus line (Fig. 2b), it enters another biphasic region (Au/Ge alloy and Ge crystal). This is where nanowire nucleation starts. Knowing the alloy volume change, it is estimated that the nucleation generally occurs at Ge weight percentage of 50% - 60%.

(III). Axial growth (Fig. 3d-f). Once the Ge nanocrystal nucleates at the liquid/solid interface, further condensation/dissolution of Ge vapor into the system will increase the amount of Ge crystal precipitation from the alloy based on the lever rule of the phase diagram. The incoming Ge species prefer to diffuse to and condense at the existing solid/liquid interface, primarily due to the fact that less energy will be involved with the crystal step growth as compared with secondary nucleation events in a finite volume. Consequently, secondary nucleation events are efficiently suppressed and no new solid/liquid interface will be created. The existing interface will then be pushed forward (or backward) to form a nanowire (Fig. 3f). This in-situ experiment unambiguously demonstrates the validity of the VLS mechanism for nanowire growth. The establishment of VLS mechanism at nanometer scale is very important for the rational control of inorganic nanowires since it provides the necessary underpinning for the prediction of metal solvents and preparation conditions.

Based on this mechanism study of the nanowire growth, it is conceivable that one can achieve controlled growth of nanowires at different levels. First of all, one can, in principle, synthesize nanowires of different compositions by choosing suitable solvents and growth temperature. A good solvent should be able to form liquid alloy with the desired nanowire material, ideally they should be able to form eutectic. Meantime, the growth temperature should be set between the eutectic point and the melting point of the nanowire material. Both physical methods (laser ablation, arc discharge, thermal evaporation) and chemical methods (chemical vapor transport and chemical vapor deposition) can be used to generate the vapor species required during the nanowire growth.

This VLS method has been exploited in the past several decades to produce 1-100 μm diameter 1D structures termed whiskers [46]. By carefully controlling the nucleation and growth, researchers now are able to produce semiconductor nanowhiskers (InAs, GaAs) using metal-organic vapor phase epitaxy (MOVPE). There have also been many reports on the VLS growth of elemental semiconductors (Si and Ge), III-V semiconductors (GaAs, GaP, InP, InAs), II-VI semiconductors (ZnS, ZnSe, CdS, CdSe), oxides (ZnO, MgO, SiO_2) [19-21,25, 47-60]. Lieber's group has developed and optimized a laser ablation based VLS process to produce semiconductor nanowires with many different compositions [21, 25]. Yang's group has applied the VLS mechanism into chemical vapor deposition/transport process to synthesize nanowires with various compositions [19, 20]. TEM studies of the product obtained after the VLS growth showed that primarily wire-like structures with remarkably uniform diameters on the order of 10 nm with lengths >1 μm were produced by these approaches. The TEM images recorded on individual nanowires further show that the nanowires consist of very uniform diameter crystalline cores. Figure 4 shows several TEM and SEM images of semiconductor ZnO nanowires prepared using this mechanism.

The synthetic framework outlined above is indeed quite general and can lead to rapid identification of catalyst materials and growth conditions. The generality of this approach is perhaps best illustrated through the rational synthesis of nanowire materials having different compositions using either laser ablation or chemical vapor transport/deposition. One example is the recent syntheses of a variety of compound semiconductor nanowires.[25] Compound semiconductors, such as GaAs and CdSe, are especially intriguing targets since their direct band gaps afford optoelectrical properties that are of considerable importance to basic science and technology. The analysis of catalyst and growth conditions can be substantially simplified by considering pseudo-binary phase diagrams for the metal and compound semiconductor of interest. For example, the pseudo-binary phase diagrams of GaAs with Au exhibit a large GaAs rich region in which liquid Au-GaAs coexists with solid GaAs. As a result, single-crystalline GaAs nanowires with diameters of a few nanometers and larger can be synthesized in large yield via laser ablation using this information to set the Au:GaAs composition and growth temperature. In many cases, the presence of catalyst nanoclusters at the nanowire ends following the termination of growth is a strong evidence supporting the VLS mechanism.

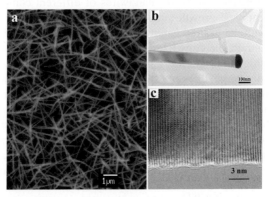

Figure 4. (a) SEM image of ZnO nanowires synthesized via the VLS mechanism (b) TEM images of the single-crystalline ZnO nanowires.

It is now possible to exploit this framework for nanowire synthesis to yield a wide range of elemental, binary, and perhaps more complex 1D nanostructures. The crucial points of this VLS approach are that (1) known equilibrium phase diagrams can be used to predict catalyst materials and growth conditions and (2) laser ablation, chemical vapor deposition and other solution methodology can be used to generate nanometer-sized diameter clusters of virtually any material, thus enabling rational growth of nanowires of many different materials. This synthetic approach represents an exciting opportunity for chemists and materials scientists and moreover, the properties and applications of these emerging 1D structures will be very rich.

2.2.2 Oxide-Assisted Nanowire Growth

In contrast to the well-established VLS mechanism, the Lee's group recently proposed a new nanowire growth route called oxide-assisted nanowire growth. They reported the synthesis of GaAs nanowires obtained by oxide-assisted laser ablation of a mixture of GaAs and Ga_2O_3[61]. No foreign metal is involved in their process. The GaAs nanowires have lengths up to tens of micrometers and diameters in the range of 10-120 nm. The nanowires have a thin oxide layer covering a crystalline GaAs core with a [111] growth direction. This oxide-assisted method for the synthesis of GaAs nanowires has the advantage of requiring neither a metal catalyst nor a template, which simplifies the purification and subsequent application of the wires.

This oxide-assisted nanowire growth mechanism has been further applied to the production of Si nanowires [62, 63]. The Lee group believed that metal catalyst (as needed in VLS mechanism) was not required for the growth of Si nanowires. In their proposed oxide-assisted growth mechanism, the vapor phase of Si_xO ($x>1$) generated by thermal evaporation or laser ablation is the key component. It was found that the precipitation, nucleation and growth of Si nanowires always occurred at the area near the cold finger, which suggested that the temperature gradient provided the external driving force for nanowire formation and growth. The nucleation of Si nanoparticles is assumed to occur on the substrate by different decompositions of Si oxide. Their TEM results suggested that these decompositions result in the precipitation of silicon nanoparticles, which are the nuclei of Si nanowires, coated by shells of silicon oxide.

2.2.3 Vapor-Solid Growth Process

Besides the vapor-liquid-solid (VLS) mechanism, the classical vapor-solid (VS) method for whiskers growth also merits attention for the growth of nanometer 1D materials [46]. In this process, vapor is first generated by evaporation, chemical reduction or gaseous reaction. The vapor is subsequently transported and condensed onto a substrate. The VS method has been used to prepare oxide, metal whiskers with micrometer diameters. The requirements for 1D crystal growth, such as the presence of a dislocation at the vapor-solid interface, are still a matter of controversy in 1D VS growth. However for most whisker growth, the control of supersaturation is a prime consideration, because there is good evidence that the degree of supersaturation determines the prevailing growth morphology. The relative supersaturation associated with the principal growth forms (whiskers, bulk crystal, powders) have been documented extensively. A low supersaturation is required for whisker growth whereas a medium supersaturation supports bulk crystal growth. At high supersaturation, powders are formed by homogeneous nucleation in the gas phase. The size of the whiskers can be controlled by supersaturation, nucleation sizes and the growth time, etc. In principle, it is possible to synthesize 1D nanostructures using the VS process if one can control its nucleation and subsequent growth process.

One such example is the recent reports of synthesis of oxide nanowires of zinc, tin, indium, cadmium, magnesium, and gallium. Among these VS studies, Yang et al. reported the synthesis of MgO, Al_2O_3, ZnO, SnO_2 nanowires via a carbon-thermal reduction process(Figure 5) [64, 65]. More recently, Wang et al. reported the synthesis of oxide nanobelts by simply evaporating the commercial metal oxide powders at high temperatures [66-68]. The as-synthesized oxide

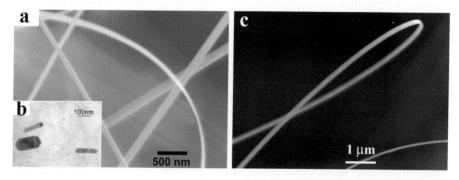

Figure 5. SEM (**a, c**) and cross sectional TEM (**b**) images of SnO2 nanobelts obtained by evaporation and condensation of SnO at 1000 °C.

nanobelts are pure, structurally uniform, and single crystalline, and most of them are free from defects and dislocations. They have a rectangular cross section with typical widths of 30 to 300 nanometers, width-to-thickness ratios of 5 to 10, and lengths of up to a few millimeters. The belt-like morphology appears to be a distinctive and common structural characteristic for the family of semiconducting oxides with cations of different valence states and materials of distinct crystallographic structures. These nanobelts could be an ideal system for fully understanding dimensionally confined transport phenomena in functional oxides and building functional devices along individual nanobelts.

2.2.4 Other Gas Phase Processes

Another important class of semiconductor nanowires is the metal silicide systems reported by the Williams's group [69,70]. In their preparation process, submonolayer amounts of Er deposited onto Si(001) react with the substrate to form epitaxial nanowires of crystalline $ErSi_2$. The $ErSi_2$ nanowires are <1 nm high, a few nanometers wide, close to a micron long, and crystallographically aligned to Si <110> directions. The growth of uniaxial structures occurs because $ErSi_2$ and Si have a good lattice match along one Si <110> crystallographic axis (- 1.3%) but a significant mismatch along the perpendicular Si <110> axis (+ 6.5%).

2.3 Overview of the Nanowire Synthesis

It can be stated that in just a matter of few years, tremendous progress has already been made in the synthesis of various highly crystalline semiconductor nanowires. Below we tabulate the nanowire systems synthesized so far with brief description of their corresponding synthetic methods. Carbon nanotubes and metal nanowires are excluded since the focus of this review is on non-carbon based inorganic semiconductor nanowires.

Table 1. Inorganic nanowire systems and their preparation methods.

Nanowire Material	Synthesis Method	References
Si	Laser Assisted VLS Growth	[21]
	Chemical Vapor Deposition of $SiCl_4$, VLS	[71,72]
	Thermal Evaporation	[73, 74, 75]
	Laser Ablation of SiO	[76]
	Solution-phase Growth, SLS	[40]
Ge	Chemical Vapor Transport, VLS	[19]
	Laser Assisted Catalytic Growth, VLS	[25]
B	Vapor transport, VLS	[55]
	Sputtering	[77]
Bi (nanotubes)	Solution-phase Process	[78]
Se/Te	Solution-phase Approach	[45]
GaN	VLS Growth with Ga and NH_3 as Reactants	[49, 79, 80, 81]
	Carbon nanotubes templating	[23]
	Oxide Assisted Growth	[82]

Table 1 (*Continued*)

	Hot Filament Chemical Vapor Deposition	[83]
	Silica-assisted Catalytic Growth	[84,85]
	Laser Assisted Catalytic Growth, VLS	[25, 50]
AlN	Aluminum Chloride Assisted Growth	[86]
	Silica-assisted Catalytic Growth	[84]
	Carbon nanotubes templating	[87]
InN	Single precursor decomposition	[38, 88]
Si_3N_4	Thermal reduction Si/SiO_2 in NH_3	[89]
Ge_3N_4	Thermal reduction Ge/SiO_2 in NH_3	[90]
GaP	Laser Assisted Catalytic Growth, VLS	[25, 91]
	Oxide Assisted Growth	[61]
InP	Laser Assisted Catalytic Growth, VLS	[25]
	Solution-liquid-solid Growth	[38, 39]
GaAs	Laser Assisted Catalytic Growth, VLS	[25]
	Oxide Assisted Growth	[61]
	Solution-liquid-solid Growth	[38, 39]
	Metal-organic Vapor-phase Epitaxy,VLS	[56,58,59]
InAs	Laser Assisted Catalytic Growth, VLS	[25]
	Solution-liquid-solid Growth	[38,39]
	Metal-organic Vapor-phase Epitaxy, VLS	[56,58,59]
$Al_xGa_{1-x}As$	Solution-liquid-solid Growth	[38,39]
$GaAs_{0.6}P_{0.4}$	Laser Assisted Catalytic Growth, VLS	[25]
$InAs_{0.5}P_{0.5}$	Laser Assisted Catalytic Growth, VLS	[25]
SiC	Reduction-Carbonization Process	[92]
	Thermal Evaporation	[93]
	Laser Ablation	[94]
	Reaction between C Nanotubes and SiO	[36, 95]

Table 1 (*Continued*)

B_xC	Fe catalyzed growth	[96]
ZnO	Chemical Vapor Transport, VLS	[17, 20]
	Thermal Evaporation of Oxide	[64-68, 97-99]
CdO	Thermal Evaporation of Oxide	[64-68]
SnO_2	Thermal Evaporation of Oxide	[64-68]
Ga_2O_3	Thermal Evaporation of Oxide	[64-68,100]
In_2O_3	Thermal Evaporation of Oxide	[64-68]
TiO_2	anodic oxidative hydrolysis of $TiCl_3$ in porous alumina	[101, 102]
GeO_2	Laser ablation	[103]
PbO_2	Thermal evaporation	[66-68]
ZnS	Laser Assisted Catalytic Growth	[25]
	Liquid Crystal Templated Growth	[104,105]
PbS	Polymer matrix templating	[106]
CdS	Laser Assisted Catalytic Growth	[25]
	Solution-phase Approach	[107-109]
	Polymer-controlled Growth	[110]
Cu_2S	Oxide assisted growth	[111, 112]
Bi_2S_3	Solvothermal Decomposition Process	[113]
CdSe	Laser Assisted Catalytic Growth, VLS	[25]
	Porous alumina templating	[114,115]
ZnSe	Solvothermal method	[116]
	Laser Assisted Catalytic Growth, VLS	[25]
Ag_2Se	Se Nanowires as template and react with $AgNO_3$	[37]
PbSe	Solution-phase Approach	[117]
CdS_xSe_{1-x}	Solvothermal Approach	[118]
Bi_2Te_3	Porous alumina templating	[30]
MgB_2	Boron Nanowires as Template	[55]
$TiSi_2$	Vapor-liquid-solid process	[119]
$ErSi_2$	Epitaxial growth on Si substrate	[69, 70]

2.4 Rational Nanowire Growth Control

2.4.1 Diameter Control

The diameter of nanowire is an important parameter. Many physical and thermodynamic properties are diameter dependent. According to the VLS mechanism, the diameter of nanowire is determined by the size of the alloy droplet, which is in turn determined by the original cluster size. By using monodispersed metal nanoclusters, nanowires with a narrow diameter distribution can be synthesized. The Yang group has utilized this strategy to grow uniform Si nanowires in a chemical vapor deposition system [47]. Uniform nanowires with 20.6±3.2, 24.6±4.0, 29.3±4.5 and 60.7±6.2 nm in diameter were grown using Au clusters with sizes of 15.3±2.4, 20.1±3.1, 25.6±4.1 and 52.4±5.3 nm, respectively. Similar results have also been observed in GaP nanowires prepared by laser ablation method [48]. The Korgel group has successfully used this strategy to prepare monodispersed Si nanowire of 4-5 nm in diameter with their supercritical preparative method [40].

2.4.2 Orientation Control

Controlling the growth orientation is important for many of the proposed application of nanowires. By applying the conventional epitaxial crystal growth technique into the VLS process, it is possible to achieve precise orientation control during the nanowire growth. This technique, vapor-liquid-solid epitaxy (VLSE), is particularly powerful in controlled synthesis of nanowire arrays.

Nanowires generally have preferred growth directions. For example, Si nanowires prefer to grow along <111> direction, ZnO nanowires prefer to grow along <001> direction. One strategy to grow vertically aligned nanowires is to properly select the substrate and to control the reaction conditions, so that the nanowires grow epitaxially on the substrate. Take Si as an example, if (111) Si wafer is used as substrate, Si nanowires will grow epitaxially and vertically on the substrate and form nanowire arrays [120, 121]. Another VLSE example is the ZnO nanowires grown epitaxially on a-plane (110) sapphire substrate (Figure 6) [17].

ZnO nanowires have wurtzite structure with lattice constant a=3.24 Å and c=5.19 Å and prefer to grow along <001> direction. ZnO nanowire can grow epitaxially on (110) plane of sapphire, because ZnO a axis and sapphire c axis are related by a factor of 4 (mismatch less than 0.08% at room temperature). Figure 6 shows vertical ZnO nanowire arrays on a-plane sapphire substrate. Their diameters range from 70-120 nm and lengths can be adjusted between 2-10 microns. Previously, InAs, GaAs nanowhiskers have been oriented on Si substrate using a fairly complicated metalorganic vapor-phase epitaxy technique [56, 58,59].

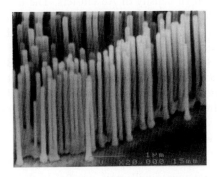

Figure 6. ZnO nanowire arrays on sapphire substrate. [Reprinted with permission from Ref. 17, Copyright American Association for Advancement of Science, 2001].

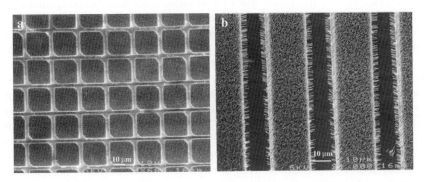

Figure 7. Site-selective growth of ZnO nanowire arrays on a-plane sapphire substrates. The pattern of nanowire array is defined by the initial Au thin film pattern fabricated by photolithography.

2.4.3 Positional Control

It is apparent from the VLS nanowire growth mechanism that the positions of nanowires can be controlled by the initial positions of Au clusters or thin films. Various lithographical techniques including for example, soft lithography, e-beam and photolithography can be used to create patterns of Au thin film for the subsequent semiconductor nanowire growth. Figure 7 shows the SEM images of ZnO nanowires grown from the line and square Au patterns on a-plane sapphire substrate. It is clear that nanowires grow vertically only from the region that is coated with Au and form the designed patterns of ZnO nanowire array [120,121].

During the nanowire array growth, generally gold thin film was deposited as the solvent/initiator for the nanowire growth. Upon heat-up, these gold thin films will self-aggregate into high density of Au clusters. The diameters and the density of these clusters are determined by the thickness of the thin film and the growth temperature. Thus it is possible to control the nanowire areal density by modifying the thin film thickness. Another approach to control the areal density of the nanowire array is to use the solution-made Au clusters. By dispersing different amount/density of Au clusters on the sapphire substrate, it is possible to obtain nanowire arrays with different densities. It is now possible to synthesize, for example, ZnO nanowire arrays with areal density spanning from 10^6 to 10^{10} cm^{-2}.

2.4.4 Nanowire Network Growth

As a comparison, when the epitaxial nanowire growth conditions are not satisfied, most of the nanowires will grow parallel to the surface, i.e., nanowires "crawl" along the substrate [120,121]. This surface-parallel growth is actually important for their potential applications in nanoscale electronics, which requires precise placement of individual wires on substrate with desired configuration. It's critical to have the capability to define the starting and ending points during the nanowire growth in order to form a surface nanowire network. While the starting point of nanowires can be defined by the positions of metal particles as outlined in the previous section, it's necessary to create certain anisotropic chemical environment so that nanowires can be guided to grow along a certain preset direction on the substrate surface. A feasible method is to create a local vapor pressure gradient near the nanowire. Nanowire grows faster along the direction with higher vapor pressure. This local vapor pressure gradient can be created by the formation of liquid alloy droplet between the metal solvent (at high temperature) and the nanowire material. This surface patterning strategy can be readily applied to ZnO system [20]. In addition, using strong electrical and magnetic field, it should be also possible to align nanowires on substrate during their growth.

2.4.5 Nanowire Heterostructures

The success of semiconductor integrated circuits has been largely hinged upon the capability of heterostructure formation through carefully controlled doping and interfacing. In fact, the 2-dimensional (2D) semiconductor interface is ubiquitous in optoelectronic devices such as light emitting diode, laser diodes, quantum cascade lasers, and transistors. Heterostructure formation in 1D nanostructures is equally important for their potential applications as efficient light emitting sources and better thermoelectrics. While there are a number of well-developed techniques (e.g., molecular beam epitaxy) for the fabrication of thin film heterostructures and superlattices, a general synthetic scheme for heterojunction and superlattice formation in 1D nanostructures with well-defined coherent interfaces was only recently introduced. Previous studies on semiconductor nanowires or nanotubes have invariably dealt with homogeneous systems with a few exceptions of heterostructure formation including heterojunctions between carbon nanotubes and silicon/carbide nanowires [122, 123]. and p-n junction on individual carbon nanotubes or GaAs/Ga$_{1-x}$In$_x$As nanowires [124-126]. Recently, a sequential electrochemical method was reported to synthesize metal bar-coded microrods [127]. This method, however, yields polycrystalline products with less than ideal interfaces. Yang et al. have recently developed a hybrid pulsed laser ablation/chemical vapor deposition (PLA-CVD) process for the synthesis of semiconductor nanowires with periodic longitudinal heterostructures [128]. In this process, Si and Ge vapor sources are independently controlled and alternately delivered into the VLS nanowire growth system. As a result, single-crystalline nanowires with Si/SiGe superlattice structure are obtained.

Figure 8 shows a scanning transmission electron microscopy (STEM) image of two nanowires in bright-field mode. Along the wire axes, dark stripes appear periodically, which originate from the periodic deposition of the SiGe alloy and Si segments. The electron scattering cross section of the Ge atom is larger than that of Si. Consequently, the SiGe alloy block appears darker than the pure Si block. The chemical composition of the darker area is examined using energy-dispersive X-ray spectroscopy (EDS), which shows a strong Si peak and apparent Ge doping (~12 wt % Ge). The periodic modulation of Ge doping is further confirmed by scanning a focused electron beam along the nanowire growth axis and tracking the change of X-ray signal from Si and Ge atoms in the wires (Figure 8b). Both Si and Ge X-ray signals show periodic modulation, and their intensities are anticorrelated: wherever the X-ray signal from Ge shows a maximum, the signal from Si shows a minimum, which confirms the formation of Si/SiGe superlattice along the wire axis.

This hybrid PLA-CVD method can be used to prepare various other heterostructures on individual nanowires in a "custom-made" fashion since the vapor source supplies can be readily programmed. It will enable the creation of various functional devices (e.g., p-n junction, coupled quantum dot structure, and heterostructured bipolar transistor) on single nanowires. These nanowires could be used as important building blocks for constructing nanoscale electronic circuits and light emitting devices. As an example, superlattice nanowires with reduced phonon transport and high electron mobility are believed to be better thermoelectrics. Using similar approaches, two other groups have been successfully prepared GaAs/GaP, InAs/InP heterostructured nanowires [129,130].

Besides the vertical heterojunction within a single nanowire, it is further possible to create lateral junctions on a single nanowire. The synthetic approach generally uses the existing nanowires as templates or substrates for conformal or selective film deposition. For example, Yang et al. have used tin dioxide nanoribbons as substrates for thin-film growth using pulsed laser deposition (PLD). Various oxides (e.g. TiO$_2$, transition metal doped TiO$_2$ and ZnO) have been deposited on these 1-dimensional nanoscale substrates. Electron microscopy and X-ray diffraction studies clearly demonstrate that these functional oxides grow epitaxially on the side surfaces of the substrate nanoribbons with sharp structural and compositional interfaces, and so form a unique class of bi-layer nanoribbons

(Figure 9) with significantly enhanced functionality [131]. Not surprisingly, if one uses chemical vapor deposition, a core-sheath nanowire structure could be prepared [132].

3 Hierarchical Assembly: Integration of Nanowires into Functional Networks

Integration of nanowire building blocks into complex functional structure in a predictable and controlled way represents a major scientific challenge in the nanowire research community. Basically, there are two possible routes. One is to form nanowire superstructures through direct one-step growth process, as illustrated above in the sections of the controlled growth of Si and ZnO nanowire networks and arrays. The other possibility is to develop suitable hierarchical assembly techniques to put nanowire building blocks together into functional structure. Atomic force microscope has been used to push or deposit nanotubes into desired configuration [133]. The shortcoming of this method is that it's time-consuming and is not a parallel process.

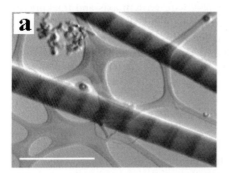
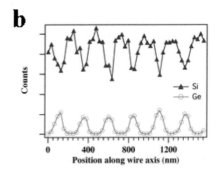

Figure 8 (a) STEM image of two nanowires in bright field mode. The scale bar is 500 nm. (b) Line profile of EDS signal from Si and Ge components along the nanowire growth axis. [Reprinted with permission from Ref. 128, Copyright American Chemical Society, 2002].

The Yang Group has developed a simple and parallel method dubbed microfluidic assisted nanowire integration (MANI) process [134]. The microchannels are formed between a poly(dimethylsiloxane) (PDMS) micromold and a flat Si/glass substrate. The microchannels have variable height of 1-4 microns, width of 1-10 microns and length of 5-10 mm. This technique has been successfully applied for the alignment of $Mo_3Se_3^-$ molecular wires, conducting polymer nanowires and carbon nanotubes. Take the $Mo_3Se_3^-$ molecular wires as example, a droplet of the wire solution/suspension was placed at the open end of the microchannels; the liquid fills the channels under capillary effect. After the evaporation of the solvent and the lift-off of the PDMS mold, bundles of molecular wires (10-100 nm in diameter) were aligned along the edges of the microchannels and formed parallel array as shown in figure 10a. After patterning the first layer of nanowires, the process can be repeated to deposit multilayer nanowires and form complex structures. By rotating the microchannel 90° during the second application, it is possible to fabricate arrays of nanowire cross-junctions in a well-controlled and reproducible fashion as shown in figure 10a inset. This method provides a general and rational approach for the hierarchical assembly of 1D nanomaterials into well-defined functional networks. The methodology has recently been extensively used in the assembly of other inorganic nanowires into functional nano-devices (Fig. 10b) [135].

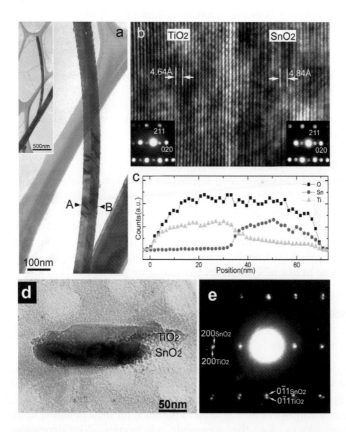

Figure 9. (a) TEM image of a $TiO_2@SnO_2$ nanotape; Inset is its low magnification image. (b) HRTEM image of the atomically sharp $TiO_2@SnO_2$ interface. 4.64 Å and 4.84 Å correspond to the lattice spacing between the (010) planes in TiO_2 and SnO_2 rutile structures. Insets are electron diffractions taken on the two sides of the interface along the same [10 2̄] zone axes. (c) Compositional line profile across the $TiO_2@SnO_2$ interface as outlined in **a** (from A to B). (e) Selected area electron diffraction pattern recorded at the cross section area in (**d**). The SnO_2 and TiO_2 layer are in the same zone axis and have the same orientation, indicating perfect epitaxy. [Reprinted with permission from Ref. 131, Copyright American Chemical Society, 2002].

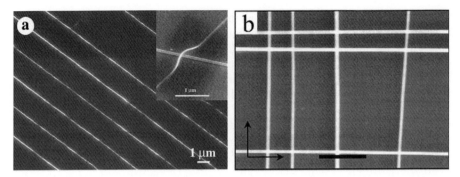

Figure 10. (a). SEM image of the aligned Mo$_3$Se$_3^-$ molecular wire bundles on Si wafer. Inset shows the cross-junction formed through multiple layer alignment process. [Reprinted with permission from Ref. 134, Copyright American Chemical Society, 2000]. (b). Typical SEM images of crossed arrays of InP nanowires obtained in a two-step assembly process with orthogonal flow directions for the sequential steps. Flow directions are highlighted by arrows in the images. The scale bar corresponds to 2 μm. [Reprinted with permission from Ref. 135, Copyright American Association for Advancement of Science, 2001].

Electrical field induced nanowire alignment represents another powerful assembly technique. Lieber et al. have fabricated metal electrode arrays on which bias is applied to generate strong electrical filed to align Si and InP nanowires dispersed in between the electrodes [136].

Although these assembly techniques have some success to align inorganic nanowires, they still have the following shortcomings: (1) the large pitch between nanowires imparted by the microchannels or fabricated electrodes makes it less likely to assemble high density nanowire arrays; (2) these techniques have limited yield and patternable substrate area. Recognizing these limitations, new methodologies such as Langmuir-Blodgett technique is currently being developed for the purpose of assembly high density nanowire assemblies [137].

4 Physical Properties of Nanowires

4.1 Thermal Stability

Compared with bulk materials, low-dimensional nanoscale materials, with their large surface area and possible quantum confinement effect, exhibit distinct electric, optical, chemical and thermal properties. Thermal stability of the semiconductor nanowires is of critical importance for their potential implementation as building blocks for the nanoscale electronics. Size-dependent melting-recrystallization process of the carbon-sheathed semiconductor Ge nanowires has been recently studied by the Yang group using an in-situ high temperature transmission electron microscope [138, 139].

Ge nanowires with diameter from 10-100 nm were made via VLS mechanism. These nanowires were further coated with 1-5 nm thick carbon sheath to confine the molten Ge and prevent the formation of liquid droplet at high temperature. Two distinct features have been observed during the melting and recrystallization process. One is the significant melting point depression, which is inversely proportional to the radius of nanowire. Another is the large hysteresis during the melting-recrystallization cycle. Take the melting and recrystallization process of a Ge nanowire with diameter of 55 nm and length of 1 micron as example, the melting starts from two ends at around 650 °C (the melting point of bulk Ge is 930 °C) and moves towards the center of the wire. At 848 °C, the whole wire melts. During cooling process, the recrystallization happens at much lower temperature (558 °C) than the initial melting temperature. Utilizing the low melting points of the nanowires encapsulated in nanotubes, the Yang group has demonstrated the capability to manipulate individual nanowires including for example, in-situ cutting, interconnection and welding. Figure 11 shows a sequential TEM images during the in-situ interconnection between two nanowires.

A related study is the recent observation of silicon-based nanostructures with different morphologies and microstructures at different formation and annealing temperature.[140] The synthetic method is thermal evaporation of Si/SiO_2 mixture in a alumina tube. It was observed that besides Si nanowires, many other kinds of Si-based nanostructures such as octopuslike, pinlike, tadpolelike, and chainlike structures were also formed. The formation and annealing temperature was found to play a dominant role on the formation of these structures. It is demonstrated that a control over the temperature can precisely control the morphologies and intrinsic structures of the silicon-based nanomaterials. This is an important step toward design and control of nanostructures using the knowledge of nanowire thermal stability [140].

The marked reduction in melting temperature for the semiconductor nanowires in nanotubes has several important implications. First, the optimum annealing temperature for preparing of high quality defect free nanowires can be expected to be a small fraction of the bulk annealing temperature. It is possible to conduct nanowire zone-refining at a modest temperature with the current simple configuration. Second, the capability of cutting, linking and welding nanowires at relatively modest temperature may provide a new approach for integrating these 1D nanostructures into functional devices and circuitry. Finally, as the dimension of the wires is reduced to nanometer length scale, the chemical and thermal stability of the new devices may be limited, which should be considered during the implementation of nanoscale electronics.

4.2 Optical Properties

4.2.1 Photoluminescence and Stimulated Emission

Due to the quantum confinement effect, the nanowires exhibit distinct optical propertied when their size is below certain critical dimension. For example, the absorption edge of the Si nanowires synthesized by Korgel et al. was strongly blue-shifted from the bulk indirect band gap of 1.1 eV. Sharp discrete absorbance features and relatively strong "band edge" photoluminescence (PL) were also observed. These optical properties likely result from quantum confinement effects, although they cannot rule out the possibility of additional surface states as well [40,141]. The <110> oriented nanowires exhibited distinctly molecular-type transitions. The <100> oriented wires exhibited a significantly higher exciton energy than the <110> oriented wires, Korgel et al. are among the first to show that the tunability of the lattice orientation in the silicon nanowires can lead to different optical properties.

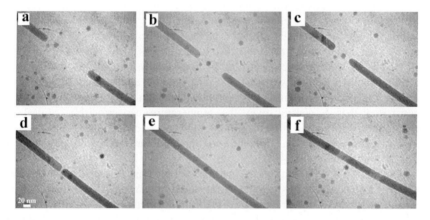

Figure 11. Sequential TEM images of linking two Ge nanowires within a carbon nanotubes. [Reprinted with permission from Ref. 139, Copyright Wiley-VCH, 2001].

In addition, Lieber et al. have characterized the fundamental PL properties of individual, isolated indium phosphide (InP) nanowires to define their potential for optoelectronics [142]. Polarization-sensitive measurements reveal a striking anisotropy in the PL intensity recorded parallel and perpendicular to the long axis of a nanowire. The order-of-magnitude polarization anisotropy was quantitatively explained in terms of the large dielectric contrast between these freestanding nanowires and surrounding environment, as opposed to quantum confinement effects. This intrinsic anisotropy has been used to create polarization-sensitive nanoscale photodetectors that may prove useful in integrated photonic circuits, optical switches and interconnects, near-field imaging, and high-resolution detectors.

More importantly, room temperature ultraviolet lasing has been successfully demonstrated within ZnO nanowires by the Yang group [17]. ZnO is a wide band-gap (3.37 eV) compound semiconductor that is suitable for blue optoelectronic applications with ultraviolet lasing action being reported in disordered particles and thin films. In addition, ZnO has exciton binding energy as high as 60 meV, significantly larger than that of ZnSe (22 meV) and GaN (25 meV), which indicates that the excitons in ZnO are thermally stable at room temperature.

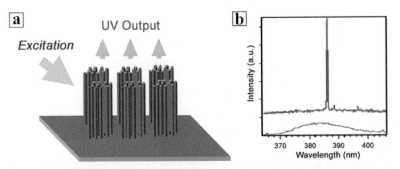

Figure 12. Experimental set-up for measuring PL and lasing spectra on nanowire arrays (a) and (b). emission spectra from vertical ZnO nanowire arrays on a-place sapphire substrate below (bottom trace) and above the lasing threshold (upper trace). The pump powers for these spectra are 20, 150 kW.cm^{-2}, respectively. The spectra are offset for easy comparison.

In the lasing experiments, the ZnO nanowire arrays were optically pumped by the fourth harmonic of Nd:YAG laser at room-temperature to measure the power-dependent emission (Fig. 12a). Figure 12b shows the evolution of the emission spectra as the pump power is increased. At low excitation intensity, the spectrum consists of a single broad spontaneous emission peak. This spontaneous emission is generally ascribed to the recombination of excitons through an exciton-exciton collision process where one of the excitons radiatively recombines to generate a photon. When the excitation intensity exceeds a threshold (~40 kW/cm^2), sharp peaks emerge in the emission spectra. Above the threshold, the integrated emission intensity increases rapidly with the pump power. The narrow line width and the rapid increase of emission intensity indicate that stimulated emission takes place in these nanowires. The observed lasing action in these nanowire arrays without any fabricated mirror suggests these single-crystalline, well-facetted nanowires can indeed function as natural resonance cavities.

The nanowire lasing has been further confirmed with the optical characterization of single ZnO nanowire by near field scanning microscopy (NSOM) [143]. (Figure 13). The nanowires were excited with short pulses (<1 ps) of 4.35 eV photons (285 nm) in order to induce photoluminescence (PL) and lasing.

The nanowire emission was collected by a chemically etched fiber optic probe held in constant-gap mode by the feedback electronics of the NSOM. The NSOM collected topographic and optical information simultaneously on both forward and reverse scans. Lasing was clearly observed from a nanowire that was situated at an angle to the quartz substrate, shown in Figure 14. The intense signal collected near the ends of the nanowire demonstrates the waveguided and confinement of the nanowire lasing emission to a cone-shaped region near the end faces.

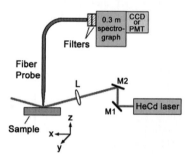

Figure 13. Schematic illustration of NSOM setup for characterizing the PL and lasing properties of individual nanowires.

It should be emphasized that the concept of using well-cleaved nanowires as natural optical cavities should be applicable to many other different semiconductor systems such as GaN [144], GaAs and CdSe. The results obtained in ZnO nanowire system suggest the feasibility of nanoscale surface-emitting lasers operating at ultraviolet or other wavelengths when the material of the nanowire cavity is altered. In addition, by creating *pn* junctions in these individual nanowires, one should be able to test the possibility of making electron ejection UV/blue lasers out of individual nanowires. Such miniaturized nanowire nanolasers will find applications in nano-photonics and microanalysis.

Figure 14. A NSOM image of a lasing ZnO nanowire. The lasing signal is waveguided to the two ends of the nanowire. [Reprinted with permission from Ref. 143, Copyright American Chemical Society, 2001].

4.2.2 Nonlinear Optical Mixing in Single Zinc Oxide Nanowires

In addition to the interesting photoluminescence and lasing properties of these nanowires, their nonlinear optical properties suggest other important potential applications as frequency converters or logic/routing elements in nanoscale optoelectronic circuitry. Coherent nonlinear optical phenomena, such as second- and third-harmonic generation (SHG and THG, respectively), depend explicitly on the crystal lattice structure of the medium, which could yield a very high (nearly 100%) polarization dependence. In addition, the temporal response of non-resonant harmonic generation is similar to the pulsewidth of the incident laser, in some cases ≈ 20 fs, while incoherent processes are at least 2-4 orders of magnitude slower. Moreover, non-resonant SHG is essentially independent of wavelength below the energy band gap of semiconductor materials, most often including the 1.3-1.5 μm wavelength region typically used in optical fiber communications. A material of particular interest is zinc oxide (ZnO). Studies of microcrystalline ZnO thin films have revealed a large second-order nonlinearity, characterized by $\chi^{(2)}$, which determines the efficiency of a material as a converter of optical frequencies via several processes (e.g., SHG and sum- and difference- frequency generation (SFG, DFG)).

To examine the nonlinear optical properties of the ZnO nanowires, Yang and Saykally et al. employed oblique collection mode NSOM, in which the sample is illuminated in the far-field, as it is preferred for nonlinear near-field imaging because of its suitability for high incident pulse intensity experiments [145]. Fig. 15 shows two series of SHG images, illustrating the input polarization and nanowire orientation dependence of the SHG. This dependence arises from the two independent, non-vanishing components of $\chi^{(2)}$ observed in SHG for ZnO, $\chi^{(2)}_{zxx}$ and $\chi^{(2)}_{zzz}$. In Fig. 15a-b, two wires are situated approximately normal to each other, with an s-polarized incident beam in Fig. 15a and p-polarization in Fig. 15b. The polarization ratio (SHG$_{s\text{-inc}}$ /(SHG$_{p\text{-inc}}$+ SHG$_{s\text{-inc}}$)) for wire 1 is 0.90. Wire 2 has an average signal that is 2.5 times that of wire 1.

To quantitatively analyze the SHG polarization effect, polarization traces were taken on several wires from Fig. 15a-b, d-e. The near-field probe was maintained above each wire, and the input polarization was rotated as the SHG signal was monitored (Fig. 15, c and f). The theoretical traces were computed. The polarization data were fit to theory, and nearly all the wires tested (diameters 80-100 nm) exhibited a ratio $\chi^{(2)}_{zzz}/\chi^{(2)}_{zxx}$ of approximately 2.0 - 2.3. These ratios can be compared with a value of 3.0 for bulk crystalline ZnO.

Using a reference material, one can determine the absolute magnitude of each component of $\chi^{(2)}$ for individual nanowires. $\chi^{(2)}_{zxx}$ and $\chi^{(2)}_{zzz}$ were determined by measuring the nanowire SHG with respect to a zinc selenide (ZnSe) disk (at 1.4 μm excitation). Using the ZnSe reference (78 pm/V), $\chi^{(2)}_{zzz}$ for a single ZnO wire = 5.5 pm/V and $\chi^{(2)}_{zxx}$ = 2.5 pm/V. The $\chi^{(2)}_{zzz}$ value is considerably lower than the reported bulk value (18 pm/V) but in relatively good agreement with values reported for ZnO thin films (4-10 pm/V). In addition, this nanowire SHG was found to be primarily wavelength independent ($\lambda_{SHG} > 400$ nm) and relatively efficient, with a larger $\chi^{(2)}_{eff} \geq 5.5$ pm/V than beta-barium borate (BBO, $\chi^{(2)}_{eff} \approx 2.0$ pm/V) a commonly-used doubling crystal.

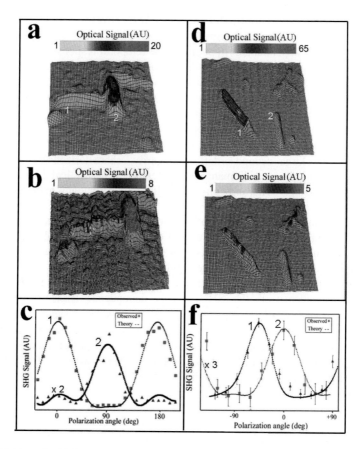

Figure 15. Nanowire SHG polarization dependence. **(a)** Combined topographical and SHG image of two wires at angle of approximately 90°. Image size is $(13 \ \mu m)^2$ and maximum topographic height is 130 nm. Beam is s-polarized, incident from the right. **(b)** SHG of same region as in (A) with p-polarized incident beam. **(c)** Polarization-dependent SHG data and theoretical predictions taken from the wires labeled in (A). The theoretical curves are calculated for the SHG signal and $\chi^{(2)}_{eff}$ for a hexagonal crystal. **(d)** Image of $(16 \ \mu m)^2$ region showing several nanowires. Maximum topographic height is 120 nm. The polarization $\alpha = -45°$. **(e)** Image of the same region as in (D) with polarization $\alpha = 45°$. **(f)** Polarization traces taken from the wires labeled 1 and 2 in (C) and theoretical simulations. Reprint with permission from [145], copyright American Chemical Society, 2002.

The nonlinear optical properties demonstrated here suggest that ZnO nanowires could be effectively employed as frequency converters or logic components in nanoscale optoelectronics.

4.2.3 Photoconductive Oxide Nanowires as Nanoscale Optoelectronic Switches

Among all possible nano-devices, switches are critical for important applications like memory and logic. Electrical switching on nanometer and molecular level has been predominantly achieved through proper electrical gating configuration, as exemplified by nanotube transistors. Yang et al. have demonstrated that it is possible to create highly sensitive electrical nanowire switches by exploring the photoconductivity of the individual semiconductor nanowires. They found that the conductance of the ZnO nanowire is extremely sensitive to ultraviolet light exposure. The light induced insulator-to-conductor transition allows us reversibly switching the nanowires from OFF to ON states, an "optical gating" phenomena as compared to commonly used electrical gating.

Four-probe measurement of individual ZnO nanowires indicates that these nanowires are essentially insulating in dark with a resistivity above 3.5 MΩcm^{-1}. When the nanowires were exposed to UV light with wavelengths below 400 nm, it was found that the nanowire resistivity instantly decreases by typically 4 to 6 orders of magnitude [146]. In addition to the high sensitivity of the nanowire photoconductor, they also exhibit excellent wavelength selectivity. Figure 16a shows the evolution of the photocurrent when a nanowire was exposed first to highly intense light at 532 nm (Nd:YAG, 2nd harmonics) for 200 s and then to UV-light at 365 nm. There is no photoresponse at all to the green light while exposure to less intense UV-light shows the typical change of conductivity of 4 orders of magnitude. Measurements of the spectral response show that our ZnO nanowires indeed have a cut-off wavelength of 385 nm which is expected from the bandgap of ZnO.

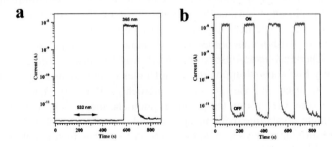

Figure 16. (a) Sensitivity of the photoresponse of a ZnO nanowire to light exposure at wavelengths of 532 nm and 365 nm. (b) Reversible switching of a ZnO nanowire between low and high conductivity states. Reprint with permission from [146], copyright Wiley-VCH, 2002.

It has been unambiguously established that oxygen chemisorption plays a profound role in enhancing photosensitivity of bulk or thin film ZnO. It is believed that a similar photoresponse mechanism could be applied to the nanowire system with additional consideration of high surface area of the nanowire, which could further enhance the sensitivity of the device. It was generally believed that the photoresponse of ZnO consists of two parts: a solid-state process where an electron and a hole are created (hv $\rightarrow$ h$^+$ + e$^-$) and a two-step process involving oxygen species adsorbed on the surface. In the dark, oxygen molecules adsorb on the oxide surface as a negatively charged ion by capturing free electrons of the n-type oxide semiconductor ($O_2(g)$ + e$^-$ $\rightarrow$ O_2^-(ad)) thereby creating a depletion layer with low conductivity near the nanowire surface. Upon exposure to UV-light, photo-generated holes migrate to the surface and discharge the negatively charged adsorbed oxygen ions (h$^+$ + O_2^-(ad) $\rightarrow$ $O_2(g)$) through surface electron-hole recombination. Meantime, photo-generated electrons destruct the depletion layer, as a result, the conductivity of the nanowire increase significantly.

The characteristics of the ZnO photoconducting nanowires indicates that they could be good candidates for optoelectronic switches, i. e. the insulating state as "OFF" in the dark and the conducting state as "ON" when exposed to UV-light. Figure 16b plots the photoresponse as a function of time while the UV-lamp was switched on and off. It is evident that the nanowires can be reversibly switched between the low conductivity state and the high conductivity state. With further optimization of the nanowire composition, e. g. through proper doping, the photocurrent decay time (i. e. the response time) could be reduced to μs level. These highly sensitive photoconducting nanowires could serve as very sensitive UV-light detector in many applications such as microanalysis and missile plume detection, as well as fast switching devices for nanoscale optoelectronics applications where ON and OFF states can be addressed optically.

4.2.4 Room Temperature NO_2 Photochemical Sensing

Another major area of application for nanowires and nanotubes is likely to be the sensing of important molecules, either for medical or environmental health purposes. The ultrahigh surface to volume ratios of these structures make their electrical properties extremely sensitive to surface-adsorbed species, as recent work has shown with carbon nanotubes, functionalized silicon nanowires and metal nanowires. Chemical nanosensors are interesting because of their potential for detecting very low concentrations of biomolecules or pollutants on platforms small enough to be used *in vivo* or on a microchip. Recently Yang et al. have demonstrated the first room temperature photochemical NO_2 sensors based on individual single-crystalline oxide nanowires and nanoribbons [147].

Tin dioxide is a wide bandgap (3.6 eV) semiconductor. For *n*-type SnO_2 single crystals, the intrinsic carrier concentration is primarily determined by deviations from stoichiometry in the form of equilibrium oxygen vacancies, which are predominantly atomic defects. The electrical conductivity of nanocrystalline

SnO_2 depends strongly on surface states produced by molecular adsorption that result in space-charge layer changes and band modulation. NO_2, a combustion product that plays a key role in tropospheric ozone and smog formation, acts as an electron-trapping adsorbate on SnO_2 crystal faces and can be sensed by monitoring the electrical conductance of the material. Because NO_2 chemisorbs strongly on many metal oxides, commercial sensors based on particulate or thin-film SnO_2 operate at 300-500°C to enhance the surface molecular desorption kinetics and continuously "clean" the sensors. The high temperature operation of these oxide sensors is not favorable in many cases, particularly in an explosive environment. Yang et al. have found that the strong photoconductive nature of individual single crystalline SnO_2 nanoribbons makes it possible to achieve equally favorable adsorption-desorption behavior *at room temperature* by illuminating the devices with ultraviolet (UV) light of energy near the SnO_2 bandgap. The active desorption process is thus photo-induced molecular desorption.

The resolution limit achieved by these oxide nanoribbons fell between 2-10 ppm. Figure 17 shows the conductance response of one nanosensor cycled between pure air and 3 ppm NO_2. Even with the low signal/noise ratio, current steps can be clearly distinguished as the NO_2 was turned on and off. This behavior was stable for >20 cycles without appreciable drift and with response times of less than one minute.

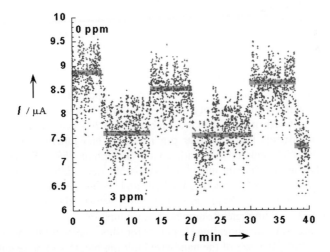

Figure 17. Cycling a nanosensor near its resolution limit under 365 nm light. NO_2 concentrations are indicated. Horizontal bars are signal averages. The average signal difference for the three cycles is 16%. Bias is 0.5 V. Reprint with permission from [147], copyright Wiley-VCH, 2002.

Individual oxide nanoribbons are small, fast and sensitive devices for detecting ppm-level NO_2 at room temperature in UV light. The advantages of low-temperature, potentially drift-free operation make these nanoribbons/wires good candidates for miniaturized, ultra-sensitive gas sensors in many applications. Single molecule chemical detection using nanowires may be within reach in near future.

4.3 Electronic Properties

Miniaturization in electronics through improvements in established top-down fabrication techniques is approaching the point where fundamental issues are expected to limit the dramatic increases in computing speed. Semiconductor nanowires have recently been used as building blocks for assembling a range of nanodevices including FETs, p-n diodes, bipolar junction transistors, and complementary inverters [148-152]. Major part of the work has been done by the Lieber group along this direction.

In contrast to carbon nanotubes, these nanowire devices can be assembled in a predictable manner because the electronic properties and sizes of the nanowires can be precisely controlled during synthesis and methods are being developed for their parallel assembly. The Lieber group [148,149] has explored the possibility of assembling various devices at the nanometer scale using their n- and p-type Si, InP nanowires. It is believed that the "bottom-up" approach to nanoelectronics has the potential to go beyond the limits of the traditional "top-down" manufacturing techniques. As the critical dimension of an individual device becomes smaller and smaller, the electron transport properties of their components become an important issue to study. As for semiconductors, recent measurements on a set of nanoscale electronic devices indicated that GaN nanowires as thin as 17.6 nm could still function properly as a semiconductor.

Another issue related to the electronic applications of chemically synthesized nanowires is the assembly of these building blocks into various device architectures. It is worth noting that the Lieber group has been able to assemble semiconductor nanowires into cross-bar p-n junctions and junction arrays having controllable electrical characteristics with a yield as high as 95%. These junctions have been further used to create integrated nanoscale field-effect transistor arrays with nanowires as both the conducting channel and gating electrode. In addition, OR, AND, and NOR logic-gate structures with substantial gain have been configured and tested to implement some basic computation.

There are several appealing features for this "bottom-up" approach to nanoelectronics. First, the size of the nanowire building blocks can be readily tuned to sub-100 nm and smaller, which should lead to high density of devices on a chip. Second, the material systems for the nanowires are essentially unlimited, which should give researchers great flexibility to select the right materials for the desired device functionality. For example, the Lieber group has recently demonstrated several GaN nanowire based nanodevices which would be of interest for their high-power/high temperature electrical applications. It is obvious that great progress has been made along the direction of using nanowire building blocks for various device applications. Nevertheless, one has to admit that to ultimately achieve the goal of "bottom-up" manufacturing in the future will still require substantial work, including, for example, the development of 3-dimensional hierarchical assembly processes, as well as the improvement of material synthesis [149].

4.4 Mechanical Properties

The mechanical properties of small, rodlike materials are of considerable interest. For example, small whiskers can have strengths considerably greater than those observed in corresponding macroscopic single crystals, an effect that is attributed to a reduction in the number of structural defects per unit length that lead to mechanical failure. The Lieber group used atomic force microscopy to determine the mechanical properties of individual, structurally isolated silicon carbide (SiC) nanowire that were pinned at one end to molybdenum disulfide surfaces [14]. The bending force was measured versus displacement along the unpinned lengths. Continued bending of the SiC nanowires ultimately led to fracture. They calculated Young's modulus of 610-660 Gpa based on their AFM measurement. These results agree well with the 600 GPa value predicted theoretically for [111]-oriented SiC and the average values obtained previously for micrometer-diameter whiskers. The large Young's modulus values determined for SiC nanowires make these materials obvious candidates for the reinforcing element in ceramic, metal, and polymer matrix composites.

4.5 Field Emission Properties

It is well-known that nanotubes and nanowires with sharp tips are promising materials for applications as cold cathode field emission devices. Field emission characteristics of the β-SiC and Si nanowires have been investigated using current-voltage measurements. The silicon carbide nanorods exhibited high electron field emission with high stability. Both Si and SiC nanowires exhibit well-behaved and robust field emission. The turn-on fields for Si and SiC nanowires were 15 and 20 V μm^{-1}, respectively [153,154] and current density of 0.01 mA cm^{-2} which are comparable with those for other field emitters including carbon nanotubes and diamond. Along with the ease of preparation, these silicon carbide and Si nanowires are believed to have potential application in electron field emitting devices.

5 Conclusions and Outlook

In this article, we have given an overview of work directed toward the rational growth and novel properties of semiconductor nanowires. Using methodologies based on solution and gas phase crystal growth, single crystalline nanowires of a wide range of elemental and compound materials have been prepared. The current capability of growth control represents a significant step toward the applications of nanowires as building blocks in nanoscale electronics and photonics. Significant challenges still exist in this synthetic direction including for example the monodispersity, interface control in 1D nanostructures and their ordered assemblies and composites.

The investigation of these 1D nanostructures will continuously be exciting and highly rewarding. Many interesting properties can be investigated, such as electron transport, optical, photoconductivity, thermoelectricity as well as their chemical properties. Although much progress has been made in the characterization of optical and electrical properties of nanowires as well as some prototype nanoscale electrical device demonstration, there is still great need in the examination of the fundamental effects of size and dimensionality on their properties. In summary, with the easy access to many different compositions and possible electronic structures, these non-carbon based inorganic semiconductor nanowires surely represent a broad class of nanoscale building blocks for fundamental studies and many technological applications.

References

1. Handbook of Nanostructured Materials and Nanotechnology (Eds. Nalwa H. S.), Academic Press, 2000.
2. Shalaev V.M. and Moskovits M., Nanostructured materials: clusters, composites, and thin films (Eds. Washington, DC), American Chemical Society, 1997.
3. Edelstein A.S.and Cammarata R.S., Nanomaterials : synthesis, properties, and applications (Eds. Bristol, UK; Philadelphia, PA), Institute of Physics, 1996.
4. Bawendi M.G., Steigerwald M.L. and Brus L.E., The quantum mechanics of larger semiconductor clusters ("quantum dots"), *Annu. Rev. Phys. Chem.* **41**(1990) pp.477-496.
5. Murray C.B., Kagan C.R. and Bawendi M.G., Synthesis and characterization of monodisperse nanocrystals and close-packed nanocrystal assemblies, *Annu. Rev. Mater. Sci.* **30**(2000) pp.545-610.
6. Likharev K.K. and Claeson T., *Sci. Am.*, *June*, 80 (1992).
7. Likharev K.K., Correlated discrete transfer of single electrons in ultrasmall tunnel junctions, *IBM J. Res. Develop.* **32**(1988) pp.144-158.

8. Markovich G., Collier C.P., Henrichs S.E., Remacle F., Levine R.D. and Heath J.R., Architectonic quantum dot solids, *Acc. Chem. Res.* **32**(1999) pp.415-423.
9. Klimov V.I., Mikhailovsky A.A., Xu S., Malko A., Hollingsworth J.A., Leatherdale C.A., Eisler H.J., and Bawendi M.G., Optical gain and stimulated emission in nanocrystal quantum dots, *Science* **290**(2000) pp.314-317.
10. Rueckes T., Kim K., Joselevich E., Tseng G.Y., Cheung C.L., and Lieber C.M., Carbon nanotube-based nonvolatile random access memory for molecular computing, *Science* **289** (2000) pp.94-97.
11. Klein D.L.,Roth R., Lim A.K.L., Alivisatos A.P. and McEuen P.L., single-electron transistor made from a cadmium selenide nanocrystal, *Nature* **389**(1997) pp.699-701.
12. Louie S.G., Electronic properties, junctions, and defects of carbon nanotubes, *Op. Appl. Phys.* **80**(2001) pp.113-145.
13. Hu J., Odom T.W. and Lieber C.M., Chemistry and Physics in One Dimension: Synthesis and Properties of Nanowires and Nanotubes, *Acc. Chem. Res.* **32**(1999), 435-445.
14. Wong E.W., Sheehan P.E., and Lieber C.M., Nanobeam mechanics: elasticity, strength, and toughness of nanorods and nanotubes, *Science* **277**(1997) pp.1971-1975.
15. Duan X.F., Huang Y., Cui Y., Wang J.F. and Lieber C.M., Indium phosphide nanowires as building blocks for nanoscale electronic and optoelectronic devices, *Nature* **409**(2001) pp.66-69.
16. Zhang Z., Sun X., Dresselhaus M.S., Ying J.Y., and Heremans J., Electronic transport properties of single-crystal bismuth nanowire arrays, *Phys. Rev.* B **61**(2000) pp.4850-4861.
17. Huang M., Mao S., Feick H., Yan H., Wu Y., Kind H., Weber E., Russo R., and Yang P., Room-temperature ultraviolet nanowire nanolasers, *Science* **292**(2001) pp.1897-1899.
18. Handbook of Microlithography, Micromachining, and Microfabrication, (Eds. P. Rai-choudhury), SPIE press, IEEE, 1999.
19. Wu Y. and Yang P., Germanium nanowire growth via simple vapor transport, *Chem. Mater.* **12**(2000) pp.605-607.
20. Huang M.H., Wu Y., Feick H., Ngan T., Webber E. and Yang P., Catalytic growth of zinc oxide nanowires by vapor transport, *Adv. Mater.* **13**(2001) pp.113-116.
21. Morales A.M. and Lieber C.M., A laser ablation method for the synthesis of crystalline semiconductor nanowires, *Science* **279**(1998) pp.208-211.
22. Martin C.R., Nanomaterials: a membrane-based synthetic approach, *Science* **266**(1994) pp.1961-1966; Almawlawi D., Liu C.Z. and Moskovits M., Nanowires formed in anodic oxide nanotemplates, *J. Mater. Res.* **9**(1994) pp.1014-1018.

23. Han W., Fan S., Li W. and Hu Y., Synthesis of gallium nitride nanorods through a carbon nanotube-confined reaction, *Science* **277**(1997) pp.1287-1289.
24. Trentler T.J., Hickman K.M., Geol S.C., Viano A.M., Gibbons P.C. and Buhro W.E., Solution-liquid-solid growth of crystalline III-V semiconductors: an analogy to vapor-liquid-solid growth, *Science* **270**(1995) pp.1791-1794.
25. Duan X. and Lieber C.M., General synthesis of compound semiconductor nanowires, *Adv. Mater.***12**(2000) pp.298-302.
26. Zach M.P., Ng K.H. and Penner R.M., Molybdenum nanowires by electrodeposition, *Science*, **290**(2000) pp.2120-2123; Song J., Wu Y., Messer B., Kind H. and Yang P., Metal nanowire formation using $Mo_3Se_3^-$ as reducing and sacrificing templates, *J. Am. Chem. Soc.* **123**(2001) pp.10397-10398.
27. Tremel W., Inorganic nanotubes, *Angew. Chem. Int. Ed.* **38**(1999) pp.2175-2179; Tenne R., Inorganic nanoclusters with fullerene-like structure and nanotubes, *Prog. Inorg. Chem.* **50**(2001) pp.269-315.
28. Wu C. G. and Bein T., Conducting carbon wires in ordered, nanometer-sized channels, *Science* **266**(1994) pp.1013-1015.
29. Huczko A., Template-based synthesis of nanomaterials, *Appl. Phys. A-Mater.* **70**(2000) pp.365-376.
30. Prieto A. L., Sander M. S., Martin-Gonzalez M.S., Gronsky R., Sands T. and Stacy A. M., Electrodeposition of Ordered Bi_2Te_3 Nanowire Arrays, *J. Am. Chem. Soc.* **123**(2001) pp.7160-7161.
31. Huber T. E., Graf M. J., Foss C. A. and Constant P., Processing and characterization of high-conductance bismuth wire array composites, *J. Mater. Res.* **15**(2000) pp.1816-1821.
32. Huang M. H. , Choudrey A. and Yang P., Ag nanowire formation within mesoporous silica, *J. Chem. Soc. Chem. Commun.* **12**(2000) pp.1063-1064.
33. Ulrich R., Du Chesne A., Templin M. and Wiesner U., Nano-objects with controlled shape, size, and composition from block copolymer mesophases, *Adv. Mater.* **11**(1999) pp.141-146.
34. Liu Z., Sakamoto Y., Ohsuna T., Hiraga K., Terasaki O., Ko C. H., Shin H. J. and Ryoo R., TEM studies of platinum nanowires fabricated in mesoporous silica MCM-41, *Angew. Chem. Int. Ed.* **39**(2000) pp.3107-3110.
35. Leon R., Margolese D., Stucky G. and Petroff P. M., Nanocrystalline Ge filaments in the pores of a meso-silicate, *Phy. Rev.B* **52**(1995) pp.2285-2288.
36. Dai H. J., Wong E. W., Lu Y.Z., Fan S. S. and Lieber C. M., Synthesis and characterization of carbide nanorods, *Nature* **375**(1995) pp.769-772.
37. Gates B., Wu Y., Yin Y., Yang P. and Xia Y., Single-crystalline nanowires of Ag_2Se can be synthesized by templating against nanowires of trigonal Se, *J. Am. Chem. Soc.* **123**(2001) pp.11500-11501.

38. Dingman S. D., Rath N.P., Markowitz P. D., Gibbons P. C. and Buhro W. E., Low-temperature, catalyzed growth of indium nitride fibers from azido-indium precursors, *Angew. Chem. Int. Ed.* **39**(2000) pp.1470-1472.
39. Trentler T. J., Goel S.C., Hickman K. M., Viano A. M., Chiang M. Y., Beatty A. M., Gibbons P. C. and Buhro W. E., Solution-Liquid-Solid Growth of Indium Phosphide Fibers from Organometallic Precursors: Elucidation of Molecular and Nonmolecular Components of the Pathway, *J. Am. Chem. Soc.* **119**(1997) pp.2172-2181; Lourie O. R., Jones C. R., Bartlett B. M., Gibbons P. C., Ruoff R. S. and Buhro W. E., CVD Growth of Boron Nitride Nanotubes, *Chem. Mater.* **12**(2000) pp.1808-1810.
40. Holmes J. D., Johnston K. P., Doty R. C. and Korgel B. A., Control of thickness and orientation of solution-grown silicon nanowires, *Science*, **287**(2000) pp.1471-1473.
41. Manna L., Scher E. C. and Alivisatos A. P., Synthesis of Soluble and Processable Rod-, Arrow-, Teardrop-, and Tetrapod-Shaped CdSe Nanocrystals, *J. Am. Chem. Soc.* **122**(2000) pp.12700-12706.
42. Chao C., Chen C. and Lang Z., Simple Solution-Phase Synthesis of Soluble CdS and CdSe Nanorods, *Chem. Mater.* **12**(2000) pp.1516-1518.
43. Liu Y. F., Zeng J. H., Zhang W. X., Yu W. C., Qian Y. T., Cao J. B. and Zhang W. Q., Solvothermal route to Bi_3Se_4 nanorods at low temperature, *J. Mater. Res.* **16**(2001) pp.3361-3365.
44. Jiang X., Xie Y., Lu J., Zhu L. Y., He W. and Qian Y. T., Simultaneous In Situ Formation of ZnS Nanowires in a Liquid Crystal Template by γ-Irradiation, *Chem. Mater.* **13**(2001) pp.1213-1218.
45. Gates B., Yin Y. and Xia Y., A solution-phase approach to the synthesis of uniform nanowires of crystalline selenium with lateral dimensions in the range of 10-30 nm, *J. Am. Chem. Soc.* **122**(2000) pp.12582-12583.
46. *Whisker Technology*, Levitt A. P. (Eds. Wiley-interscience), New York, 1970.
47. Wu Y. and Yang P., Direct observation of vapor-liquid-solid nanowire growth, *J. Am. Chem. Soc.* **123**(2001) pp.3165-3166.
48. Gudiksen M. S. and Lieber C. M., Diameter-selective synthesis of semiconductor nanowires, *J. Am. Chem. Soc.* **122**(2000) pp.8801-8802.
49. Chen C. C., Yeh C. C., Chen C. H., Yu M. Y., Liu H. L., Wu J. J., Chen K. H., Chen L. C., Peng J. Y. and Chen Y. F., Catalytic Growth and Characterization of Gallium Nitride Nanowires, *J. Am. Chem. Soc.* **123**(2001) pp.2791-2798.
50. Duan X. F. and Lieber C. M., Laser-Assisted Catalytic Growth of Single Crystal GaN Nanowires, *J. Am. Chem. Soc.* **122**(2000) pp.188-189.
51. Shi W. S., Peng H. Y., Zheng Y. F., Wang N., Shang N. G., Pan Z. W., Lee C. S. and Lee S. T., Synthesis of large areas of highly oriented, very long silicon nanowires, *Adv. Mater.* **12**(2000) pp.1343-1345.
52. Tang C. C., Fan S. S., de la Chapelle M. L., Dang H. Y. and Li P., Synthesis of gallium phosphide nanorods, *Adv. Mater.* **12**(2000) pp.1346-1348.

53. Gu G., Burghard M., Kim G. T., Dusberg G. S., Chiu P.W., Krstic V., Roth S. and Han W. Q., Growth and electrical transport of germanium nanowires, *J. Appl. Phys.* **90**(2001) pp.5747-5751.
54. Pan Z. W., Lai H. L., Au F. C. K., Duan X. F., Zhou W. Y., Shi W. S., Wang N., Lee C. S., Wong N. B., Lee S. T. and Xie S. S., Oriented silicon carbide nanowires: synthesis and field emission properties, *Adv. Mater.* **12**(2000) pp.1186-1190.
55. Wu Y., Messer B. and Yang P., Superconducting MgB_2 nanowires, *Adv. Mater.* **13**(2001) pp.1487-1489.
56. Yazawa M., Kohuchi M., Muto A. and Hiruma K., Semiconductor nanowhiskers, *Adv. Mater.* **5**(1993) pp.577-580.
57. Liu J., Zhang X., Zhang Y. J., He R. R. and Zhu J., Novel synthesis of AlN nanowires with controlled diameters, *J. Mater. Res.* **16**(2001) pp.3133-3138.
58. Hiruma K., Yazawa M., Katsuyama T., Ogawa K., Haraguchi K., Koguchi M. and Kakibayashi H., Growth and optical properties of nanometer-scale GaAs and InAs whiskers, *J. Appl. Phys.* **77**(1995) pp.447-462.
59. Shimada T., Hiruma K., Shirai M., Yazawa M., Haraguchi K., Sato T., Matsui M. and Katsuyama T., Size, position and direction control on GaAs and InAs nanowhisker growth, *Superlattices and Microstructures* **24**(1998) pp.453-458.
60. Duan X. F., Wang J. F. and Lieber C. M., Synthesis and optical properties of gallium arsenide nanowires, *Appl. Phys. Lett.* **76**(2000) pp.1116-1118.
61. Lee S. T., Wang N., Zhang Y. F. and Tang Y. H., Oxide-assisted semiconductor nanowire growth, *MRS Bulletin* **24**(1999) pp.36-42.
62. Shi W. S., Peng H. Y., Wang N., Li C. P., Xu L., Lee C. S., Kalish R. and Lee S. T., Free-standing single crystal silicon nanoribbons, *J. Am. Chem. Soc.* **123**(2001) pp.11095-11096.
63. Lee S. T., Wang N. and Lee C. S., Semiconductor nanowires: synthesis, structure and properties, *Mater. Sci. Eng.* **A 286**(2000) pp.16-23.
64. Yang P. and Lieber C.M., Nanostructured high-temperature superconductors: Creation of strong-pinning columnar defects in nanorod/superconductor composites, *J. Mater. Res.* **12**(1997) pp.2981-2996; Yang P., Lieber C. M., US Patent 5897945.
65. Yang P. and Lieber C. M., Nanorod-superconductor composites: a pathway to materials with high critical current densities, *Science* **273**(1996) pp.1836-1840.
66. Wang Z. L., Gao R. P., Pan Z. W. and Dai Z. R., Nano-scale mechanics of nanotubes, nanowires, and nanobelts, *Adv. Eng. Mater.* **3**(2001) pp.657-661.
67. Pan Z. W., Dai Z. R. and Wang Z. L., Nanobelts of semiconducting oxides, *Science* **291**(2001) pp.1947-1949.
68. Dai Z. R., Pan Z. W. and Wang Z. L., Ultra-long single crystalline nanoribbons of tin oxide, *Solid State Commun.* **118**(2001) pp.351-354.

69. Chen Y., Ohlberg D. A. A. and Williams R. S., Epitaxial growth of erbium silicide nanowires on silicon(001), *Mater. Sci. & Eng.* B **87**(2001) pp.222-226.
70. Chen Y., Ohlberg D. A. A., Medeiros-Ribeiro G., Chang Y. A. and Williams R. S., Self-assembled growth of epitaxial erbium disilicide nanowires on silicon (001), *Appl. Phys. Lett.* **76**(2000) pp.4004-4006.
71. Zhang Y. J., Zhang Q., Wang N. L., Yan Y. J., Zhou H. H. and Zhu J., Synthesis of thin Si whiskers (nanowires) using $SiCl_4$, *J. Cryst. Growth* **226**(2001) pp.185-191.
72. Westwater J., Gosain D.P., Tomiya S. and Usui S., Growth of silicon nanowires via gold/silane vapor-liquid-solid reaction, *J. Vac. Sci. Technol.* **B15**(1997) pp.554-557.
73. Gole J. L., Stout J. D., Rauch W. L. and Wang Z. L., Direct synthesis of silicon nanowires, silica nanospheres, and wire-like nanosphere agglomerates, *Appl. Phys. Lett.* **76**(2000) pp.2346-2348.
74. Gu Q., Dang H. Y., Cao J., Zhao J. H. and Fan S. S., Silicon nanowires grown on iron-patterned silicon substrates, *Appl. Phys. Lett.* **76**(2000) pp.3020-3021.
75. Shi W. S., Peng H. Y., Zheng Y. F., Wang N., Shang N. G., Pan Z. W., Lee C. S. and Lee S. T., Synthesis of large areas of highly oriented, very long silicon nanowires, *Adv. Mater.* **12**(2000) pp.1343-1345.
76. Tang Y. H., Zhang Y. F., Wang N., Shi W. S., Lee C. S., Bello I. and Lee S. T., Si nanowires synthesized from silicon monoxide by laser ablation, *J. Vac. Sci. & Tech.* B **19**(2001) pp.317-319.
77. Cao L. M., Zhang Z., Sun L. L., Gao C. X., He M., Wang Y. Q., Li Y. C., Zhang X. Y., Li G., Zhang J. and Wang W. K., Well-aligned boron nanowire arrays, *Adv. Mater.* **13**(2001) pp.1701-1704.
78. Li Y. D., Wang J. W., Deng Z. X., Wu Y. Y., Sun X. M., Yu D. P. and Yang P. D., Bismuth nanotubes. A rational low-temperature synthetic route, *J. Am. Chem. Soc.* **123**(2001) pp.9904-9905.
79. Zhang J., Peng X. S., Wang X. F., Wang Y. W. and Zhang L. D., Micro-Raman investigation of GaN nanowires prepared by direct reaction Ga with NH3, *Chem. Phys. Lett.* **345**(2001) pp.372-376.
80. Chen C. C. and Yeh C. C., Large-scale catalytic synthesis of crystalline gallium nitride nanowires, *Adv. Mater.* **12**(2000) pp.738-741.
81. He M. Q., Zhou P. Z., Mohammad S. N., Harris G. L., Halpern J. B., Jacobs R., Sarney W. L. and Salamanca-Riba L., Growth of GaN nanowires by direct reaction of Ga with NH_3, *J. Crys. Growth* **231**(2001) pp.357-365.
82. Shi W. S., Zheng Y. F., Wang N., Lee C. S. and Lee S. T., Microstructures of gallium nitride nanowires synthesized by oxide-assisted method, *Chem. Phys. Lett.* **345**(2001) pp.377-380.
83. Peng H. Y., Zhou X. T., Wang N., Zheng Y. F., Liao L. S., Shi W. S., Lee C. S. and Lee S. T., Bulk-quantity GaN nanowires synthesized from hot filament chemical vapor deposition, *Chem. Phys. Lett.* **327**(2000) pp.263-270.

84. Tang C. C., Fan S. S., Dang H. Y., Li P. and Liu Y. M., Simple and high-yield method for synthesizing single-crystal GaN nanowires, *Appl. Phys. Lett.* **77**(2000) pp.1961-1963.
85. Tang C. C., Fan S. S., de la Chapelle M. L. and Li P., Silica-assisted catalytic growth of oxide and nitride nanowires, *Chem. Phys. Lett.* **333**(2001) pp.12-15.
86. Haber J. A., Gibbons P. C. and Buhro W. E., Morphological Control of Nanocrystalline Aluminum Nitride: Aluminum Chloride-Assisted Nanowhisker Growth, *J. Am. Chem. Soc.* **119**(1997) pp.5455-5456.
87. Zhang Y. J., Liu J., He R. R., Zhang Q., Zhang X. Z. and Zhu J., Synthesis of Aluminum Nitride Nanowires from Carbon Nanotubes, *Chem. Mater.* **13**(2001) pp.3899-3905.
88. Devi A., Parala H., Rogge W., Wohlfart A., Birkner A. and Fischer R. A., Growth of InN whiskers from single source precursor, *J. de Phys.* iv **11**(2001) pp.577-584.
89. Zhang Y. J., Wang N. L., He R. R., Liu J., Zhang X. Z. and Zhu J., A simple method to synthesize Si_3N_4 and SiO_2 nanowires from Si or Si/SiO_2 mixture, *J. Cryst. Growth* **233**(2001) pp.803-808.
90. Gao Y. H., Bando Y. and Sato T., Nanobelts of the dielectric material Ge_3N_4, *Appl. Phys. Lett.* **79**(2001) pp.4565-4567.
91. Shi W. S., Zheng Y. F., Wang N., Lee C. S. and Lee S. T., Synthesis and microstructure of gallium phosphide nanowires, *J. Vac. Sci.&Tech.* B **19**(2001) pp.1115-1118.
92. Hu J. Q., Lu Q. Y., Tang K. B., Deng B., Jiang R. R., Qian Y. T., Yu W. C., Zhou G. E., Liu X. M. and Wu J. X., Synthesis and characterization of SiC nanowires through a reduction-carburization route, *J. Phys. Chem.* B **104**(2000) pp.5251-5254.
93. Wang Z. L., Dai Z. R., Gao R. P., Bai Z. G. and Gole J. L., Side-by-side silicon carbide-silica biaxial nanowires: synthesis, structure, and mechanical properties, *Appl. Phys. Lett.* **77**(2000) pp.3349-3351.
94. Shi W. S., Zheng Y. F., Peng H. Y., Wang N., Lee C. S. and Lee S. T., Laser ablation synthesis and optical characterization of silicon carbide nanowires, *J. Am. Ceram. Soc.* **83**(2000) pp.3228-3230.
95. Pan Z. W., Lai H. L., Au F. C. K., Duan X. F., Zhou W. F., Shi W. S., Wang N., Lee C. S., Wong N. B., Lee S. T. and Xie S. S., Oriented silicon carbide nanowires: synthesis and field emission properties, *Adv. Mater.* **12**(2000) pp.1186-1190.
96. McIlroy D. N., Zhang D., Kranov Y. and Norton M. G., Nanosprings, *Appl. Phys. Lett.* **79**(2001) pp.1540-1542.
97. Wang Y. W., Zhang L. D., Wang G. Z., Peng X. S., Chu Z. Q. and Liang C. H., Catalytic growth of semiconducting zinc oxide nanowires and their photoluminescence properties, *J. Cryst. Growth* **234**(2002) pp.171-175.

98. Kong Y. C., Yu D. P., Zhang B., Fang W. and Feng S. Q., Ultraviolet-emitting ZnO nanowires synthesized by a physical vapor deposition approach, *Appl. Phys. Lett.* **78**(2001) pp.407-409.
99. Li J. Y., Chen X. L., Li H., He M. and Qiao Z. Y., Fabrication of zinc oxide nanorods, *J. Cryst. Growth* **233**(2001) pp.5-7.
100. Wu X. C., Song W. H., Huang W. D., Pu M. H., Zhao B., Sun Y. P. and Du J. J., Crystalline gallium oxide nanowires: intensive blue light emitters, *Chem. Phys. Lett.* **328**(2000) pp.5-9.
101. Zhang X. Y., Yao B. D., Zhao L. X., Liang C. H., Zhang L.D. and Mao Y. Q., Electrochemical fabrication of single-crystalline anatase TiO_2 nanowire arrays, *J. Electrochem. Soc.* **148**(2001) pp.398-400.
102. Lei Y., Zhang L. D. and Fan J. C., Fabrication, characterization and Raman study of TiO_2 nanowire arrays prepared by anodic oxidative hydrolysis of TiCl3, *Chem. Phys. Lett.* **338**(2001) pp.231-236.
103. Tang Y. H., Zhang Y. F., Wang N., Bello I., Lee C. S. and Lee S. T., Germanium dioxide whiskers synthesized by laser ablation, *Appl. Phys. Lett.* **74**(1999) pp.3824-3826.
104. Jiang X., Xie Y., Lu J., Zhu L. Y., He W. and Qian Y. T., Simultaneous In Situ Formation of ZnS Nanowires in a Liquid Crystal Template by γ-Irradiation, *Chem. Mater.* **13**(2001) pp.1213-1218.
105. Li Y., Wan J. H. and Gu Z. N., Synthesis of ZnS nanowires in liquid crystal systems, *Mol. Crys. Liq. Crys.* **337**(1999) pp.193-196.
106. Wang S. H. and Yang S. H., Preparation and Characterization of Oriented PbS Crystalline Nanorods in Polymer Films, *Langmuir* **16**(2000) pp.389-397.
107. Yan P., Xie Y., Qian Y. T. and Liu X. M., A cluster growth route to quantum-confined CdS nanowires, *Chem. Commun.* 14(1999) pp.1293-1294.
108. Xu D. S., Xu Y. J., Chen D. P., Guo G. L., Gui L. L. and Tang Y. Q., Preparation and characterization of CdS nanowire arrays by dc electrodeposit in porous anodic aluminum oxide templates, *Chem. Phys.Lett.* **325**(2000) pp.340-344.
109. Routkevitch D., Bigioni T., Moskovits M. and Xu J. M., Electrochemical fabrication of CdS nanowire arrays in porous anodic aluminum oxide templates, *J. Phys. Chem.* **100**(1996) pp.14037-14047.
110. Zhan J. H., Yang X. G., Wang D. W., Li S. D., Xie Y., Xia Y. N. and Qian Y. T., Polymer-controlled growth of CdS nanowires, *Adv. Mater.* **12**(2000) pp.1348-1357.
111. Wang N., Fung K. K., Wang S. and Yang S., Oxide-assisted nucleation and growth of copper sulfide nanowire arrays, *J. Cryst. Growth* **233**(2001) pp.226-232.
112. Wang S. H., Yang S. H., Dai Z. R. and Wang Z. L., The crystal structure and growth direction of Cu_2S nanowire arrays fabricated on a copper surface, *Phys. Chem. Chem. Phys.* **3**(2001) pp.3750-3753.

113. Yu S. H., Shu L., Yang J. A., Han Z. H., Qian Y. T. and Zhang Y. H., A solvothermal decomposition process for fabrication and particle sizes control of Bi2S3 nanowires, *J. Mater. Res.* **14**(1999) pp.4157-4162.
114. Peng X. S., Zhang J., Wang X. F., Wang Y. W., Zhao L. X., Meng G. W. and Zhang L. D., Synthesis of highly ordered CdSe nanowire arrays embedded in anodic alumina membrane by electrodeposition in ammonia alkaline solution, *Chem. Phys. Lett.* **343**(2001) pp.470-474.
115. Xu D. S., Shi X. S., Guo G. L., Gui L. L. and Tang Y. Q., Electrochemical Preparation of CdSe Nanowire Arrays, *J. Phys. Chem. B* **104**(2000) pp.5061-5063.
116. Wang W. Z., Geng Y., Yan P., Liu F. Y., Xie Y. and Qian Y. T., Synthesis and characterization of MSe (M = Zn, Cd) nanorods by a new solvothermal method, *Inorg. Chem. Commun.* **2**(1999) pp.83-85.
117. Wang W. H., Geng Y., Qian Y. T., Ji M. R. and Liu X. M., A novel pathway to PbSe nanowires at room temperature, *Adv. Mater.* **10**(1998) pp.1479-1481.
118. Yu S. H., Yang J., Han Z. H., Yang R. Y., Qian Y. T. and Zhang Y. H., Novel Solvothermal Fabrication of CdS_xSe_{1-x} Nanowires, *J. Solid State Chem.* **147**(1999) pp.637-640.
119. Kamins T. I., Stanley Williams R., Basile D. P., Hesjedal T. and Harris J. S., Ti-catalyzed Si nanowires by chemical vapor deposition: Microscopy and growth mechanisms, *J. Appl. Phys.* **89**(2001) pp.1008-1016.
120. Wu Y., Yan H., Huang M., Messer B., Song J. and P. Yang, Inorganic semiconductor nanowires: rational growth, assembly, and novel properties, *Chemistry, Euro. J.* **8**(2002) pp.1260-1268.
121. Wu Y., Yan H. and Yang P., Semiconductor nanowire array: potential substrates for photocatalysis and photovoltaics, *Topics in Catalysis* **19**(2002) pp.197-202.
122. Hu J., Ouyang M., Yang P. and Lieber C. M., Controlled growth and electrical properties of heterojunctions of carbon nanotubes and silicon nanowires, *Nature* **399**(1999) pp.48-51.
123. Zhang Y., Ichihashi T., Landree E., Nihey F. and Iijima S., Heterostructures of single-walled carbon nanotubes and carbide nanorods, *Science* **285**(1999) pp.1719-1722.
124. Zhou C., Kong J., Yenilmez E. and Dai H., Modulated chemical doping of individual carbon nanotubes *Science* **290**(2000) pp.1552-1555.
125. Haraguchi K., Katsuyama T. and Hiruma K., Polarization dependence of light emitted from GaAs p-n junctions in quantum wire crystals, *J. Appl. Phys.* **75**(1994) pp.4220-4225.
126. Markowitz P. D., Zach M. P., Gibbons P. C., Penner R. M. and Buhro W. E., Phase Separation in $Al_xGa_{1-x}As$ Nanowhiskers Grown by the Solution-Liquid-Solid Mechanism, *J. Am. Chem. Soc.* **123**(2001) pp.4502-4511.

127. Nicewarner-Pena S., Freeman R. G., Reiss B. D., He L., Pena D. J., Walton I. D., Cromer R., Keating C. D. and Natan M. J., Submicrometer metallic barcodes, *Science* **294**(2001) pp.137-141.
128. Wu Y., Fan R. and Yang P., Block-by-Block Growth of Single-Crystalline Si/SiGe Superlattice Nanowires, *Nano Lett.* **2**(2002) pp.83-86.
129. Gudiksen M. S., Lauhon L. J., Wang J., Smith D. C. and Lieber C. M., Growth of nanowire superlattice structures for nanoscale photonics and electronics, *Nature* **415**(2002) pp.617-620.
130. Bjoerk M. T., Ohlsson B. J., Sass T., Persson A. I., Thelander C., Magnusson M. H., Deppert K., Wallenberg L. R., Samuelson, L., One-dimensional Steeplechase for Electrons Realized, *Nano. Lett.* **2**(2002) pp.87-89.
131. He R., Law M., Fan R., Kim F. and Yang P., Functional Bimorph Composite Nanotapes, *Nano. Lett.* **2**(2002) pp.1109-1112.
132. Lauhon L. J., Gudiksen M. S., Wang C. L. and Lieber C. M., Epitaxial core-shell and core-multishell nanowire heterostructures, *Nature* **420**(2002) pp.57-61.
133. Cheung C. L., Hafner J. H., Odom T. W., Kim K. and Lieber C. M., Growth and fabrication with single-walled carbon nanotube probe microscopy tips, *Appl. Phys. Lett.* **76**(2000) pp.3136-3138.
134. Messer B., Song J. H. and Yang P., Microchannel networks for nanowire patterning, *J. Am. Chem. Soc.* **122**(2000) pp.10232-10233.
135. Huang Y., Duan X. F., Wei Q. Q. and Lieber C. M., One-dimensional nanostructures into functional networks, *Science* **291**(2001) pp.630-633.
136. Duan X. F., Huang Y., Cui Y., Wang J. F. and Lieber C. M., Indium phosphide nanowires as building blocks for nanoscale electronic and optoelectronic devices, *Nature* **409**(2001) pp.66-69.
137. Kim F., Kwan S., Arkana J. and Yang P., Langmuir-Blodgett nanorod assembly, *J. Am. Chem. Soc.* **123**(2001) pp.4360-4361.
138. Wu Y. and Yang P., Germanium / carbon core-sheath nanostructures, *Appl. Phys. Lett.* **77**(2000) pp.43-45.
139. Wu Y. and Yang P., Melting and welding semiconductor nanowires in nanotubes, *Adv. Mater.* **13**(2001) pp.520-523.
140. Pan Z. W., Dai Z. R., Xu L., Lee S. T. and Wang Z. L., Temperature-Controlled Growth of Silicon-Based Nanostructures by Thermal Evaporation of SiO Powders, *J. Phys. Chem. B* **105**(2001) pp.2507-2514; Peng H. Y., Pan Z. W., Xu L., Fan X. H., Wang N., Lee C. S. and Lee S. T., Temperature dependence of Si nanowire morphology, *Adv. Mater.* **13**(2001) pp.317-320.
141. Brus L., Luminescence of Silicon Materials: Chains, Sheets, Nanocrystals, Nanowires, Microcrystals, and Porous Silicon, *J. Phys. Chem.* **98**(1994) pp.3575-3581.
142. Wang J. F., Gudiksen M. S., Duan X. F., Cui Y. and Lieber C. M., Highly polarized photoluminescence and photodetection from single indium phosphide nanowires, *Science* **293**(2001) pp.1455-1457.

143. Johnson J.C., Yan H., Schaller R.D., Haber L., Saykally R. J. and Yang P., Single Nanowire Lasers, *J. Phys. Chem.* **105**(2001) pp.11387-11390.
144. Johnson J., Choi H. J., Knutsen K. P., Schaller R. D., Saykally R. J. and Yang P., Single gallium nitride nanowire lasers, *Nature Materials* **1**(2002) pp.106-110.
145. Johnson J. C., Yan H., Schaller R. D., Peterson P. B., Yang P. and Saykally R. J., Near-Field Imaging of Nonlinear Optical Mixing in Single Zinc Oxide Nanowires, *Nano. Lett.* **2**(2002) pp.279-283.
146. Kind H., Yan H., Messer B., Law M. and Yang P., Nanowire ultraviolet photodetectors and optical switches, *Adv. Mater.* **14**(2002) pp.158-160.
147. Law M., Kind H., Kim F., Messer B. and Yang P., Photochemical sensing of NO_2 with SnO_2 nanoribbon nanosensors at room temperature, *Angew. Chem. Int. Ed.* **41**(2002) pp.2405-2408.
148. Cui Y. and Lieber C. M., Functional nanoscale electronic devices assembled using silicon nanowire building blocks, *Science* **291**(2001) pp.851-853.
149. Huang Y., Duan X. F., Cui Y., Lauhon L. J., Kim K. H. and Lieber C. M., Logic gates and computation from assembled nanowire building blocks, *Science* **294**(2001) pp.1313-1317.
150. Yu J. Y., Chung S. W. and Heath J. R., Silicon Nanowires: Preparation, Device Fabrication, and Transport Properties, *J. Phys. Chem.* B **104**(2000) pp.11864-11870.
151. Cui Y., Duan X. F., Hu J. T. and Lieber C. M., Doping and Electrical Transport in Silicon Nanowires, *J. Phys. Chem.* B **104**(2000) pp.5213-5216.
152. Chung S. W., Yu J. Y. and Heath J. R., Silicon nanowire devices, *Appl. Phys. Lett.* **76**(2000) pp.2068-2070.
153. Pan Z. W., Lai H. L., Au F. C. K., Duan X. F., Zhou W. Y., Shi W. S., Wang N., Lee C. S., Wong N. B., Lee S. T. and Xie S. S., Oriented silicon carbide nanowires: synthesis and field emission properties, *Adv. Mater.* **12**(2000) pp.1186-1190.
154. Zhou X. T., Lai H. L., Peng H. Y., Au F. C. K., Liao L. S., Wang N., Bello I., Lee C. S. and Lee S. T., Thin β-SiC nanorods and their field emission properties, *Chem. Phys. Lett.* **318**(2000) pp.58-62.

HARNESSING SYNTHETIC VERSATILITY TOWARD INTELLIGENT INTERFACIAL DESIGN: ORGANIC FUNCTIONALIZATION OF NANOSTRUCTURED SILICON SURFACES

LON A. PORTER, JR., JILLIAN M. BURIAK
Purdue University, Department of Chemistry, 1393 Brown Laboratories of Chemistry, West Lafayette, IN 47907-1393, USA

Nanostructured semiconductors continue to attract intense research interest. Porous silicon has received the greatest consideration primarily due to its light emitting properties, which make it a promising candidates for use in optoelectronic, sensor array, light harvesting, and/or biocomposite applications. Because interfacial characteristics have proven to play such a dominant role in the materials properties of high surface area systems, control over the surface chemistry is critical. Only recently over the past decade, however, has the chemical reactivity and stability of porous silicon surfaces been explored. In particular, research efforts have labored to yield an improved understanding of the fundamental reactivity of non-oxidized group IV semiconductor surfaces, providing a diversity of facile methods utilized to prepare and characterize organic monolayers bound directly through E–C bonds (where E = Si or Ge). Each of these methods, described herein, provide access to well-defined, functional monolayers through wet, bench-top reactions ammenable to the majority of scientists and engineers. The following chapter serves to spotlight recent advances in the preparation of functional monolayers from hydride-terminated, porous silicon and germanium substrates.

1 Introduction

Nanostructured materials of varying composition continue to entice the materials research community with the promise of innumerable practical applications as well as advancing an understanding of their fundamentals [1,2,3]. The exploration of size and structure influences on materials properties such as dielectric constant, conductivity, and luminescence has developed into a burgeoning new subdiscipline of materials science with hundreds of papers published every year [4]. Nowhere is this trend more prevalent than in the study of micro- and mesoporous materials [5,6]. These framework solids offer unparalleled surface area-to-volume ratios with morphologies that can usually be selectively tailored to fine tolerances, utilizing facile methodologies. Synthetic strategies routinely achieve complex subarchitectures, sending pore sizes plummeting into the nanoscale regime, where the ratio of surface atoms to bulk increases to significant proportions. It is in this size regime that the surface properties of these materials become of paramount significance. The chemical properties of the interfacial atoms directly influence the stability, reactivity, and overall utility of a material in a given application [7]. Although the interfacial regions have proven to play such a dominant role in the materials properties of these systems, only recently has their chemical reactivity and stability been intricately explored.

While zeolites and related materials have remained the most prominent group of porous materials [8,9], porous semiconductors have received a great amount of attention over the course of the past decade, due in part to their unique electronic structures and light emission [10,11]. Of the group(IV) semiconductors, silicon in particular has enjoyed decades of intense interest as a result of its useful bulk properties in IC systems and widespread utilization in modern electronics, thus serving as the foundation of a multi-billion dollar industry [12,13,14]. Thus far, simple insulating layers of oxide have passivated these silicon surfaces [15]. While effective at insulating the bulk semiconductor and protecting it from ambient conditions, the silicon/silicon dioxide system [16] is the exception, rather than the rule. Germanium and the compound semiconductors do not form ordered native oxides capable of surface passivation [17]. As these alternative semiconductors, including gallium arsenide, indium phosphide, and cadmium selenide, begin to find specialized applications due to their wide and often direct energy band gaps [18], an ever-increasing amount of attention has been focused upon their surface chemistry. While ultrahigh vacuum (UHV) techniques offer pristine surfaces for dry processing, the power, ease and efficiency of solution phase synthetic organic/organometallic chemistry can be harnessed for the preparation of functional monolayers, tailor-made to meet the demands of specialized, high-performance applications [19]. In order to attain such a goal, a fundamental understanding of the surface chemistry of semiconductors must be vigorously pursued. In this chapter, we focus upon the high surface area silicon material, nanocrystalline porous silicon.

2 Porous Silicon

Porous silicon [20], a rather unique variant of crystalline silicon, is comprised of a network of one-dimensional nanowires and zero-dimensional nanocrystallites [21]. A thin layer of porous silicon is prepared on commercially available Si(100) wafers, via a variety of methodologies [22]. These include both chemical [23] and electrochemical [24] etching processes, which may be tailored to precisely modulate the feature size and morphology of the resulting porous silicon layer (Figure 1) [25,26,27,28]. Such samples may exhibit extreme surface areas, in excess of nearly 600 m^2/cm^3 [22]. Porous silicon may be prepared to contain nanocrystallites as small as three nanometers, each consisting of only about one thousand silicon atoms each. For features of this size, nearly one half of the silicon atoms reside at the surface. Such samples provide ideal substrates for the design and characterization of functional interfaces, owing to the immense ratio of surface atoms to those that comprise the bulk. Porous silicon has also attracted a great deal of attention due to its electro- [29], chemo- [30], and photo-luminescent properties [22], which can be finely tuned by modulation of etching conditions [31]. Samples of porous silicon have been prepared to yield quantum efficiencies in excess of five percent, whereas bulk silicon remains an extremely poor light emitter, about 6 orders of magnitude weaker [32]. Furthermore, the preparatory etching protocols allow for the facile incorporation of porous silicon into well-established VLSI and MEMS fabrication processes [33,34,35]. These properties make porous silicon an attractive candidate for use in the fabrication of exotic new devices [36,37], ranging from optoelectronics displays [38] and sensors to biocompatible materials [6,39,40,41,42,43,44,45,46,47,48,49,50].

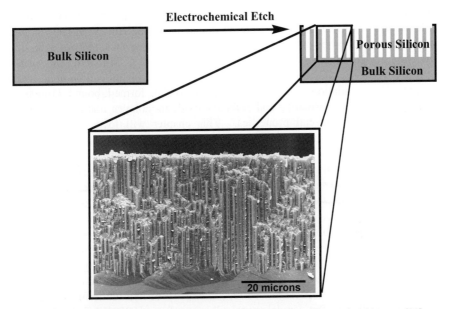

Figure 1. Cross-sectional scanning electron micrograph of porous silicon subarchitecture [21].

3 Synthetic Routes to Surface Functionalization

3.1 The Native Hydride Surface

One of the most substantial obstacles preventing the widespread application of porous silicon in specialized devices results from the inherent instability of its native interface [51]. The surface termination, produced in the presence of the HF etchant, consists of a surface layer of silicon-terminated bonds. Hydride-terminated porous silicon is only metastable with respect to oxidation under ambient conditions, thereby limiting its synthetic utility over extended periods of time. Greater than ninety-nine percent of the silicon atoms comprising the surface are bound to either one, two, or three hydrogen atoms in order to fulfill their tetravalency. These ≡SiH,

=SiH$_2$, and –SiH$_3$ moieties exist in a diverse variety of environments and local orientations, resulting from the inherent nature of this porous material (Figure 2) [22]. Freshly prepared hydride-terminated porous silicon, is virtually "oxide-free" and serves as an attractive precursor to more highly functionalized interfaces. Silicon-terminated and silicon-silicon bonds exist as chemical handles, through which surface chemistry may be carried out. The literature provides a vast, yet diverse degree of synthetic and mechanistic studies involving the functionalization of solution and gas phase silicon-terminated and silicon-silicon molecular species [52]. A great many of these organosilicon transformations have been implemented in the functionalization of hydride-terminated porous silicon [53,54], whereas recent work has provided reactions that have no molecular precedent. In such reactions, the electronic properties of the underlying bulk silicon play an important role, thus differentiating these reactions from those in molecules. Much of the surface chemistry of porous silicon revolves around formation of silicon-carbon bonds, with no concomitant oxidation. The organic monolayers thus formed, bound through Si-C linkages, avoid the insulating and defective oxide monolayer, and allow for good control over the interfacial properties. This chapter will focus exclusively on chemistries resulting in a stable silicon-carbon bond forming event on the porous silicon surface.

While detrimental to device applications, the metastability of the native hydride-terminated surface has its advantages: The surface is stable enough to be manipulated in the presence of air for short periods of time (on the order of tens of minutes), and yet reactive enough to functionalize. The short term inertia allows utilization of well-established Schlenk and glove box handling methods [55]. These methods, commonplace in the modern laboratory, give access to porous silicon as a model system for the design of functional interfaces to the large majority of chemists, engineers, and materials scientists. In addition, from a fundamental perspective, in many instances the surface chemistry of hydride-terminated porous silicon mirrors that of flat, single crystal silicon surfaces [19]. While UHV studies served to pioneer the understanding of silicon surface chemistry [56], the organosurface chemistry of silicon at ambient conditions has been studied only over the past decade. This resulted from the inherent difficulties associated with the characterization of monolayer coverages of thin organic films. While flat, crystalline surfaces produce interfaces that exhibit a truly exceptional degree of structural homogeneity, they are limited by the means available for their characterization. Flat interfaces are typically bound to the more traditional methods of surface analysis and low sensitivities. These porous silicon and germanium substrates, however, offer extreme surface areas and IR permeability, thus offering routes to more conventional characterization methods such as solid-state nuclear magnetic resonance spectrometry, transmission-mode Fourier transform infrared spectroscopy, and Raman spectroscopy [19]. The end result yields two

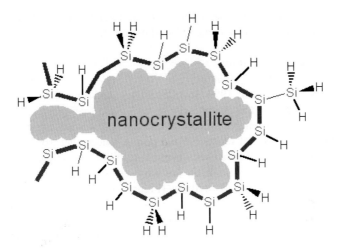

Figure 2. Cartoon portraying a hydride-terminated porous silicon nanocrystallite exhibiting SiH, SiH$_2$, and SiH$_3$ surface species.

complementary substrates for the design and characterization of functional interfaces.

3.2 Contrasting Si-C Bonds Versus an Oxide Interface

Oxidation processes have served as the traditional method of electronically and chemically passivating silicon components, and this resulting oxide has been successfully employed in numerous approaches toward monolayer formation on porous silicon [57,58,59,60,61,62,63,64,65,66]. The resulting silicon-oxygen surface linkage falls short of providing an ideal system for utilization in practical device applications. The highly electronegative nature of the oxygen atom serves to polarize the silicon, thus rendering it susceptible to nucleophilic attack. For instance, silicon-oxygen bonds are thermally labile [67] and known to undergo hydrolysis upon exposure to ambient conditions. Therefore, there is a need for methods resulting in monolayers with increased stability. The silicon-carbon bond appears ideal because it is much less polarized in comparison to the silicon-oxygen bond and is substantially more kinetically stable. From a thermodynamic perspective, the Si-C bond is about as stable as a C-C bond [19]. It has been shown that aliphatic monolayers, anchored to Si(111) and Si(100) through silicon-carbon linkages, remain intact at temperatures approaching 615 K under vacuum [68]. Furthermore, oxide functionalization methods are not practical routes to surface functionalization for other semiconductor systems. Germanium and the compound semiconductors do not form ordered native oxides capable of surface passivation [69]. For example, germanium oxide, unlike SiO$_2$, is soluble in water [70]. The chemical versatility of the organosilicon interface, its stability and clear-cut interface makes it desirable for many fundamental studies and applications.

3.3 Electrochemical Alkylation

Experiments conducted in 1996 by Chazalviel and coworkers pioneered the formation of a stable organic monolayer on an "oxide-free" porous silicon surface [71]. Bound directly through a silicon-carbon linkage, this was an attempt aimed at the deliberate preparation of functionalized porous silicon that would exhibit a high resistance to oxidation and enhanced chemical stability. Freshly etched, hydrogen-terminated samples of porous silicon are exposed to methyl lithium or a methyl Grignard reagent under anodic conditions (Figure 3). Electrochemical activation of the silicon-terminated surface, utilizing porous silicon as a semiconductor electrode serves to facilitate the reaction. The reaction, as proposed by Chazalviel and coworkers, involves the consumption of surface hydride through a transmetallation reaction resulting in the formation of silicon-methyl groups, with either MgHX or LiH as reaction by-products although these salts were not characterized.

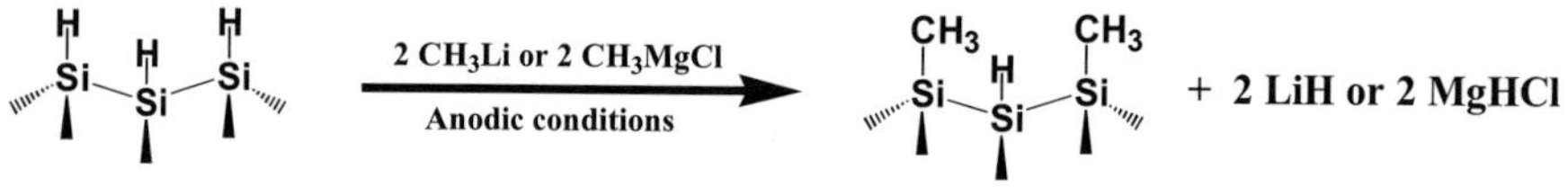

Figure 3. Electrochemical grafting of methyllithium or methyl Grignard onto hydride-terminated porous silicon.

The methyl termination was chosen by the authors in order to produce a robust, silicon-carbon linkage of low polarity. Furthermore, the methyl group was selected due to steric considerations. Prior studies by Chazalviel indicated that any silicon-terminated groups that remained after functionalization were prone to subsequent oxidation. In order to maximize the conversion of hydride, it was believed that a termination exhibiting the least steric hindrance would produce optimal conversion rates. Although transmission IR analysis confirmed the preparation of methyl-terminated porous silicon, complete substitution of surface hydride was not obtained. Prevention of concomitant silicon oxidation also proved to be somewhat of a challenge, requiring rigorous exclusion of air and moisture throughout the procedure for optimal results. The preparation of a methyl-terminated surface served both as an excellent proof-of-concept, and as the impetus to jump-start the field of Si-C bond formation on porous silicon surfaces. The authors have since expanded the variety of surfaces prepared through this method to include alkenyl, alkynyl, and aryl functionalities [72].

Electrochemical routes to silicon-carbon bond formation are not limited to organometallic reagents. Work by Sailor and coworkers has demonstrated that the electrochemical reduction of organo halides results in monolayer formation on porous silicon [73]. Hydride-terminated porous silicon samples are immersed in dry, deoxygenated solutions of the selected organo halide in acetonitrile, containing LiBF$_4$ as an electrolyte (Figure 4). After brief application of a low cathodic current (~2 min), good coverage of the organic group results Owing to the stability of the silicon-carbon bond, the resulting organic thin films proved to be stable against oxidation under ambient conditions, as well as displaying excellent resistance to organic solvents and ethanolic HF rinses. The formation of this silicon-carbon bond is further supported by transmission IR analysis; a substantial decrease in the integrated intensity of the silicon-terminated stretching vibrations provides indirect evidence for the consumption of the hydride terminations. In addition, emergence of the silicon-carbon stretching mode for surfaces prepared through the electrochemical reduction of methyl iodide makes the reaction product unequivocal [74].

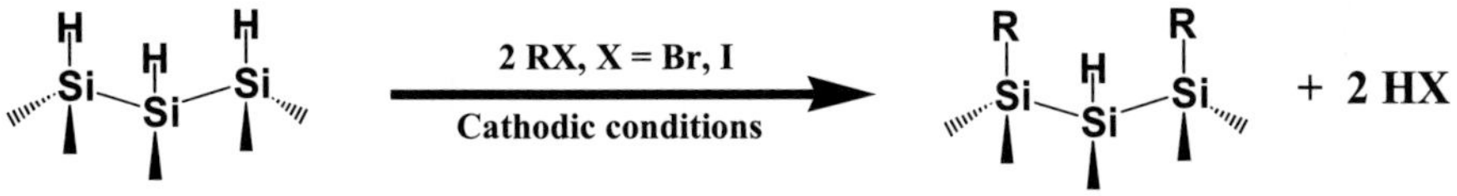

$$\text{Si-Si-Si(H,H,H)} \xrightarrow[\text{Cathodic conditions}]{2\ RX,\ X = Br,\ I} \text{Si-Si-Si(R,H,R)} + 2\ HX$$

Figure 4. Electrochemical grafting of organo halides on porous silicon.

The authors suggest that the electrochemical reduction proceeds with the initial formation of either an alkyl or benzyl radical, as dictated by the employed organo halide (Figure 5). The ensuing radical species abstracts H· radicals from the hydride-terminated silicon surface, resulting in the generation of silicon radicals. These silicon radicals on the surface proceed to react directly with available carbon radicals, leading to formation of a silicon-carbon bond. An alternate mechanistic pathway involves the formation of silyl anions at the surface of porous silicon, which participate in nucleophilic attack on organo halide species in solution (Figure 6). Subsequent electrochemical reduction of surface silicon radicals, generated from the previously noted reaction with carbon radicals, to silyl anions provides one plausible route to anion formation. The formation of 'long-lived' silicon radicals, however, might lead to Si-X bond formation through abstraction of the halogen radical from the organo halide. Since little Si-X bond formation is observed, the authors suggest that any halogenated surface termination may be reduced to silicon anions under cathodic conditions. Ultimately, nucleophilic attack on the organo halide species by the silyl anion, regardless of its origin, would subsequently result in a silicon-carbon surface linkage. Finally, the authors also realize the possibility of attack of the weak silicon-silicon bonds on the surface. Carbon radicals, which may become further reduced to carbanions, could react with the porous silicon surface in a nucleophilic fashion to facilitate the cleavage of surface silicon-silicon bonds (Figure 7, and vide infra). Although the authors maintain that only a minor fraction of carbon radicals would be subsequently reduced to generate carbanion species, it serves to highlight the diverse nature of silicon surface chemistry. The variety of feasible mechanistic pathways indicates the underlying need for further fundamental investigations aimed at exploring the complicated nature of porous silicon surface chemistry.

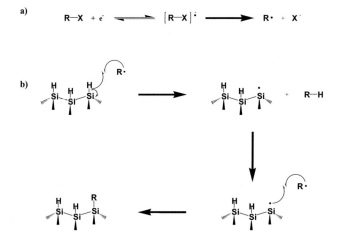

Figure 5. Radical mechanism for the grafting of organo halides on porous silicon. a) Initial electrochemical reduction of the organo halide to yield an alkyl or aryl radical. b) Subsequent generation of surface silicon radicals, followed by silicon-carbon bond formation.

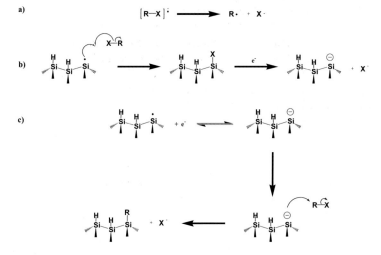

Figure 6. Nucleophilic mechanism for electrochemical grafting of organo halides on porous silicon a) Electrochemical reduction of organo halide to yield an alkyl or aryl radical and a halide anion b) Subsequent generation of surface silyl anion via halide release c) Formation of silyl anion from the silicon radical, and subsequent attack of organo halide, leading to Si-C bond formation.

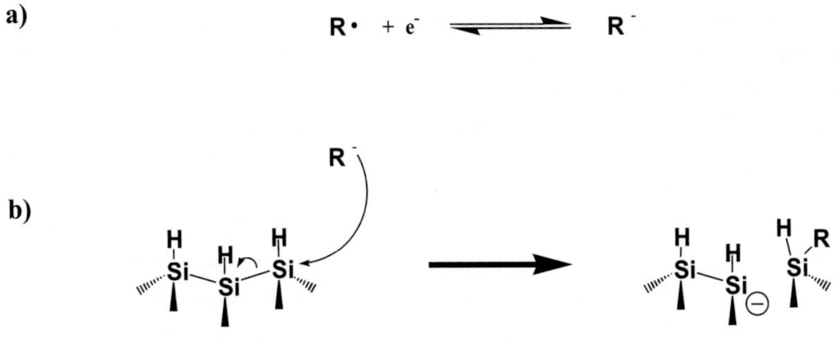

Figure 7. Nucleophilic attack of Si-Si bonds by carbanions produced through electrochemical reduction of organo halides on porous silicon a) Reduction of an alkyl or aryl radical to generate a carbon nucleophile. b) Subsequent nucleophilic attack to generate the silicon-carbon bond.

In terms of the intrinsic photoluminescence of the porous silicon, functionalization utilizing alkyl iodide species results in nearly complete quenching due to iodide contamination, which serves as an efficient surface trap for the excitons responsible for emission. Attempts made to recover photoluminescence through removal of these surface contaminants have proven unsuccessful. If, however, organo bromides are employed in surface functionalization, samples may be rinsed with ethanolic HF, thereby restoring nearly ninety percent of the original photoluminescence intensity.

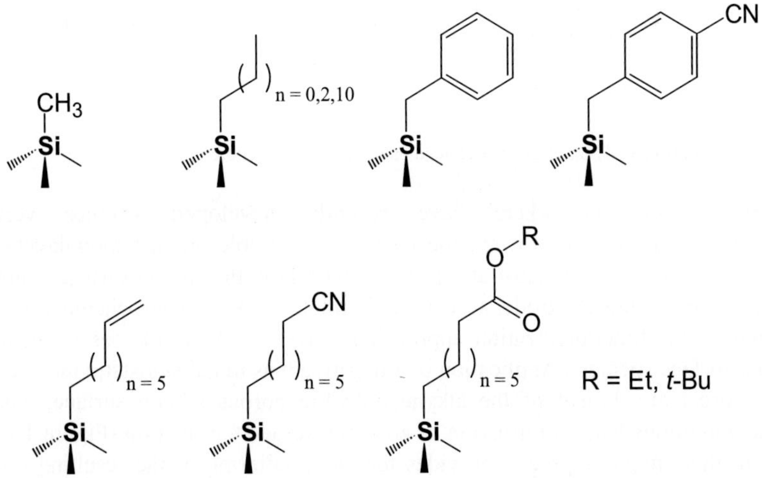

Figure 8. Examples of monolayers prepared from the electrochemical grafting of organo halides on porous silicon.

Electrochemical reduction of organo halides provides a fast and facile route to the efficient preparation of a variety of organic interfaces. The methodology is applicable to both p- and n-type porous silicon, and may take full advantage of the plethora of mono and bifunctional organo halides available, both commercially and synthetically. The authors successfully demonstrated the versatility of this technique by preparing monolayers of a number of compatible functional groups (Figure 8). Furthermore, utilizing classic synthetic organic techniques, a simple protecting group was employed to access surfaces that could not be prepared directly. By employing a ω-ester-terminated organo halide, a stable hydrophilic interface was prepared. Subsequent cleavage of the ω-ester with trifluoroacetic acid in methylene chloride yielded the ω-carboxylic acid-terminated surface (Figure 9). Such manipulations clearly demonstrate that alkyl monolayers, bound through stable silicon-carbon bonds exhibit the durability required to undergo further chemical transformations.

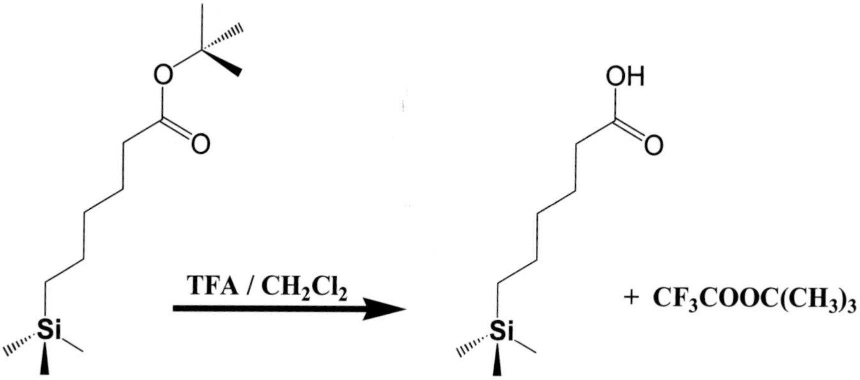

Figure 9. Ester cleavage executed on a functionalized porous silicon surface.

Buriak and coworkers have recently developed another versatile electrochemical approach toward the formation of stable organic monolayers onto oxide-free porous silicon substrates [75]. Once again, the silicon surface is utilized as a semiconducting electrode, serving to facilitate a direct electrochemical grafting reaction. This functionalization approach involves terminal alkynes to yield two distinct surface species. Application of a negative bias in galvanostatic mode results in the direct attachment of the alkyne onto the porous silicon surface, whereas anodic conditions lead to formation of an alkyl surface termination (Figure 10). As a result, this unique approach provides for facile tailoring of the resulting organic monolayer; quite literally, at the flick of a switch, the surface scientist is provided with a choice of two surface terminations, prepared from a single starting molecule.

Cathodic electrografting (CEG) of alkynes proceeds through a quick reaction (~2 min) resulting in a stable monolayer, bound to the underlying porous silicon substrate through a silicon-carbon bond (Figure 11). Infrared analysis of the resulting surfaces reveals a substantial decrease in the integrated intensity of the silicon-terminated stretching region, as well as the absence of the terminal $v(\equiv$C-H$)$ mode. This indirect evidence is further supported by the observation of a sharp $v(C\equiv C)$ signal ~2160 cm^{-1}, corresponding to a silylated acetylene linkage. For instance, in the case of a dodecynyl-terminated surface, the $v(C\equiv C)$ is observed at 2176 cm^{-1} which compares well with the 2176 cm^{-1} feature seen in 1-trimethylsilyldodecyne. Note that the $v(C\equiv C)$ of 1-dodecyne appears at 2120 cm^{-1}. The fact that the monolayers are bonded to the porous silicon through a Si-C≡CR bond suggest that direct electonic communication between the organic fragment (like molecular wires) and the underlying silicon may be possible. The resulting alkynyl monolayers tolerate boiling pH 10 KOH solutions for 15 minutes, conditions under which a hydride-terminated sample rapidly dissolves. Rinsing with 1:1 48% HF (aq)/EtOH solutions results in no changes in the transmission FTIR spectrum, indicating the covalent nature of the bonding to the surface.

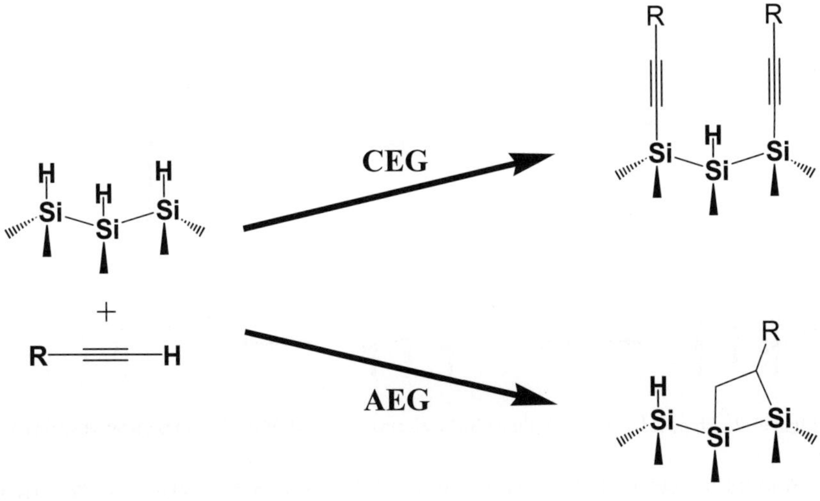

Figure 10. Electrografting of alkynes on porous silicon. CEG = <u>c</u>athodic <u>e</u>lectro<u>g</u>rafting, and AEG = <u>a</u>nodic <u>e</u>lectro<u>g</u>rafting.

The authors propose that cathodic electrografting is initiated through the formation of silyl anions, generated by the reduction of surface hydride. This anion then serves to abstract the proton from the weakly acidic alkyne, thus regenerating the silicon-terminated bond. The resulting acetylide anion is then free to participate in direct nucleophilic attack of a surface silicon-silicon bond, yielding the direct grafting of the alkyne and propagating the silyl anion to further facilitate surface functionalization (Figure 12). This mechanism is supported by the fact that the reaction is suppressed by a proton source (0.1 M HCl in ether), which quenches the silyl anion, prohibiting any surface functionalization from taking place.

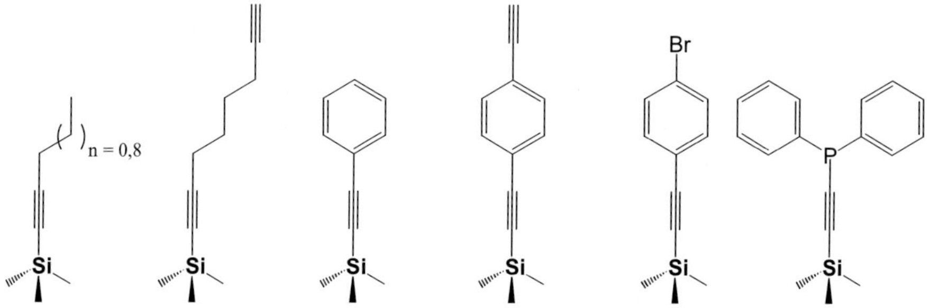

Figure 11. Examples of monolayers prepared through cathodic electrografting (CEG) of alkynes on porous silicon.

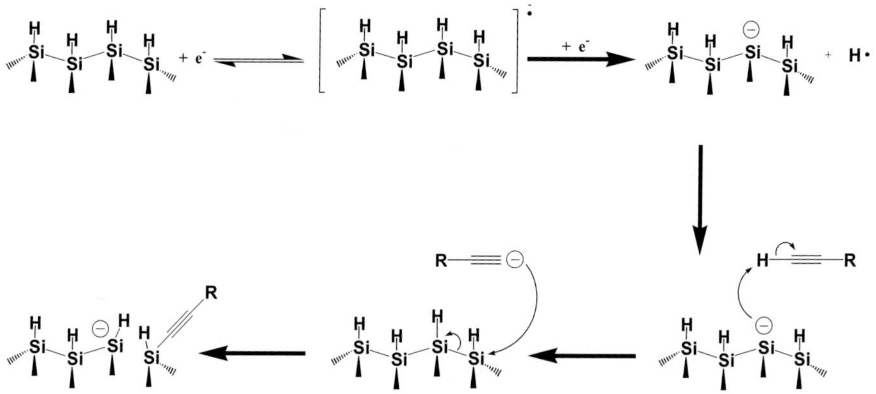

Figure 12. Proposed mechanism for cathodic electrografting (CEG) of alkynes on porous silicon.

Anodic electrografting (AEG) similarly produces densely packed organic monolayers on porous silicon. The reaction conditions are identical to CEG, with the exception of the polarity of the surface bias. Whereas CEG produces direct grafting of alkynyl functionalities, the AEG reaction results in a completely saturated, aliphatic termination; no $\nu(C=C)$ or $\nu(C\equiv C)$ modes are observed by transmission FTIR. The authors, therefore, propose that bis-hydrosilylation of the alkyne is taking place, although mono-hydrosilylation followed by cationic oligomerization cannot be ruled out. Surfaces prepared by AEG of aliphatic alkynes were subjected to boiling solutions of chloroform (20 minutes) and pH 10 KOH solutions (10 minutes), and no changes in their IR spectra were observed. This evidence serves to support the presence of covalent surface functionalization as opposed to simple chemical physisorption, although the authors do not rule out the presence of oligomeric species trapped within the porous silicon nanoarchitecture.

To account for the formation of the bis-hydrosilylated surface product, the authors propose a surface-initiated cationic hydrosilylation mechanism (Figure 13) which is based upon related electrochemical studies carried out on trialkylsilanes in solution [76]. Under anodic conditions, the surface silicon-terminated is rendered cationic, and is then attacked by the alkyne in solution, forming the Si-C bond and a β-silyl alkenyl carbocation. Abstraction of a neighboring hydride yields the neutral alkenyl group, and a proximal silicon-based positive charge. This site is then attacked by the surface bound olefin, leading to the second hydrosilylation step. Formation of the final surface bis-hydrosilylation product is completed with a hydride transfer from an adjacent Si-H bond. It is possible that cationic oligomerization could occur via attack of the intermediate carbocations by an alkyne.

Photoluminescence studies of the surfaces prepared through CEG and AEG were carried out. Most of the intrinsic photoluminescence of samples prepared

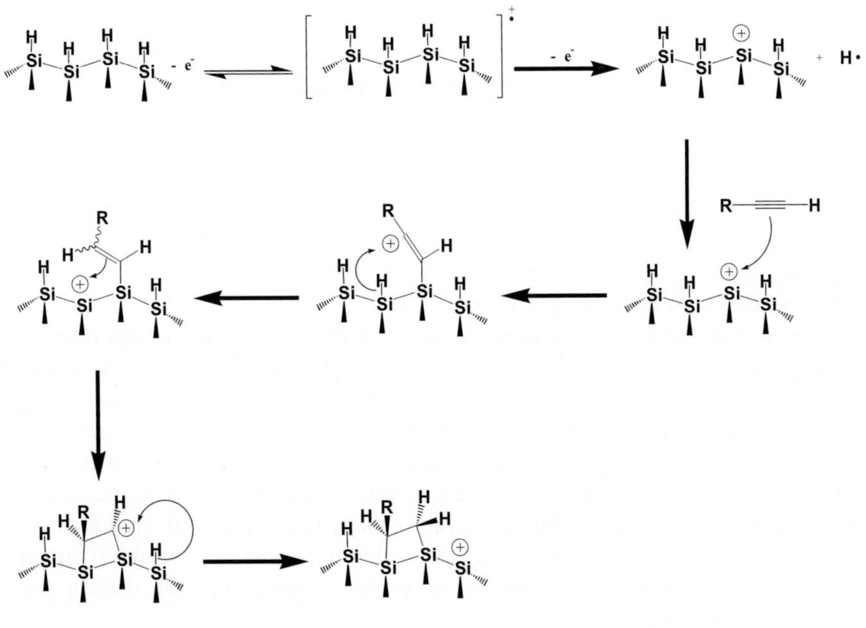

Figure 13. Proposed mechanism for bis-hydrosilylation observed during anodic electrografting (AEG) of alkynes on porous silicon.

through CEG exhibited significant loss of light emission, with aliphatic alkynyl derivatization retaining five to fifteen percent of the original photoluminescence intensity, and arynyl terminations leading to complete quenching. Conjugated species, when functionalized onto porous silicon, provide a site for nonradiative recombination of the exciton [77]. AEG functionalized surfaces exhibit more intense photoluminescence with little quenching of the native light emission, owing to the aliphatic nature of their surface attachment, which do not appear to favor exciton non-radiative recombination.

3.4 Metal Complex Mediated Hydrosilylation

Homogeneous catalysis utilizing late transition metal species, such as platinum, palladium, and rhodium have proven quite effective in facilitating solution phase hydrosilylation reactions [78]. At first glance, it would appear that these efficient catalysts would serve as excellent candidates for mediating the hydrosilylation of alkenes and alkynes onto hydride-terminated porous silicon. In such a pursuit, Buriak and coworkers attempted to utilize rhodium and palladium hydrosilylation catalysts in the preparation of functionalized porous silicon surfaces [79]. Exposure of Wilkinson's catalyst, $RhCl(PPh_3)_3$, to freshly etched porous silicon samples, in the presence of an alkyne resulted in the formation of a vinyl-terminated surface. Transmission FTIR confirmed the presence of a silicon bound vinyl group, along with a decrease in the integrated intensity of the surface silicon-terminated stretching modes. Unfortunately, a substantial amount of surface oxidation resulted from this approach even if rigorous precautions were maintained to exclude the presence of air and moisture. Furthermore, surfaces prepared in this manner appeared "blackened," presumably through metal deposition which resulted in compete photoluminescence quenching, with no possibility for recovery. Palladium complexes, such as $PdCl_2(PEt_3)_2$ and $Pd(OAc)_2/1,1,3,3$-tetramethylbutylisocyanide, employed in an attempt to insert alkynes directly into surface silicon-silicon bonds, suffered from these same complications.

Faced with the highly problematic nature of hydrosilylation through transition metal complexes, Buriak and coworkers employed Lewis acid catalysis as a viable alternative. The Lewis acid, ethylaluminum dichloride, an inexpensive commercially available reagent, had been reported in the organic literature to facilitate alkyne hydrosilylation with soluble, molecular trialkyl substituted silanes with high conversions under extremely gentle conditions to yield *cis* alkenyl products, exclusively [80]. The literature also reports aluminum trichloride to be a highly effective olefin hydrosilylation catalyst but due to its poor solubility in non-polar solvents, $EtAlCl_2$ was chosen to avoid complications of dealing with multiphasic reactions.

Lewis acid mediated hydrosilylation proved to be an efficient approach to the preparation of organic monolayers on hydride-terminated porous silicon [81]. A diverse variety of alkynes and alkenes were utilized, owing to the mild nature of the reaction, to yield the corresponding vinyl and alkyl surfaces, respectively (Figure 14). The experimental setup is simple and straightforward, with the entire procedure carried out at room temperature. Under inert atmosphere conditions (N_2 or Ar), a freshly etched sample of porous silicon is exposed to a few microliters of a commercially prepared 1.0 M hexane solution of $EtAlCl_2$, followed by the addition of 3-50 microliters of the alkyne or alkene of interest. Hydrosilylation of alkynes is completed in under an hour, whereas alkenes require reaction times of approximately twelve hours to reach completion. The reaction is then quenched with tetrahydrofuran and the surface subsequently rinsed with copious amounts of ethanol to remove any residual aluminum catalyst.

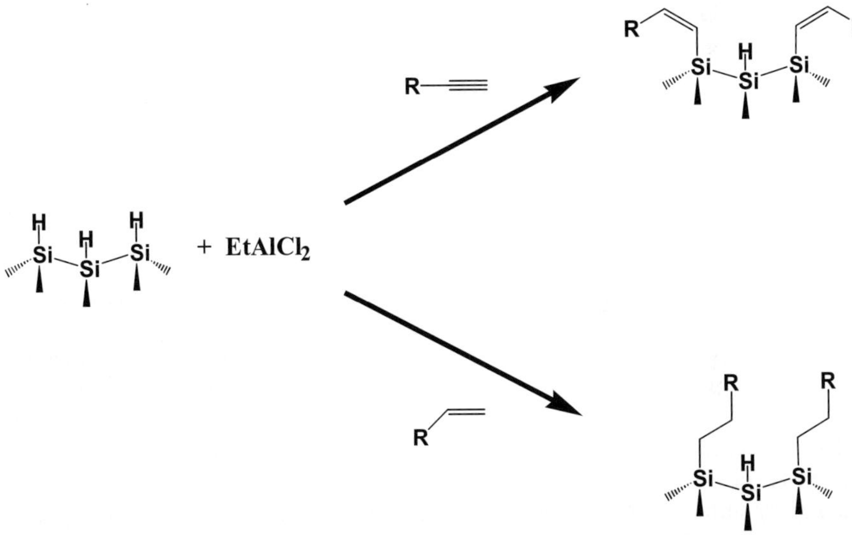

Figure 14. Lewis acid mediated hydrosilylation of alkynes and alkenes on hydride-terminated porous silicon, leading to alkenyl and alkyl moieties, respectively.

Monolayer formation via Lewis acid catalysis is supported by both consumption of silicon-terminated termini, and the concomitant appearance of C-H stretching modes, as determined by transmission FTIR analysis. Furthermore, solid-state ^{13}C-NMR studies of 1-pentyne functionalized, freestanding porous silicon samples clearly demonstrate the resulting pentenyl termination. The ^{13}C-NMR porous silicon spectrum was compared to a solution-phase, molecular analog, the Lewis acid mediated hydrosilylation product of 1-pentyne with tris(trimethylsilyl)silane [81]. Both the surface and molecular products contain the identical pentenyl moiety, bound through a covalent Si-C bond to a silicon atom substituted silicon center. The solution ^{13}C-NMR spectrum of the molecular hydrosilylation product correlated very well to that on the free-standing porous silicon wafer, indicating identical organic fragments.

Chemical means were employed to further elucidate the structure of the resulting functionalized surfaces. Hydrosilylation of 1-dodecyne produces a surface bound vinyl group, which results in a sharp v(C=C) signal in the transmission FTIR spectrum at 1595 cm^{-1}. Exposing this functionalized surface to a 0.8 M borane solution in THF, under nitrogen, leads to the expected hydroboration reaction of the surface–bound olefin functionality (Figure 15). After quenching the reaction in air and extensive rinsing with THF, FTIR reveals complete disappearance of the v(C=C) signal along with the advent of new signals corresponding to the v(B-O) modes. This same result was verified through the use of solid-state ^{13}C-NMR of a surface functionalized with 1-pentyne. Following hydroboration, the olefinic peaks attributed to the alkenyl carbon centers disappear and as before, the result was reproduced with the solution-phase analog.

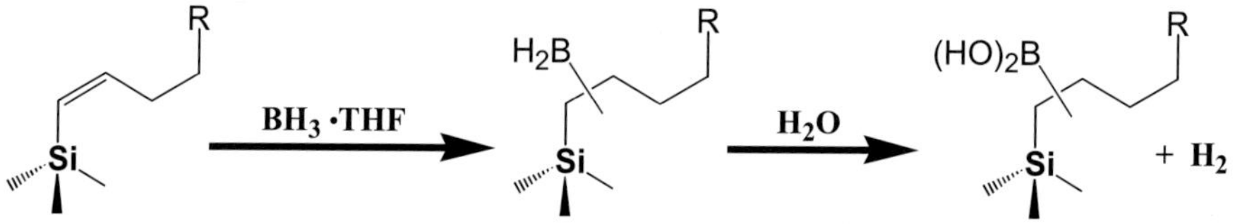

Figure 15. Hydroboration of the surface-bound alkenyl group carried out on a porous silicon surface.

The mechanism for Lewis acid mediated hydrosilylation on porous silicon is most likely similar to that of its solution-phase analog since the reaction products are similar (Figure 16). The alkyne species first coordinates to the ethylaluminum dichloride, leading to a highly reactive intermediate which needs to react rapidly with Si-H, or else oligomerization occurs. The electron deficient carbon is then attacked by surface hydride, thus generating a silyl cation and the aluminate species. Recombination of the silylium cation and the alumimum-bound carbanion leads to the neutral product.

The effect of surface functionalization on the photoluminescence of the resulting surfaces was investigated. Aliphatic and unconjugated vinyl groups result in ~80% quenching of the intrinsic photoluminescence of the porous silicon. As previously noted, conjugated surface terminations, such as a styrenyl group, lead to efficient nonradiative exciton recombination, completely quenching photoluminescence [77]. This was also observed for surfaces prepared by this method. Samples functionalized with phenylacetylene, leading to a Si-CH=CH$_2$Ph group, resulted in a complete loss of photoluminescence intensity, which was entirely independent of any substitutions, either electron donating or withdrawing, made to the phenyl moiety.

The authors successfully demonstrated that hydrosilylation utilizing EtAlCl$_2$, tolerates a wide range of different alkynes and alkenes (Figure 17). Secondary ionization mass spectrometry (SIMS) clearly indicated that the hydrosilylation reaction takes place in a homogeneous manner throughout the porous structure, and is not limited exclusively to the exterior sites of porous silicon. Furthermore, Brunauer-Emmet-Teller (BET) and Barrett-Joyner-Halenda (BJH) nitrogen adsorption/desorption analysis confirmed that the hydrosilylation reaction conditions were indeed so mild that no significant degradation in pore size nor surface area was observed. These analytical results were verified through chemical stability tests, which highlighted the excellent stability of these surfaces to highly demanding chemical environments. Surfaces demonstrated stern resistance to soaking in HF and boiling in aerated solutions of aqueous KOH (pH 10) for over four hours with no significant degradation, as monitored by FTIR. Even boiling pH 14 solutions can be tolerated for 5 minute periods.

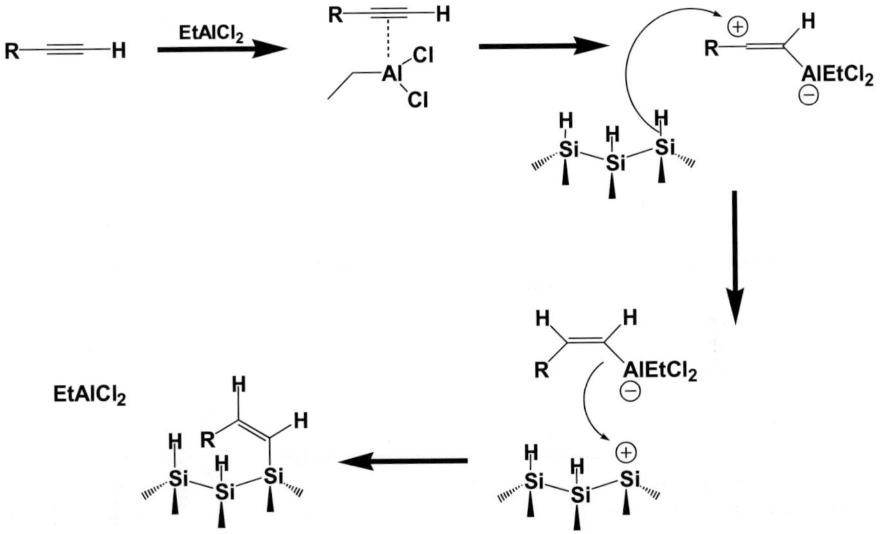

Figure 16. Possible mechanism for Lewis acid mediated hydrosilylation of alkynes on hydride-terminated porous silicon.

3.5 Alkylation Involving Carbon Nucleophiles

Alkylation and arylation reactions using carbanion nucleophiles were investigated by the Sailor [82] and Laibinis [83] groups, who found that an applied potential was not required for a reaction to proceed on porous silicon (Figure 18). Both groups of investigators independently developed similar methods to chemically modify porous silicon surfaces, resulting in stable silicon-carbon bond formation. The straightforward procedure is as follows: the desired organolithium or Grignard reagent is introduced onto a freshly prepared sample of hydride-terminated porous silicon, and allowed to react for a period of approximately two hours at room temperature. Because a silyl anion is generated (Si⁻M⁺), it must be quenched with an electrophilic reagent before exposure to air to avoid oxidation. The appropriate electrophilic quenching agent is subsequently applied and the sample thoroughly rinsed. Transmission FTIR analysis provides evidence for the formation of a dense organic monolayer.

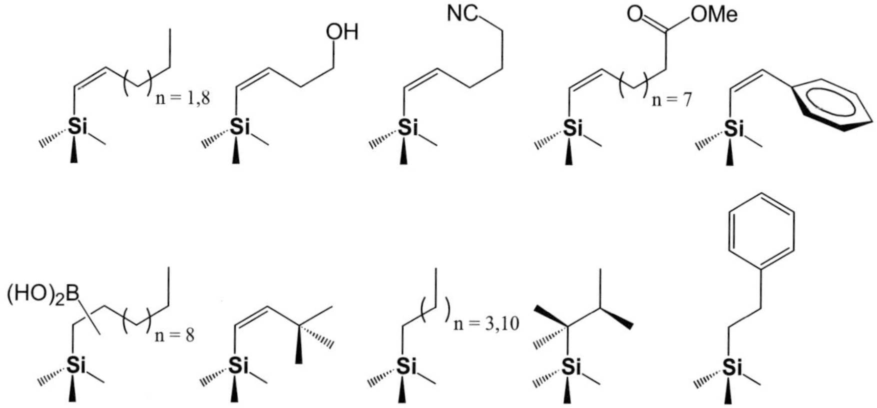

Figure 17. Examples of monolayers prepared through Lewis acid mediated hydrosilylation alkynes and alkenes on hydride-terminated porous silicon

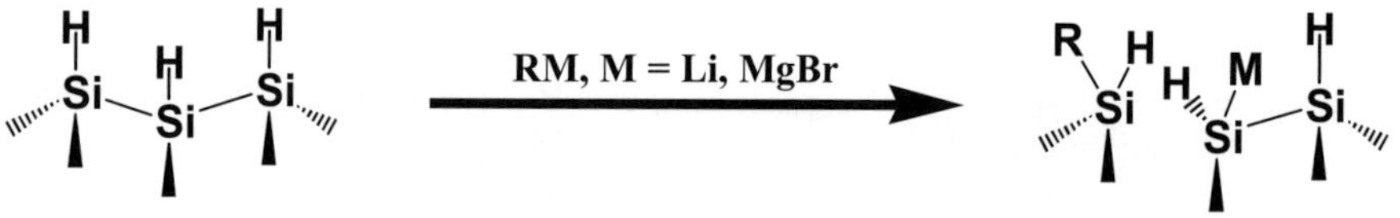

Figure 18. Alkylation of hydride-terminated porous silicon through the use of carbanion nucleophiles.

Detailed mechanistic studies attributed the resulting alkylation to nucleophilic attack directed upon weak surface silicon-silicon bonds (Figure 19). These bonds are known to be nearly 75 kJ mol^{-1} weaker than the silicon-terminated bond (Si-Si 215-250 kJ mol^{-1}, Si-H 323 kJ mol^{-1}), therefore making them more prone to cleavage. Nucleophilic attack by the carbanion species on the Si-Si bond results in the formation of a silyl anion, which in all probability exists as the metallated Si⁻M⁺ species (M = Li, MgX). This highly reactive intermediate surface species undergoes rapid oxidation in the presence of air and moisture. If, however, an electrophile such as a proton (ie, trifluoroacetic acid, TFA) or an acyl chloride is added, the silyl anion is neutralized. A variety of acyl halides have been demonstrated to react readily with the silyl anion surface intermediates in the to form "mixed" monolayers, including acetyl, heptanoyl, 3-(chloromethyl)benzoyl, and 4-butylbenzoyl chloride Negligible oxidation was observed by FTIR, and the surfaces resist any change upon immersion in an HF medium. This approach offers a facile route to hybrid organic monolayers with properties that may be tailored to study interactions on more intricate surfaces.

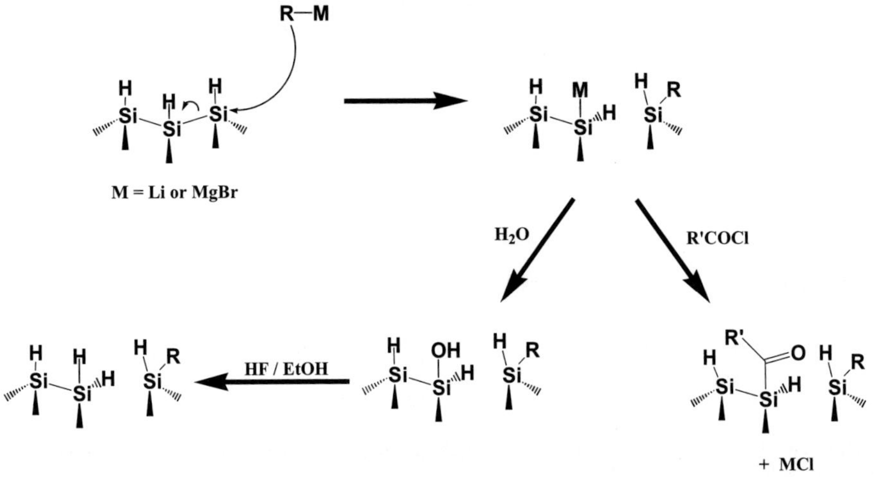

Figure 19. Mechanism of the carbanion reaction on porous silicon surfaces, and subsequent reaction with an acyl chloride reagent, leading to mixed monolayers.

In terms of the effects of this chemistry on the instrinsic photoluminescence of porous silicon, conjugated phenylacetylenyl groups (formed from the reaction with lithium phenylacetylide) result in almost complete quenching [82], again pointing to the effect of an aromatic ring conjugated through either an alkynyl or alkenyl group to the surface [77]. No degree of HF rinsing could regain the photoluminescence. Methyl, *n*-butyl, phenyl, and 4-fluorophenyl-terminated surfaces resulted in intermediate values of photoluminescence quenching, perhaps providing a chemical method of tuning photoluminescent efficiencies.

3.6 Hydrosilylation Involving Hydride Abstraction Agents

Lambert and co-workers successfully demonstrated catalytic solution phase hydrosilylation of 1,1'-diphenylethene with trialkylsilanes, through the utilization of the triphenylcarbenium cation, a hydride abstracting agent (Figure 20) [84]. Via Corey hydride transfer [85], the triphenylcarbenium cation removes the hydride from the silane, forming the silylium cation. Nucleophilic attack of the alkene species on this reactive species results in the formation of a β-silyl carbocation that

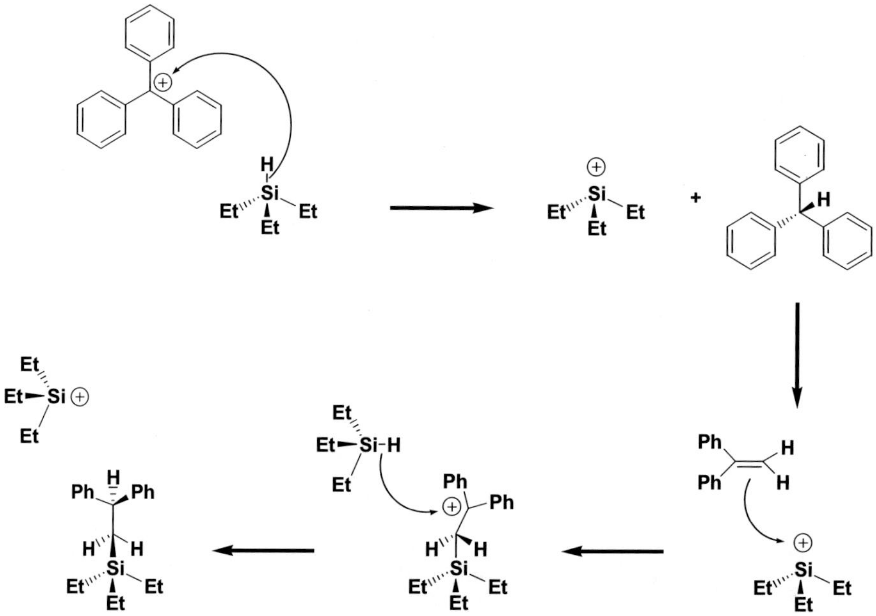

Figure 20. Solution phase hydrosilylation of 1,1'-diphenylethene with trialkylsilanes catalyzed by the triphenylcarbenium cation.

is quenched through a hydride abstraction step from excess silane in solution, yielding the final hydrosilylation product.

This efficient manner of silicon-carbon bond formation was subsequently utilized by Buriak and co-workers to facilitate the organic functionalization of hydride-terminated porous silicon in an analogous fashion [86]. In an inert atmosphere, a fresh sample of hyride-terminated porous silicon is exposed to a dilute solution of triphenylcarbenium salt and an equal volume of neat alkene or alkyne. The reaction proceeds quickly at room temperature and is completed within three hours. Following functionalization, the sample is removed from the inert atmosphere and washed with copious amounts of tetrahydrofuran, ethanol, methylene chloride, and pentane. The formation of densely packed organic monolayers were supported through the use of FTIR and solid-state ^{13}C NMR.

A surface hydrosilylation mechanism for porous silicon functionalization that is analogous to the one presented by Lambert (Figure 21) is believed to be in operation. In order to rule out the possibility of a parallel radical-mediated reaction pathway, the reaction was run in the presence of radical quenchers. Carbocation-mediated hydrosilylation of alkenes and alkynes on hydride-terminated porous silicon in the presence of an equivalent 2,4,6-tri-*tert*-butylphenol (with respect to the

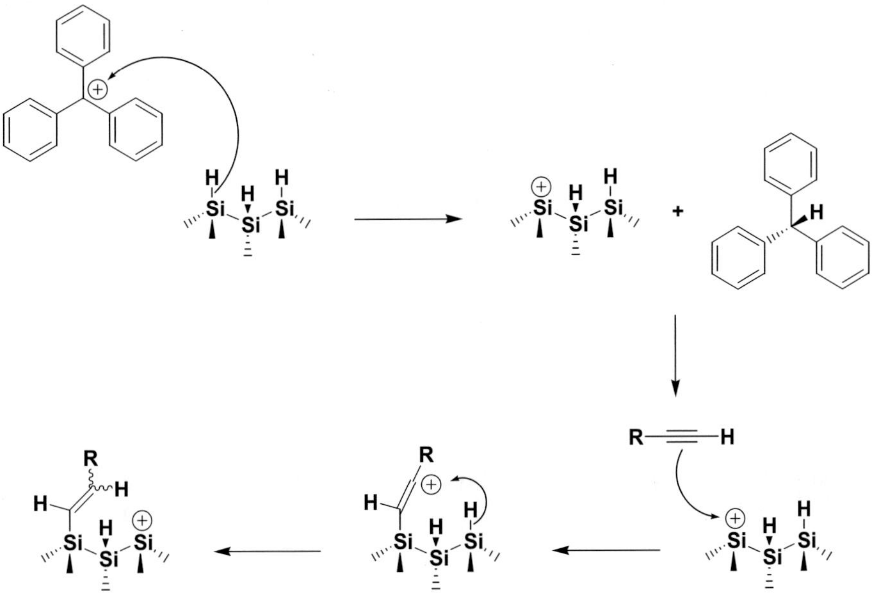

Figure 21. Carbocation mediated hydrosilylation of alkynes on hydride-terminated porous silicon.

triphenylcarbenium cation) proceeded without interference. The authors are currently conducting detailed mechanistic studies employing various hydride abstraction agents, as well as looking into solvent and anion effects.

This method of porous silicon functionalization is attractive since it is carried out quickly at room temperature with a minimum of experimental apparatus, and is sufficiently mild as to tolerate the preparation of a wide variety of functional groups (Figure 23). The functionalized porous silicon surfaces prepared through this methodology withstand oxidation under harsh chemical environments such as extended immersion times in aerated boiling water and ethanolic solutions at elevated pH levels.

3.7 Thermal Hydrosilylation

Experiments conducted by the groups of Horrocks and Houlton, and Wayner utilized thermal hydrosilylation to prepare organic monolayers on photoluminescent samples of porous silicon [87,88]. Freshly prepared samples of hydride-terminated porous silicon were immersed in an aliphatic alkene or alkyne and heated to 110-180°C for nearly twenty hours (Figure 23). After washing with toluene to remove any physisorbed species, a chemically bound alkyl or alkenyl-terminated surface remained. The resulting monolayers are densely packed and stable, based upon FTIR data and stability studies.

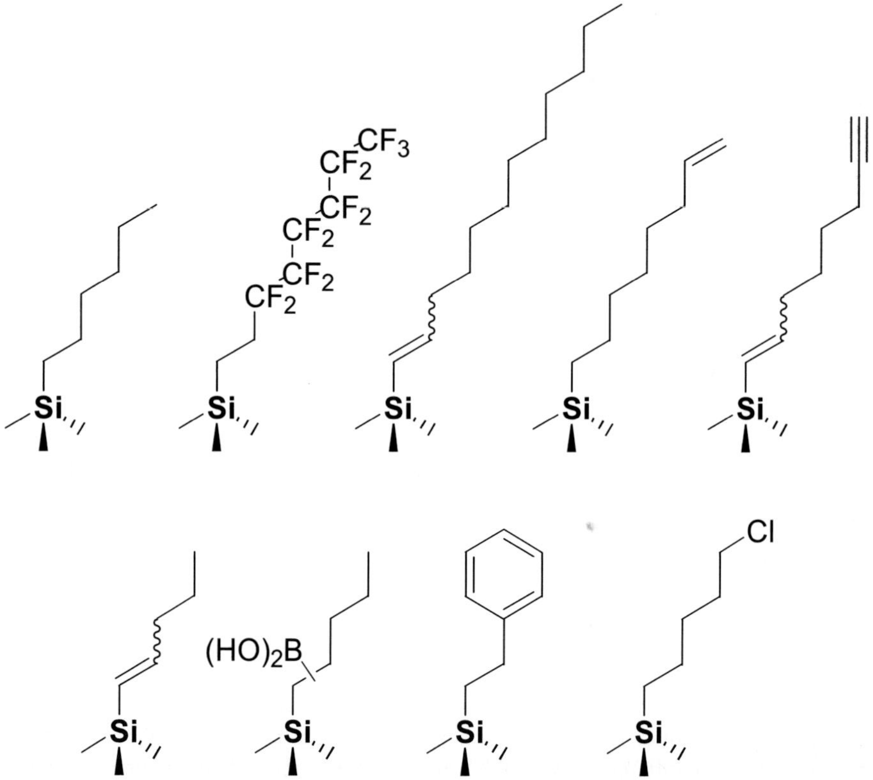

Figure 22. Examples of monolayers prepared through carbocation mediated hydrosilylation of alkynes and alkenes on hydride-terminated porous silicon.

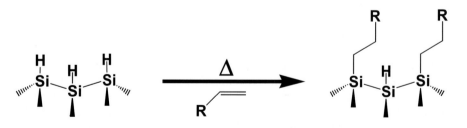

Figure 23. Thermally mediated hydrosilylation of alkenes on hydride-terminated porous silicon.

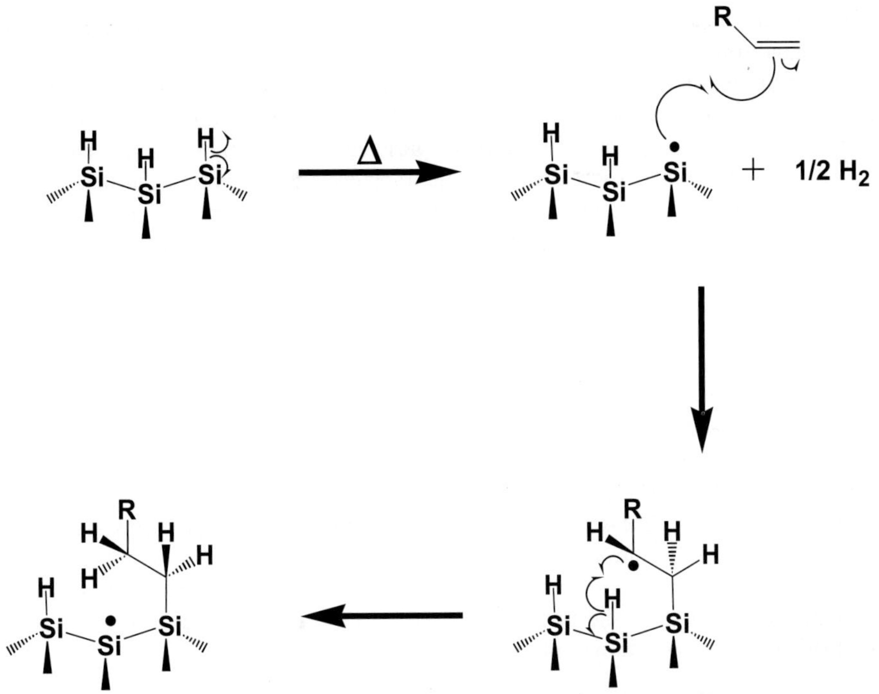

Figure 24. Mechanism of thermal hydrosilylation on hydride-terminated porous silicon.

In absence of any initiator or catalyst, it appears that hydrosilylation proceeds through a radical chain reaction, initiated by homolytic Si-H bond cleavage (Figure 24) [19]. The silicon radical on the surface then reacts with the olefin, leading to formation of a silicon-carbon bond. The rapid reactivity of olefins with silyl radicals is well documented in the molecular literature [89]. The newly formed carbon radical readily abstracts a hydrogen from an adjacent surface hydride site, or from the allylic position of unreacted olefin in solution. It has been reported that thermal alkyne hydrosilylation on porous silicon leads to the alkyl surface as opposed to the expected alkenyl product [87], although other groups have seen the alkenyl surface on both porous [81] and flat silicon [90]. Recently Buriak and co-workers have utilized this approach to prepare alkyl monolayers on porous Ge [91].

Functionalized porous silicon samples using this approach appear highly robust, withstanding the standard exposure stability tests to boiling solvents and aqueous acids and bases. Significant quenching of photoluminescence is, however, observed by some groups after monolayer formation via thermal hydrosilylation [87] although others contest this [88]. Given the complex architecture of porous silicon, it appears that the nanocrystallites may be too inherently fragile and degrade under extended reflux conditions.

The thermal reaction is simple to carry out, and a number of non-Lewis basic organic functionalities have been incorporated into organic monolayers prepared on porous silicon through this approach (Figure 25). By combining the chemical versatility and stability, subsequent organic transformations involving surface-bound reactive groups have been carried out to prepare functional interfaces. For example, Horrocks and Houlton recently reported the successful immobilization of single-stranded DNA onto the surface of porous silicon (Figure 26) [92]. Surface-immobilized DNA is promising of interest for gene chip and biosensor applications. The single-stranded DNA could not be prepared though direct surface functionalization, and instead, a molecular scaffold was employed as a chemical linkage. First, 4,4'-dimethoxytrityl-protected (DMT) ω-undecynol was functionalized onto the surface of porous silicon through thermal hydrosilylation. The protecting group was then removed and a 17-mer oligonucleotide was synthesized at the surface, utilizing a specially modified DNA synthesizer/column setup. In order to access the efficiency of this process, cleavable linkages were inserted so that gel electrophoresis could be utilized. Futhermore, double-stranded DNA was also prepared from the single-stranded surface upon hybridization. This example serves to show the myriad of surfaces yet to be explored through rational synthetic design of functional interfaces.

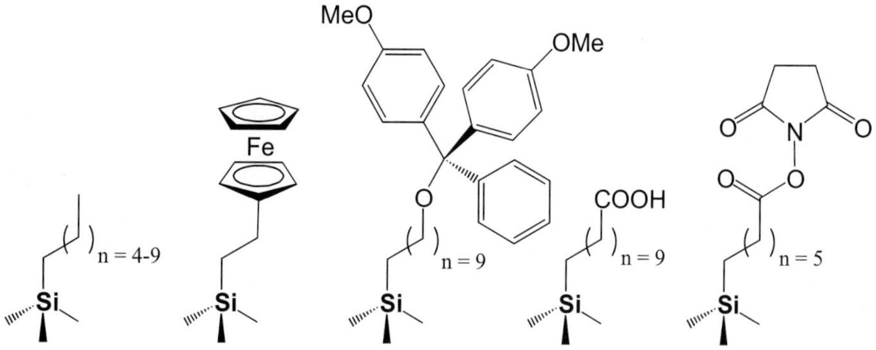

Figure 25. Examples of monolayers prepared through thermally mediated hydrosilylation on porous silicon.

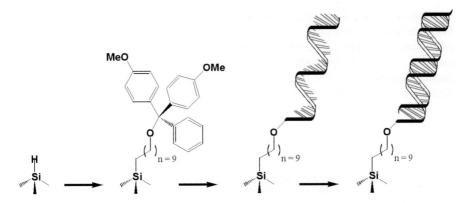

Figure 26. Porous silicon functionalized with 4,4'-dimethoxytrityl-protected (DMT) ω-undecanol was deprotected and a 17-mer oligonucleotide was synthesized at the surface, utilizing a specially modified DNA synthesizer/column setup. The double-stranded DNA was prepared through a subsequent hybridization reaction.

3.8 Photochemical hydrosilylation

Buriak and coworkers have recently demonstrated hydrosilylation of alkenes and alkynes on hydride-terminated porous silicon utilizing a simple white light source to facilitate the reaction [93]. Photoluminescent samples of porous silicon are coated with the neat alkene/alkyne, or a methylene chloride solution of the substrates, and illuminated. While an unfiltered source works well, 400 nm long pass, 600 nm short pass and IR absorbing filters were consistently utilized to ensure that the window of light, from 400-600 nm, contained no stray UV or excess thermal energy. Sample illumination, is maintained under inert conditions, at a moderate intensity of 22-44 mW cm^{-2} and results in efficient hydrosilylation at room temperature. Exposure times as short as only fifteen minutes result in the formation of an alkyl surface termination for alkenes, and an alkenyl surface termination when alkynes are utilized. These results were confirmed through FTIR and SIMS analysis, and reaction conditions proved mild enough for the incorporation of a variety of functionalities. Through simple masking procedures, the hydrosilylation can be carried out in a regiospecific fashion on the surface, since only the illuminated areas react, as shown schematically in Figure 27. Examples of patterned surfaces can be seen in Figure 28.

The authors recently completed a revealing mechanistic study of this process, bringing forth evidence for a novel, unprecedented reaction at the silicon surface. Initially, radical formation through photolytic Si-H bond cleavage provided a possible explanation, similar to the thermal hydrosilylation reaction of Figure 24. Cleavage of Si-H bonds with light of such low energy is unprecedented, since wavelengths shorter than 350 nm are required to do this in an efficient manner [94].

In addition, radical quenching agents had little effect, and studies of the molecular model, tris(trimethyl)silylsilane, showed that this molecule did not react under these conditions. The key to the reaction came about when it was discovered that the reaction is successful only on samples of porous silicon that exhibit photoluminescence. No hydrosilylation above background was observed on nonluminescent samples of porous silicon, flat hydride-terminated silicon substrates,

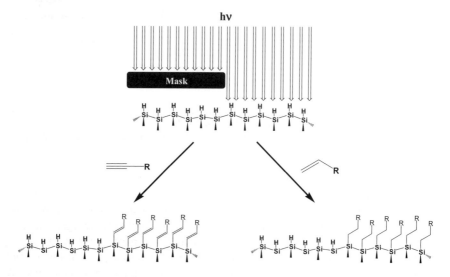

Figure 27. White light promoted hydrosilylation of alkynes and alkenes on porous silicon, utilizing a photomask to limit light exposure to predetermined domains.

Figure 28. Photographs of porous silicon samples prepared through light promoted hydrosilylation of 1-dodecene or 1-dodecyne through masking procedures. a) The dodecenyl surface appears as the darker patterned area when illuminated with a 365 nm hand-held UV lamp; the sample is glowing a characteristic orange due to the intrinsic photoluminescence of the porous silicon. The other areas of the wafer are unfunctionalized (native Si-H termination). b) A dodecyl surface appears as the slightly red-shifted area upon UV illumination. c) Sample in room light from b) after boiling in aerated, aqueous KOH (pH 12) solution for 30 seconds. The unfunctionalized porous silicon (grey area) has dissolved while the hydrosilylated surface (surface **1,** golden area) remains intact. (d) Illumination of the surface from (c) with a 365 nm hand held UV lamp. The PL of the hydrosilylated area (dodecyl-terminated surface) remains intact while most of the unfunctionalized PL is destroyed.

or molecular phase silane analogues. This observation suggested that the quantum confined excitons in the silicon nanocrystallites in porous silicon, responsible for light emission, were somehow involved in the surface chemistry. Finally, conjugated alkynes, such as phenylacetylene, 4-methylphenylacetylene, and 4-chlorophenylacetylene, yield poor levels of surface incorporation, even for extended reaction times, whereas they react very well with other hydrosilylation chemistries, including thermal and Lewis acid hydrosilylation chemistries.

These observations led to postulation of an exciton-mediated hydrosilylation reaction, involving nucleophilic attack of alkynes/alkenes at the porous silicon surface (Figure 29). Upon exposure to white light, electron-hole pairs (excitons) are formed which, a small fraction of the time, recombine in a radiative fashion, leading to the observed photoluminescence emission of porous silicon. Migration of the positively charged hole to the Si-H-terminated interface will render the surface much more susceptible to nucleophilic attack, leading to Si-C bond formation and a β-silyl carbocation. The remaining half of the exciton couple, the electron, can combine with a proximal surface hydrogen atom, giving what can formally be called a 'hydride'; combination of the hydride and carbocation leads to the final organic termination, and a neutral nanocrystallite which can undergo further reactions in this manner. When reagents known to quench the photoluminescence through energy and charge transfer pathways were employed, the hydrosilylation was essentially blocked, thus definitively tying together light emission and surface reactivity. Futhermore, hydrosilylation efficiencies for phenylacetylene and other conjugated alkenes and alkynes yielded similarly low values. Hydrosilylation of phenylacetylene results in the formation of styrenyl-terminated surface, which effectively quenches the porous silicon photoluminescence [77,82,95]. Thus, the reaction of only a few phenylacetylene molecules serves to quench the excitons, thereby preventing any additional hydrosilylation. This mechanism accounts for all of the observed reactivity pathways and provides a materials property driven reaction with no solution phase analogue

This approach toward silicon-carbon bond formation proved so mild that the underlying nanocrystallite subarchitecture remained largely intact. This is supported by the observation that alkyl functionalized porous silicon samples, prepared in this manner, maintain all of their intrinsic photoluminescence. As observed by others, monolayers containing unsaturated bonds result in lower values; utilizing the white light mediated hydrosilylation, aliphatic (unconjugated) alkenyl surface terminations resulted in only about 40% quenching of the porous silicon photoluminescent intensity. A small hypsochromic shift of approximately 10 nm is observed in the photoluminescence spectra of these samples. Futhermore, these monolayers proved to be quite robust in serving to passivate the underlying porous silicon towards chemical attack and oxidation under ambient conditions. Functionalized surfaces easily withstood immersion into HF, organic solvents, and boiling alkaline solutions. This method may serve as an excellent candidate for surface functionalization of

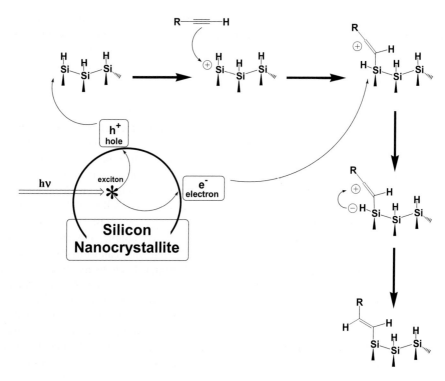

Figure 29. Proposed exciton-mediated mechanism for white light promoted hydrosilylation on hydride-terminated porous silicon.

photoluminescent samples of porous silicon for implementation into optoelectronic devices or chemical sensors.

White light mediated surface functionalization of porous silicon is ideal for photopatterning since simple masking procedures may be utilized to produce organic features on the surface down to the micron scale. Utilizing a simple and inexpensive optical apparatus, monolayer photopatterning, can be utilized to result in spatially defined domains of differing chemical functionalities (Figure 28). Employing nothing more sophisticated than a f/75 reducing lens and a transparency photomask, prepared from a high quality laser printer, features as small as 30 microns were obtained. The key to the procedure is that only the silicon-terminated areas under illumination proceed to react with the selected alkyne or alkene. The surface protected by the photomask remains hydride-terminated, which may subsequently participate in hydrosilylation with different alkenes or alkynes. The sample of porous silicon may be decorated, through this approach, to yield intricate designs of differing functionalities, limited only by imagination and lithographic resolution. This method allows for facile interfacial design, through simple apparatus and may be readily incorporated into already existing microprocessor fabrication routes.

4 Conclusions

These examples of systematic interfacial design serve to demonstrate the powerful role of surface functionalization for the fabrication and tailoring of nanostructured materials to a host of high-performance devices. This rapidly evolving field of research presents a unique interchange between fundamental surface science and technological applications. A sound fundamental understanding of surface chemistry is required for rational design and preparation of functional interfaces, yet the pursuit of exotic new materials leads to the discovery of new reactivity and mechanisms. With continued exploration, perhaps the breadth of surface functionalization chemistry will someday begin to rival the diversity and selectivity of solution phase synthesis. Certainly, the emphasis will shift from chemical passivation to the design of truly functional interfaces, capable of catalysis, molecular recognition, and much more. In the interim, intelligent interfacial design will yield remarkable new classes of electronics, sensors, and biocompatible implants. While many frontiers remain to be investigated, especially the surface chemistry of the compound semiconductors, the last decade has witnessed the infancy of a truly interdisciplinary and extremely exciting field, combining the best of chemistry, physics, and materials science.

References

1. Amorphous and Nanocrystalline Materials: Preparation, Properties, and Applications, ed. by A. Inoue and K. Hashimoto (Springer, Berlin, 2001).
2. Nanostructured Materials, ed. by J. Y. Ying (Academic, San Diego, 2001).
3. M. N. Rittner, *Amer. Ceramic Soc.*, **81** (2002) pp. 33.
4. H. Gleiter, *Nanostructured Mat.* **6** (1995) pp. 3.
5. F. Schüth and E. Schmidt, *Adv. Mater.* **14** (2002) pp. 629.
6. M. E. Davis, *Nature* **417** (2002) pp. 813.
7. Nano-Surface Chemistry, ed. by M. Rosoff (Marcel Dekker, New York, 2002).
8. Advanced Zeolite Science and Applications, ed. by J. Weitkamp, H. G. Karge, and J. C. Jansen (Elsevier, Amsterdam, 1994).
9. Introduction to Zeolite Science and Practice, ed by P. A. Jacobs, J. C. Jansen, H van Bekkum, and E. M. Flanigen (Elsevier, Amsterdam, 2001).
10. A. G. Cullis, L. T. Canham, and P. D. J. Calcott, *J. Appl. Phys.* **82** (1997) pp. 909.
11. V. Parkhututik, *J. Por. Mat.* **7** (2000) pp. 363.

12. A. Allann, D. Edenfield, W. H. Joyner, A. B. Kahng, M. Rodgers, and Y. Zorian, *Computer* **35** (2002) pp. 42.
13. C. M. Wolfe, N. Holonyak, Jr., and G. E. Stillman, Physical Properties of Semiconductors (Prentice Hall, Englewood Cliff, NJ, 1989); S. M. Sze, Semiconductor Devices, Physics, and Technology (Wiley, New York, 2002).
14. J. Page, *Smithsonian* **1** (2000) pp. 37; J. D. Plummer, M. D. Deal, and P. B. Griffin, Silicon VLSI Technology, Fundametals, Practice, and Modeling (Prentice Hall, London, 2000).
15. X. G. Zhang, Electrochemistry of Silicon and its Oxide, (Kluwer, New York, 2001).
16. Fundamental Aspects of Silicon Oxidation, ed. by Y. J. Chabal (Springer, Berlin, 2001); The Si-SiO$_2$ System, ed. by P. Balk (Elsevier, Amsterdam, 1988).
17. A. M. Green and W. E. Spicer, *J. Vac. Sci. Technol. A* **11** (1993) pp. 1061.
18. V. S. Vavilov, *Physics-Uspekhi* **37** (1994) pp. 269.
19. J. M. Buriak, *Chem. Rev.* **102** (2002) pp. 1271.
20. A. Uhlir, *Bell Syst. Techn. J.* **35** (1956) pp. 333; D. R. Turner, *J. Electrochem. Soc.* **105** (1958) pp. 402.
21. M. J. Sailor and E. J. Lee, *Adv. Mater.* **9** (1997) pp. 783.
22. Properties of Porous Silicon, ed. by L. T. Canham (INSPEC, London, 1997).
23. S. Shih, K. H. Jung, T. Y. Hsieh, J. Sarathy, J. C. Cambell, and D. L. Kwong, *Appl. Phys. Lett.* **60** (1992), pp. 1863.
24. R. L. Smith and S. D. Collins, *J. Appl Phys.* **71** (1992) pp. R1.
25. M. I. J. Beale, N. G. Chew, M. J. Uren, A. G. Cullis, and J. D. Benjamin, *Appl. Phys. Lett.* **46** (1985) pp. 86.
26. F. Ronkel and J. W. Schultze, *J. Por. Mat.* **7** (2000) pp. 11.
27. B. Hamilton, *Semicond. Sci. Technol.* **10** (1995) pp. 1187.
28. D. Hamm, J. Sasano, T. Sakka, and Y. H. Ogata, *J. Electrochem. Soc.* **149** (2002) pp. C331.
29. A. Halimaoui, C. Oules, G. Bomchil, A. Bsiesy, F. Gaspard, R. Herino, M. Ligeon, and F. Muller, *Appl. Phys. Lett.* **59** (1991) pp. 304; H. C. Choi and J. M. Buriak, *Chem. Mater.* **12** (2000) pp. 2151.
30. P. McCord, S. L. Yau, and A. J. Bard, *Science* **257** (1992) pp. 68.
31. X. Chen, D. Uttamchandani, C. Trager-Cowan, and K. P. O'Donnell, *Semicond. Sci. Technol.* **8** (1993) pp. 92.
32. K. J. Nash, P. D. J. Calcott, L. T. Canham, M. J. Kane, and D. Brumhead, *J. Lumin.* **60** (1994) pp. 297; A. G. Cullis and L. T. Canham, *Nature* **353** (1991) 335; D. J. Lockwood and A. G. Wang, *Solid State Commun.* **94** (1995) pp. 905.
33. P. Van Zant, Microchip Fabrication: A Practical Guide to Semiconductor Processing (McGraw Hill, New York, 2000); S. A. Campbell, The Science and Engineering of Microelectronic Fabrication (Oxford, New York, 2001).

34. M. J. Sailor, J. L. Heinrich, and J. M. Lauerhaas, Semiconductor Nanoclusters, ed. by P. V. Karmat and D. Meisel (Elsevier, Amsterdam, 1996) pp. 209.
35. A. Gupta, V. K. Jain, C. R. Jalwania, G. K. Singhal, O. P. Arora, P. P. Puri, R. Singh, M. Pal, and V. Kumar, *Semicond. Sci. Technol.* **10** (1995) pp. 698.
36. G. Marsh, *Materials Today* **1** (2002) pp. 36.
37. V. P. Parkhutik and L. T. Canham, *Phys. Stat. Sol. A* **182** (2000) pp. 591.
38. L. T. Canham, T. I. Cox, A. Loni, and A. J. Simons, *Appl. Surf. Sci.* **102** (1996) pp. 436.
39. M. P. Stewart and J. M. Buriak, *Adv. Mater.* **12** (2000) pp. 859.
40. S. E. Letant, S. Content, TT Tan, F. Zenhausen, and M. J. Sailor, *Sensors and Actuators B* **69** (2000) pp. 193.
41. A. M. Tinsley-Brown, L. T. Canham, M. Hollings, M. H. Anderson, C. L. Reeves, T. I. Cox, S. Nicklin, D. J. Squirrell, E. Perkins, A. Hutchinson, M. J. Sailor, and A. Wun, *Phys. Stat. Sol. A* **182** (2000) pp. 547.
42. S. Content, W. C. Trogler, and M. J. Sailor, *Chem. Eur. J.* **6** (2000) pp. 2205.
43. L.T. Canham, M. P. Stewart, J. M. Buriak, C. L. Reeves, M. Anderson, E. K. Squire, P. Allcock, and P. A. Snow, *Phys. Stat. Sol. A* **182** (2000) pp. 512.
44. H. Sohn, S. Letant, M. J. Sailor, and W. C. Trogler, *J. Am. Chem. Soc.* **122** (2000) pp. 5399.
45. L. T. Canham, *Adv. Mater.* **7** (1995) pp. 1033.
46. V. S.-Y. Lin, K. Motesharei, K.-P. S. Dancil, M. J. Sailor, and M. R. Ghadiri, *Science* **278** (1997) pp. 840.
47. J. Wei, J. M. Buriak, G. Siuzdak, *Nature* **399** (1999) pp. 243.
48. S. Fan, M. G. Chapline, N. R. Franklin, T. W. Tombler, A. M. Cassell, and H. Dai, *Science* **283** (1999) pp. 512.
49. S. C. Bayliss, and L. D. Buckberry, *Mater. World* **7** (1999) pp. 212.
50. L. T. Canham, C. L. Reeves, J. P. Newey, M. H. Houlton, T. I. Cox, J. M. Buriak, and M. P. Stewart, *Adv. Mater.* **11** (1999) pp. 1505.
51. Y. H. Ogata, T. Tsuboi, T. Sakka, and S. Naito, *J. Por. Mat.* **7** (2000) pp. 63.
52. The Chemistry of Organic Silicon Compounds, ed. by S. Patai and Z. Rappoport (John Wiley and Sons, New York, 1989).
53. J. M. Buriak, *Adv. Mater.* **11** (1999) pp. 265; J. M. Buriak, *J. Chem. Soc. Chem. Commun.* (1999) pp. 1051.
54. M. P. Stewart, E. G. Robins, T. W. Geders, M. J. Allen, H. C. Choi, and J. M. Buriak, *Phys. Stat. Sol. A* **182** (2000) pp. 109.
55. B. J. Tufts, A. Kumar, A. Bansal, and N. S. Lewis, *J. Phys. Chem.* **96** (1992) pp. 4581.
56. R. J. Hamers and Y. Wang, *Chem. Rev.* **96** (1996) pp. 1261.
57. J. H. Song and M. J. Sailor, *Comment Inorg. Chem.* **21** (1999) pp. 69.
58. M. Warntjes, C. Vieillard, F. Ozanam, and J.-N. Chazalviel, *J. Electrochem. Soc.* **42** (1995) pp. 4138.

59. K.-H. Li, C. Tsai, J. C. Campbell, M. Kovar, and J. M. White. *J. Electon. Mater.* **23** (1994) pp. 409.
60. J. A. Glass, E. A. Wovchko, and J. T. Yates, *Surf. Sci.* **338** (1995) pp. 125.
61. E. J. Lee, J. S. Ha, and M. J. Sailor, *Mater. Res. Soc. Symp. Proc.* **358** (1995) pp. 387.
62. E. J. Lee, T. W. Bitner, J. S. Ha, M. J. Shane, and M. J. Sailor, *J. Am. Chem. Soc.* **118** (1996) pp. 5375.
63. R. C. Anderson, R. S. Muller, and C. W. Tobias, *J. Electrochem. Soc.* **140** (1993) pp. 1393.
64. V. M. Dubin, C. Vieillard, F. Ozanam, and J. N. Chazalviel, *Phys. Stat. Sol. B* **190** (1995) pp. 47.
65. Y. Duvault-Herrera, N. Jaffrezic-Renault, P. Clechet, J. Serpinet, and D. Morel, *Colloids Surf.* **50** (1990) pp. 197.
66. J. T. C. Wojtyk, K. A. Morin, R. Boukherroub, and D. D. M. Wayner, *Langmuir* **18** (2002) pp. 6081; R. Boukherroub, D. D. M. Wayner, and D. J. Lockwood, *Appl. Phys. Lett.* **81** (2002) pp. 601.
67. M. Calistry-Yeh, E. J. Kramer, R. Sharma, W. Zhao, M. H. Rafailovich, J. Sokolov, and J. D. Brock, *Langmuir* **12** (1996) pp. 2747.
68. M. M. Sung, J. Kluth, O. W. Yauw, and R. Maboudian, *Langmuir*, **13** (1997) pp. 6164.
69. Semiconductor Surfaces and Interfaces, ed. by W. Mönch (Springer, Berlin, 2001).
70. F. Glockling, The Chemistry of Germanium (Academic Press, London, 1969).
71. C. Vieillard, M. Warntjes, F. Ozanam, and J.-N. Chazalviel, *Proc. Electrochem. Soc.* **95** (1996) pp. 250.; F. Ozanam, C. Vieillard, M. Warntjes, T. Dubois, M. Pauly, and J.-N. Chazalviel, *Can J. Chem. Eng.* **76** (1998) pp. 1020.
72. A. Teyssot, A. Fidelis, S. Fellah, F. Ozanam, and J. N, Chazalviel, *Electrochimica Acta* **47** (2002) pp. 2565.
73. C. Gurtner, A. W. Wun, and M. J. Sailor, *Angew. Chem. Int. Ed.* **38** (1999) pp. 1966.
74. C. A. Canaria, I. N. Lees, A. W. Wun, G. M. Miskelly, and M. J. Sailor, *Inorg. Chem. Commun.* **8** (2002) pp. 560.
75. E. G. Robins, M. P. Stewart, and J. M. Buriak, *Chem. Commun.* (1999) pp. 2479.
76. (a) A. A. Khapicheva, N. T. Berberova, E. S. Klimov, and O. Yu Okhlobystin, *Zh. Obsch. Khim.* **55** (1985) pp. 1533; (b) O. Yu Okhlobystin and N. T. Berberova, *Dokl. Akad. Nauk. SSSR*, **332** (1993) pp.599.
77. Buriak, J. M.; Allen, M. J. *J. Lumin.* **80** (1999) pp. 29.
78. B. Marciniec, and J. Gulinski, *J. Organomet. Chem.* **446** (1993) pp. 15.; Y. Ito, M. Suginome, and M. Murakami, *J. Org. Chem.* **56** (1991) pp. 1948.
79. J. M. Holland, M. P. Stewart, M. J. Allen, and J. M. Buriak, *J. Solid State Chem.* **147** (1999) pp. 251.

80. N. Asao, T. Sudo, and Y. Yamamoto, *J. Org. Chem.* **61** (1996) pp. 7654; T. Sudo, N. Asao, V. Gevorgayan, and Y. Yamamoto, *J. Org. Chem.* **64** (1999) pp. 2494.
81. J. M. Buriak, and M. J. Allen, *J. Am. Chem. Soc.* **120** (1998) pp. 1339.; J. M. Buriak, M. P. Stewart, T. W. Geders, M. J. Allen, H. Cheul Choi, J. Smith, D. Raftery, and L. T. Canham, *J. Am. Chem. Soc.* **121** (1999) pp. 11491.
82. J. H. Song, and M. J. Sailor, *J. Am. Chem. Soc.* **120** (1998) pp. 2376.; J. H. Song, and M. J. Sailor, *Inorg. Chem.* **38** (1999) pp. 1498.
83. N. Y. Kim, and P. E. Laibinis, *J. Am. Chem. Soc.* **120** (1998) pp. 4516.
84. J. B. Lambert and Y. Zhao, *J. Am. Chem. Soc.* **118** (1996) pp. 7867; J. B. Lambert, Y. Zhao, and H. Wu, *J. Org. Chem.* **64** (1999) pp. 2729.
85. J. Y. Corey, *J. Am. Chem. Soc.* **97** (1975) pp. 3237; J. B. Lambert, Y. Zhao, and S. M. Zhang, *J. Phys. Org. Chem.* **14** (2001) pp. 370.
86. J. M Schmeltzer, L. A. Porter, Jr., M. P. Stewart, and J. M. Buriak, *Langmuir* **18** (2002) pp. 2971.
87. J. E. Bateman, R. D. Eagling, D. R. Worrall, B. R. Horrocks, and A. Houlton, *Angew. Chem. Int. Ed.* **37** (1998) pp. 2683; J. E. Bateman, R. D. Eagling, B. R. Horrocks, and A. Houlton, *J. Phys. Chem. B* **104** (2000) pp. 5557.
88. R. Boukherroub, D. D. M. Wayner, D. J. Lockwood, and L. T. Canham, *J. Electrochem. Soc.* **148** (2001) pp. H91; R. Boukherroub, S. Morin, D. D. M. Wayner, and D. J. Lockwood, *Phys. Stat. Sol. A* **182** (2000) pp. 117; Boukherroub, S. Morin, D. D. M. Wayner, F. Bensebaa, G. I. Sproule, J. M. Baribeau, D. J. Lockwood, *Chem. Mater.* **13** (2001) pp. 2002; R. Boukherroub, J. T. C. Wojtyk, D. D. M. Wayner, and D. J. Lockwood, *J. Electochem. Soc.* **149** (2002) pp. H59.
89. C. Chatgilialoglu, *Acc. Chem. Res.* **25** (1992) pp. 188.
90. M. R. Linford, P. Fenter, P. M. Eisenberger, and C. E. D. Chidsey, *J. Am. Chem. Soc.* **117** (1995) pp. 3145.
91. H. C. Choi, and J. M. Buriak, *J. Chem. Soc., Chem. Commun.* (2000) pp. 1669.
92. A. R. Pike, L. H. Lie, R. A. Eagling, L. C. Ryder, S. N. Patole, B. A. Connolly, B. R. Horrocks, and A. Houlton, *Angew. Chem. Int. Ed.* **41** (2002) pp. 615.
93. M. P. Stewart, and J. M. Buriak, *Angew. Chem. Int. Ed.* **37** (1998) pp. 3257; M. P. Stewart, and J. M. Buriak, *J. Am. Chem. Soc.* **123** (2001) pp. 7821.
94. R. L. Cicero, M. R. Linford, and C. E. D. Chidsey, *Langmuir* **16** (2000) pp. 5688.
95. J. H. Song and M. J. Sailor, *J. Am. Chem. Soc.* **119** (1997) pp. 7381.

MOLECULAR NETWORKS AS NOVEL MATERIALS

WENBIN LIN*, HELEN L. NGO
Department of Chemistry, CB#3290, University of North Carolina, Chapel Hill NC 27599, USA

A brief review of the latest developments in the field of solid-state supramolecular chemistry is presented, with a particular focus on rational design of functional materials based on molecular networks. Since molecular networks are typically synthesized from discrete molecular building blocks under mild conditions, the structural integrity and functions of the building units can be retained which allow for their use as modules in the assembly of extended networks. By assembling such tunable organic and metal ion or cluster building blocks (modules) into a network using highly directional intermolecular interactions (such as hydrogen bonding and metal-ligand coordination), molecular networks with predictable topologies and prescribed functions can be readily designed. This chapter will illustrate the power of this bottom-up synthetic approach by presenting a few case studies in which highly porous solids capable of selective sorption and gas storage, heterogeneous catalytic systems, second-order nonlinear optical materials, single-crystalline nanocomposites, and novel magnetic materials have been successfully synthesized based on molecular networks.

1 Introduction

Crystal engineering, an area of research devoted to predicting and controlling the packing of molecular building units in the solid-state, has attracted much attention over the past few decades owing to its potential exploitation for the synthesis of technologically important materials. The major objective of crystal engineering is concerned with the ability to predictably synthesize solid-state supramolecular structures with prescribed functions from well-designed constituent building blocks [1,2]. Although it is still not possible to precisely predict the structure of crystalline molecular solids from a knowledge of their chemical composition [3], there has been a surge of research efforts in the crystal engineering of functional materials based on polymeric molecular networks. In such a molecular network approach, a single strong interaction such as hydrogen bonding or metal-ligand coordination is utilized to assemble supramolecular networks with desired topologies from their constituent building blocks. Because a single strong interaction dominates in such systems, other weak intermolecular interactions play a less important role in the crystal packing. By designing appropriate molecular subunits (modules), molecular networks with predictable topologies can be constructed based on dominant hydrogen bonding or metal-ligand coordination interactions. The modular nature of such an approach also allows precise engineering of a multitude of properties via systematic tuning of the molecular modules.

Since the properties of molecular networks are not only dependent upon the nature of the molecular modules but are also a direct result of the arrangement of these constituents within the crystal lattice, these solids typically are not amenable to traditional characterization techniques developed for discrete molecules. Technological advances in both experimental as well as computational methodologies in single crystal X-ray diffraction have accelerated the development of crystal engineering of molecular networks into an emerging field of interdisciplinary science. For example, CCD-based diffractometers allow the data collection to be done in hours or even minutes rather than days or weeks required for diffractometers equipped with conventional point detectors [4]. Computational advances have not only made structural solutions much less time consuming and challenging but also allowed easy visualization and presentations of complicated molecular networks. Facile retrieval of structural information from databases has also had a significant impact in the rapid development of crystal engineering of molecular networks.

This chapter will provide a brief review of the latest developments in the rational design of functional materials based on molecular networks. Since many excellent reviews have appeared in recent years in structural aspects of molecular networks [5-10], a major focus of this chapter will be placed upon the materials aspects of molecular networks. In this chapter, we will attempt to illustrate the power of this bottom-up synthetic approach by presenting a few case studies in which highly porous solids for selective sorption and gas storage, heterogeneous catalytic systems, second-order nonlinear optical materials, single-crystalline nanocomposites, and novel magnetic materials have been successfully synthesized based on molecular networks.

2 Highly Porous Metal-Organic Coordination Networks

The synthesis of metal-organic coordination networks (MOCNs) with porous structures has witnessed revolutionary growth over the past decade, and many highly porous solids based on metal-ligand coordination linkages are now available. Such MOCNs may one day extend the utility of inorganic zeolites as selective adsorbents for separations and purifications of important commodity chemicals. Traditional zeolites are microporous inorganic solids with aluminosilicate tetrahedral building units, which possess accessible internal cavities that are capable of binding guest molecules and catalyzing chemical reactions. Zeolites are used for the production of many consumer products ranging from gasoline to detergents and contribute to almost a trillion dollars of global economy annually [11]. For example, size-selective heterogeneous catalytic transformation of petrochemicals has been extensively used to improve the quality of gasoline, and thus significantly reduce the pollution of incomplete combustions of fossil fuels. However, the cavity sizes of

4 – 13 Å in traditional zeolites have precluded their applications in the transformation of molecules larger than xylenes.

Research efforts of many groups, particularly those of Yaghi et al., have led to numerous highly porous MOCNs over the past few years [11,12]. As pointed out by Yaghi et al, there exist three major challenges in the rational synthesis of highly porous MOCNs: (a) control of the orientation and stereochemistry of the building blocks in the solid state; (b) attainment of X-ray diffraction quality single-crystalline MOCNs; and (c) access to the pores within the open structures of MOCNs. These challenges have been met by combining well-designed (chosen) organic spacers and transition metal ions or metal clusters. Scheme I illustrates the construction of 2D porous structures via the addition of a triangular organic spacer to a trigonal planar metal complex (Y-shaped). Loss of the ligands or counterions (small triangles) that are associated with the metal ion allows for their incorporation into the pores. Intrinsic binding geometry of the metal ion and the functionality of the organic spacers will direct the assembly of porous MOCNs with tailorable pores of diverse sizes, shapes, and functions.

Scheme I.

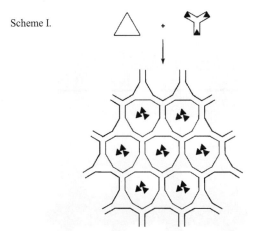

Numerous porous MOCNs based on a variety of organic spacers and transition metal ions have been synthesized according to the strategy illustrated in Scheme I and several excellent reviews on MOCNs have already appeared [5-10]. The most successful approach towards highly porous MOCNs was developed by Yaghi et al., which uses metal clusters as the secondary building units (SBUs) and thus overcomes the interpenetration problem often encountered when simple metal ions were used as the metal connecting points [7,12].

The tetrahedral μ_4-O bridged zinc acetate structure can be used as a SBU and extended into a 3D network with 1,4-benzenedicarboxylic acid (BDC) to give $Zn_4O(BDC)_3 \cdot (DMF)_8(C_6H_5Cl)$ (**MOF-5**), a simple cubic 5-connected net [13]. In this extended network of **MOF-5**, the basic zinc carboxylate SBUs with an octahedral extension occupy the nodes (vertices) of the cubic net, while finite rods of BDC ligands form the links (edges) of the net (Figure 1). A space-filling model clearly shows that the pores in **MOF-5** are a 3D channel system of 8 Å aperture and 12 Å cross section. These pores are filled with eight DMF and one chlorobenzene. The guests can be easily exchanged or/and evacuated from the pores without loss of the framework integrity.

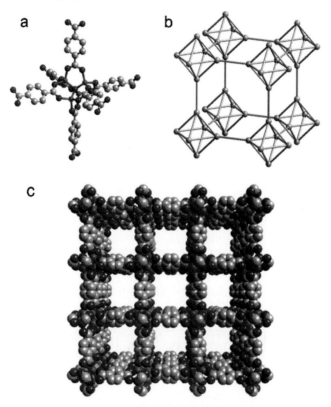

Figure 1. (a) Building unit present in crystals of $Zn_4O(BDC)_3 \cdot (DMF)_8(C_6H_5Cl)$ (**MOF-5**). (b) The primitive cubic lattice of **MOF-5**. (c) Highly porous 3D network of **MOF-5**. Coloring scheme: M, purple; O, red; C, grey (Reprinted from Ref 13 with permission from the American Association for the Advancement of Science).

Gas sorption studies indicated that **MOF-5** is a highly porous MOCN which exhibits a reversible type I isotherm (Figure 2). Such a reversible type I isotherm was also observed when $Ar_{(g)}$ and organic vapors such as CH_2Cl_2, $CHCl_3$, CCl_4, C_6H_6, and C_6H_{12} were used as the adsorbates. In comparison to zeolites which typically have a surface area of 500 m^2/g, **MOF-5** exhibits an apparent Langmuir surface area of ~2900 m^2/g.

1600
1200
800
400
0
Amount sorbed (mg/g)
0
0.2
0.4
0.6
0.8
1
P/P0
Ar
CCl4
CHCl3
N2
C6H6
C6H12

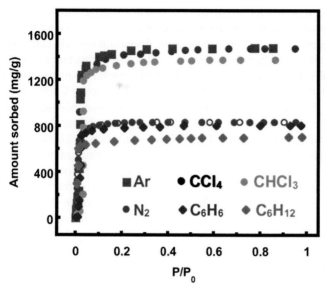

Figure 2. Gas sorption isotherms for **MOF-5**. An apparent Langmuir surface area of ~2900 m^2/g has been estimated for **MOF-5** (Reprinted from Ref 13 with permission from the American Association for the Advancement of Science).

2.1 Porous MOCNs for Selective Sorptions

Compared to microporous zeolites and recently developed mesoporous MCM-type materials, MOCNs are built with weaker metal-ligand coordination bonds and are inherently less stable. It is thus imperative to incorporate functions that are not possible with inorganic oxide materials into MOCNs. This section will attempt to illustrate potential applications of porous MOCNs in the areas of size-, shape-, and functional group-selective sorption of volatile organics, gas storage, and heterogeneous asymmetric catalysis.

Scheme II.

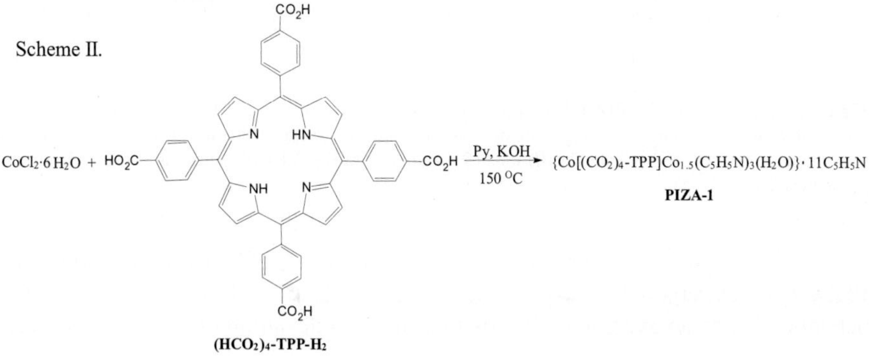

$CoCl_2 \cdot 6H_2O$ +
HO_2C
CO_2H
CO_2H
CO_2H
N
HN
NH
N
Py, KOH
150 °C
$\{Co[(CO_2)_4\text{-TPP}]Co_{1.5}(C_5H_5N)_3(H_2O)\} \cdot 11C_5H_5N$
PIZA-1
$(HCO_2)_4$-TPP-H_2

Suslick et al. reported the synthesis of a functional zeolite analog that is built from carboxylate-substituted metalloporphyrins [14]. The compound **PIZA-1** was prepared solvothermally by heating a mixture of freebase porphyrin-carboxylic acid and cobalt(II) chloride (Scheme II). Single crystal X-ray structure determinations showed that **PIZA-1** adopts a neutral 3D network structure composed of ruffled Co(III) porphyrin cores that are connected by bridging trinuclear Co(II)-carboxylate clusters (Figure 3). **PIZA-1** contains large, bi-directional oval-shaped 9×7 Å channels along the crystallographic b and c axes and another set of channels of 14×7 Å along the crystallographic a axis. These channels are filled with 44 molecules of disordered pyridine guest molecules (per unit cell). Thermal gravimetric analysis (TGA) indicated a significant (60%) weight loss attributable to loosely held solvate molecules and coordinated pyridine and water molecules. The framework stability of **PIZA-1** was confirmed by powder X-ray diffraction studies on the solvate-filled and evacuated framework solids. **PIZA-1** is thermally stable in vacuum to >250 °C for five days, and its decomposition begins during TGA at ~375 °C. Molecular modeling studies indicated that **PIZA-1** contains a large occupiable and accessible void volume of 47.9%.

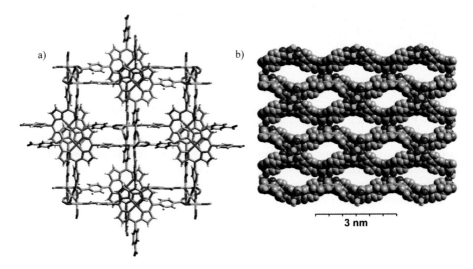

Figure 3. (a) A view of the **PIZA-1** network. (b) Space filling model of the **PIZA-1** network as viewed down the b or c axis. Grey coloring indicates carbon atom, red oxygen, blue nitrogen, green Co(II) ions in trinuclear clusters, and purple Co(III) ions (Reprinted from Ref 14 with permission from Macmillan Publishers Ltd).

The void volume of **PIZA-1** has been shown to be highly hydrophilic in nature. **PIZA-1** has extremely high capacity for repeated selective adsorption of water. The network adsorbs an average of 162 (with a standard deviation of 9) water molecules per unit cell (78 mL H$_2$O per 100 g solid host). In comparison, zeolites commonly used as desiccants do not have nearly as large a capacity: zeolite 4A and ALPO$_4$-18 molecular sieves (of pore size 3.8×3.8 Å) absorb 22 mL water per 100 g desiccant and 32.8 mL water per 100 g solid, respectively. Interestingly, **PIZA-1** also has a high affinity for amines and exhibits size- and shape-selective sorption behavior.

For the series of linear alkyl amines (C$_n$H$_{2n+1}$NH$_2$), chain length makes a substantial impact on the extent of sorption. Short-chain alkyl amines were adsorbed to the greatest extent, and as chain length increases a progressive decrease in sorption was observed. The size-selectivity was illustrated by a series of progressively larger aromatic amines with the sorption order of pyridine->aniline>3,5-dimethylaniline>4-t-butylpyridine >2,4,6-trimethylpyridine (Figure 4).

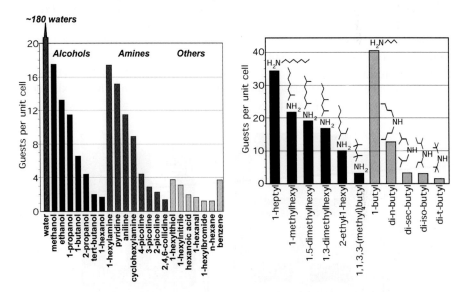

Figure 4. Size-, shape-, and functional group-selectivity of PIZA-1 as probed by thermal desorption of guest molecules. (a) Selectivity observed in a variety of chemical groups. (b) Comparison of linear, unsubstituted alkylamine and substituted alkylamine isomers (Reprinted from Ref 14 with permission from Macmillan Publishers Ltd).

Shape-selectivity was probed using the picoline series. The observed sorption has an inverse relationship to the steric hindrance of the picoline, i.e., the sorption order is 4-picoline>3-picoline>2-picoline. Shape-selectivity was also investigated by observing the sorption of the isomers of dibutyl-substituted amines: di-n-butylamine>di-iso-butylamine>di-sec-butylamine>di-$tert$-butylamines. It is apparent that as the bulky organic substituents encroach upon the hydrophilic group, guest sorption declines. This type of size- and shape-selectivity was also observed with alcohols as the adsorbates. Given the extreme insolubility of **PIZA-1** in all common solvents including dimethyl formamide, dimethyl sulphoxide and strong acids such as concentrated sulphuric and hydrochloric acids, **PIZA-1** possesses extraordinary properties as a desiccant for the drying of common organic solvents.

2.2 Porous MOCNs for Gas Storage

Kitagawa et al. first demonstrated potential applications of porous MOCNs in methane storage [15]. Methane is the major ingredient of natural gas and can be used as a fuel in the emerging fuel-cell technology. The design of microporous solids that have high and reversible methane uptake would greatly facilitate the development of methane-based fuel cells.

3D coordination networks $\{[Cu(AF_6)(4,4'\text{-bipy})_2]\cdot 8H_2O\}_n$, (A=Si, **1a**, and A=Ge, **1b**) were obtained by treating $Cu(BF_4)_2 \cdot 6H_2O$, $(NH_4)_2AF_6$ (A = Si or Ge), and 4,4'-bipy at room temperature [15, 16]. X-ray structural analyses show that **1a** and **1b** are isostructural and adopt a 3D network structure based on square grids of $[Cu(4,4'\text{-bipy})_2]_n$ and pillars of AF_6^{2-} ions (A = Si or Ge). The 3D networks of **1a** and **1b** contains open channels of dimensions of ~8×8 Å^2 along the c axis and of ~6×2 Å^2 along the a and b axes (Figure 5).

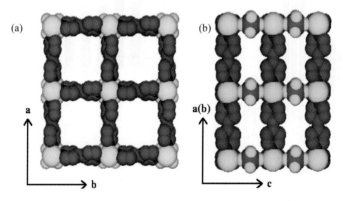

Figure 5. (a) Space-filling model of **1a** as viewed down the c axis. (b) Space-filling model of **1a** as viewed down the b (or a) axis (Reprinted from Ref 15 with permission from Wiley-VCH).

Argon gas adsorption measurements carried out at 87.3 K showed that both **1a** and **1b** exhibit a BET surface area of 1337 m^2/g and a micropore volume of 0.56 mL/g. They both have a pore size of ~8 Å, entirely consistent with the X-ray crystallographic results. Interestingly, methane adsorption measurements indicated that both **1a** and **1b** have higher methane uptake than zeolite 5A, which has the highest methane adsorption capacity in zeolites. At 25 °C and 36 atm, the density of methane adsorbed in **1a** and **1b** for micropore volume is 0.21 g/mL. In comparison, the density of the compressed methane gas at 27 °C and 280 atm is 0.16 g/mL. This

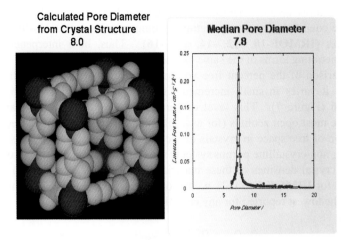

Figure 6. (left) The 3D network of **1a**. (right) Horvath-Kawazoe differential pore volume plot for **1a** (Reprinted from Ref 15 with permission from Wiley-VCH).

result strongly indicates a concentration effect by strong micropore filling in the cavities of **1a** and **1b**.

Yaghi *et al.* has extended their earlier work on **MOF-5** to synthesize a series of isoreticular MOCNs with systematically varied pore size and functionality and demonstrated their potential application in methane storage [17]. By reacting an N,N-diethylformamide solution of $Zn(NO_3)_2 \cdot 4H_2O$ and the dicarboxylic acids shown in Scheme III in a closed vessel at elevated temperatures, an isoreticular series of 16 crystalline MOCNs with the formula of $Zn_4O(link)_3(DEF)_x$ was obtained. These solids adopt the same framework topology as the prototypical **MOF-5** and are termed **IRMOFs**. All the **IRMOFs** have the expected topology of CaB_6 in which an oxide-centered Zn_4O tetrahedron is edge-bridged by six carboxylates to afford the octahedron-shaped SBU. These octahedron-shaped SBUs are further linked by the finite aromatic rods of the bridging dicarboxylic acids to form 3D cubic porous networks. In **IRMOF-2** through **IRMOF-7**, BDC links with bromo, amino, *n*-propoxy, *n*-pentoxy, cyclobutyl, and fused benzene functional groups have been incorporated into the frameworks. The above functional groups are pointing into the voids and thus drastically influence the properties of the resulting porous networks. Pore expansion can also be achieved by simply adjusting the length of the bridging dicarboxylic acids as illustrated with **IRMOF-8** through **IRMOF-16**.

Previous work by the same authors indicated that expansion of links can result in interpenetrating frameworks. Indeed, with the exception of the noninterpenetrating structure involving 2,6-NDC (**IRMOF-8**), BPDC, HPDC, PDC, and TPDC (**IRMOF-9, -11,-13, and -15**, respectively) are each reticulated as doubly interpenetrating structures. However, by carrying out the original reactions under more dilute conditions, noninterpenetrating counterparts were successfully achieved for all links (**IRMOF-10, -12, -14,** and **-16**). Thus, both interpenetrating and noninterpenetrating forms of the same extended structure were obtained.

Comparison of the percent free volume in crystals of **IRMOF-1** through **–16** shows that it varies in small increments from 55.8% in **IRMOF-5** to 91.1% in **IRMOF-16** (Figure 7). This level of percent free volume exceeds that found in some of the most open zeolites (for example, faujasite has 45 to 50% free volume). The fraction of free space in crystals of the expanded **IRMOF** series has only been achieved in non-crystalline porous systems such as SiO_2 xerogels and aerogels. The calculated crystal densities of these materials (in the absence of guests) vary from 1.00 g/cm^3 for **IRMOF-5** to 0.21 g/cm^3 for **IRMOF-16** (Figure 7). This value of density is the lowest reported for any crystalline material known to date.

Scheme III

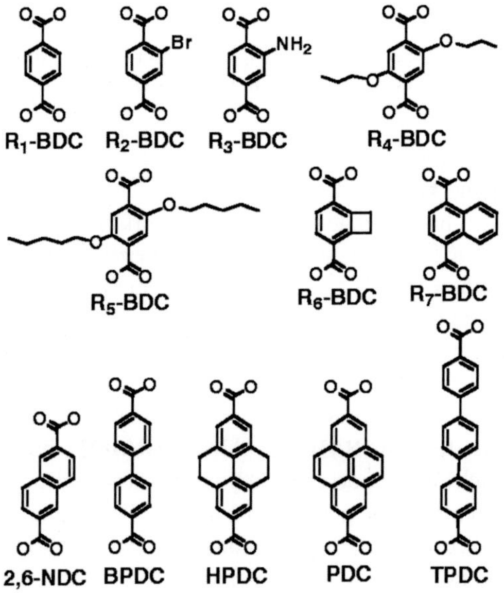

R_1-BDC R_2-BDC R_3-BDC R_4-BDC

R_5-BDC R_6-BDC R_7-BDC

2,6-NDC BPDC HPDC PDC TPDC

The gas sorption isotherms indicated that **IRMOF-6** maintains its porosity in the absence of guests. N_2 sorption at 78 K revealed a reversible type I isotherm behavior characteristic of a microporous material. The plateau was reached at relatively low pressure with no additional uptake at higher pressures. From the adsorption isotherms, Langmuir surface area and pore volume were estimated to be 2630 m^2/g and 0.60 cm^3/cm^3, respectively. Organic vapors (CH_2Cl_2, C_6H_6, CCl_4, and C_6H_{12}) were also used as the sorbates and similar type I reversible isotherms were obtained. The pore volumes obtained from organic vapor sorptions range from 0.57 to 0.60 cm^3/cm^3. These results confirm the homogeneity of the pores.

The exceptionally high surface area and pore volumes and appropriately designed aperture for **IRMOF-6** made it an ideal candidate for methane storage. Methane sorption studies indicated that **IRMOF-6** has an uptake of 155 cm^3 [(STP)/cm^3] at 298 K and 36 atm. This level of methane uptake exceeds that of other crystalline materials including zeolite 5A (87 cm^3/cm^3).

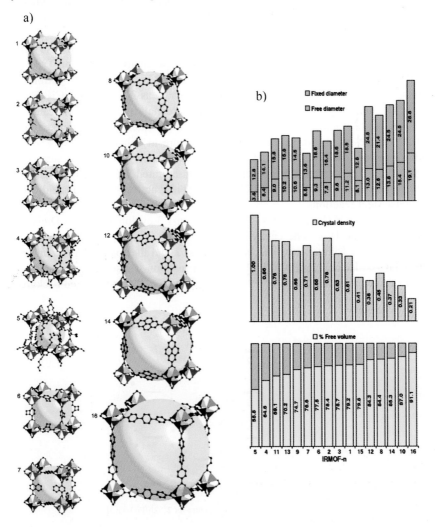

Figure 7. (a) Single crystal structures of IRMOF-n (n = 1 through 7, 8, 10, 12, 14, and 16). Coloring scheme: Zn (blue polyhedra), O (red), C (grey). (b) Bottom to top, the calculated percent free volume (yellow), crystal densities (light brown), and free diameter (green) and fixed diameter (blue), respectively, for **IRMOF-1** through **–16** (Reprinted from Ref 17 with permission from the American Association for the Advancement of Science).

2.3 Porous MOCNs for Heterogeneous Catalysis

Porous MOCNs have also been used as heterogeneous catalysts. Since MOCNs are typically synthesized under mild conditions, porous MOCNs with interesting catalytic sites can be rationally designed by incorporating bridging ligands with desired size, shape, chirality, and electronic properties. Such hybrid solids will combine both advantage of homogeneous and heterogeneous catalysts, and represent an interesting class of recyclable "artificial enzymes".

By linking anthracene-bis(resorcinol), **2**, with Ti connecting points, Aoyama *et al*. designed interesting microporous solid Lewis acid catalysts based on Ti(IV) and Al(III) centers [18]. Treatment of hydrogen-bonded **2** with $(^iPrO)_2TiCl_2$ afforded an insoluble orange-colored amorphous solid formulated as $2^{4-}\cdot 2[(^iPrO)TiCl]$ (Scheme IV). IR and ^{13}C CPMAS spectra and microanalysis results indicated that deprotonated tetraanionic species of the host (2^{4-}) are extensively networked via the O-Ti-O bridges as shown in the idealized structure in Scheme IV. N_2 adsorption studies showed that $2^{4-}\cdot 2[(^iPrO)TiCl]$ possesses a modest microporous surface area of 80 m^2/g. The presence of coordinatively unsaturated Ti(IV) centers within the microporous network has rendered $2^{4-}\cdot 2[(^iPrO)TiCl]$ active as a Lewis acid catalyst for Diels-Alder reactions.

$2^{4-}\cdot 2[(^iPrO)TiCl]$ catalyzes the acrolein-1,3-cyclohexadiene Diels-Alder reaction (Scheme IV). The catalytic activity of $2^{4-}\cdot 2[(^iPrO)TiCl]$ is much higher than those of its components, i.e., both apohost **2** and the soluble Ti^{4+} counterpart, $(^iPrO)_2TiCl_2$. The $2^{4-}\cdot 2[(^iPrO)TiCl]$-catalyzed reaction is also more stereoselective with an endo/exo product ratio of >99/1 (in comparison, the uncatalyzed background

Scheme IV

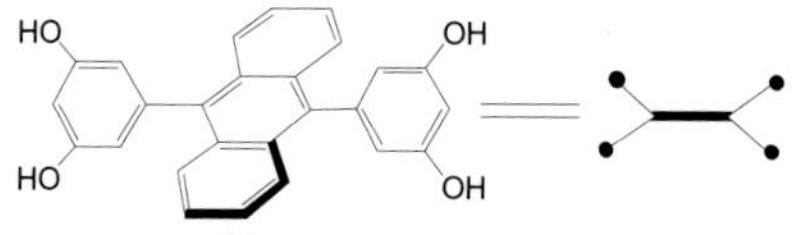

(2)

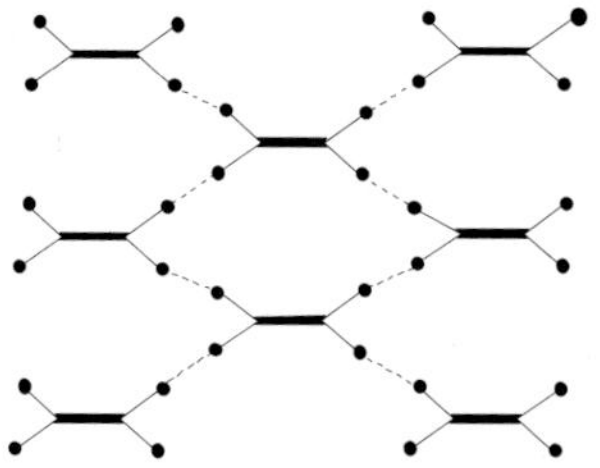

Proposed network structure for $2^{4-}\cdot 2[(^iPrO)TiCl]$

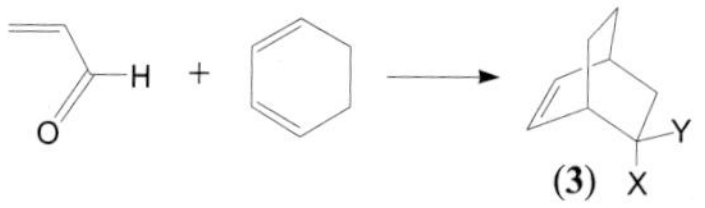

3_{endo}: X = COH, Y = H
3_{exo}: X = H, Y = COH

(3)

reaction has an endo/exo product ratio of 90/10). The solid Ti(IV) catalyst was also recovered and re-used without showing a significant deactivation. $2^{4-}\cdot2[(^iPrO)TiCl]$ was also used to catalyze Diels-Alder reactions with other dienophiles. An aluminum(III) analog with the formulation of $2^{4-}\cdot2(AlCH_3)$ was similarly prepared and exhibited a BET surface area of 240 m^2/g. $2^{4-}\cdot2(AlCH_3)$ also catalyzes the Diels-Alder reactions in a similar manner.

Heterogeneous catalysis with porous MOCNs demonstrated by Aoyama et al. is interesting, but unlikely to find practical applications. The ability to incorporate chirality into porous MOCNs will render them useful in heterogeneous asymmetric catalysis. Kim et al. demonstrated the first enantioselective catalysis with a single-crystalline, homochiral MOCN [19]. The enantiopure chiral organic building block **4** was synthesized from readily available D-tartaric acid (Scheme V), and reacted with Zn^{2+} ions to produce a homochiral open-framework solid with the formula of $[Zn_3(\mu_3\text{-}O)(4\text{-}H)_6]\cdot2H_3O\cdot12H_2O$ (referred to as D-**POST-1**). In D-**POST-1**, three zinc ions are held together with six carboxylate groups of the deprotonated chiral ligands **4** and a bridging oxo oxygen, to form a trinuclear unit, in which a threefold axis parallel to the *c* axis passes through the center of the trinuclear unit (Figure 8). The trinuclear units in **POST-1** are interconnected through three pyridyl groups of **4** to generate 2D infinite layers consisting of large edge-sharing chair-shaped hexagons with a trinuclear unit at each corner. The 2D layers stack along the *c* axis with a mean interlayer separation of 15.47 Å. Interestingly, large chiral 1D channels exist along the *c* axis, which can be best described as an equilateral triangle with a side length of ~13.4 Å. The void volume of the channels of **POST-1** is estimated to be ~47% of the total volume and is filled with 47 water molecules per unit cell. **POST-1** loses crystallinity upon removal of the solvate molecules by evacuation, but the X-ray powder diffraction pattern of **POST-1** can be regenerated upon exposing the evacuated sample to ethanol or water vapor.

Scheme V

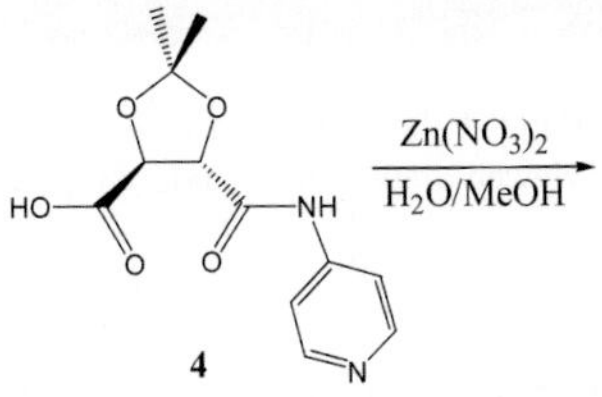

$$\xrightarrow{Zn(NO_3)_2}{H_2O/MeOH} [Zn_3(\mu_3\text{-}O)(4\text{-}H)_6]\cdot2H_3O\cdot12H_2O$$

D-POST-1

4

Three of six pyridyl groups in each trinuclear unit ($[Zn_3(\mu_3\text{-}O)(1\text{-}H)_6]^{2-}$) are coordinated to the zinc ions of three neighboring trinuclear units, the other three extrude into the channel without any interactions with the framework. Charge neutrality argument would suggest that two of the three dangling pyridyl groups have been protonated, which has been supported by elemental analysis. The protons bound to the pyridyl groups can be exchanged with alkali metal ions, which confirms the protonated nature of the two pyridyl groups. Interestingly, the pore size of **POST-1** can be varied by *N*-alkylation of the pyridyl groups. TGA results

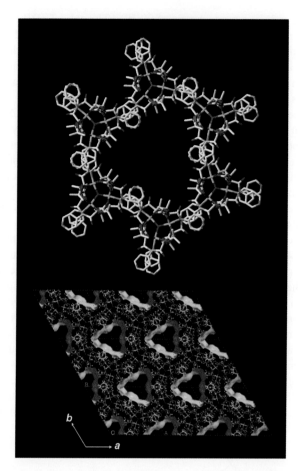

Figure 8. (top) The hexagonal framework of **POST-1** that is formed with the trinuclear SBU. (bottom) The chiral trigonal channels of **POST-1** as viewed down the c axis (Reprinted from Ref 19 with permission from Macmillan Publishers Ltd).

suggest that the pore volume of **POST-1** shrinks by 14% and 60% upon N-alkylation with iodomethane and 1-iodohexane, respectively.

The dangling pyridyl groups in **POST-1** were used to catalyze transesterification reactions. While the reaction of **5** and ethanol in the presence of **POST-1** in carbon tetrachloride for 55 h at 27 °C produced ethyl acetate in 77% yield, no or little transesterification occurs without **POST-1** or with the N-methylated **POST-1**, respectively. Transesterification of **5** with bulkier alcohols such as isobutanol, neopentanol and 3,3,3-triphenyl-1-propanol occurs with a much slower rate under otherwise identical reaction conditions. Such size selectivity suggests that the

Scheme VI

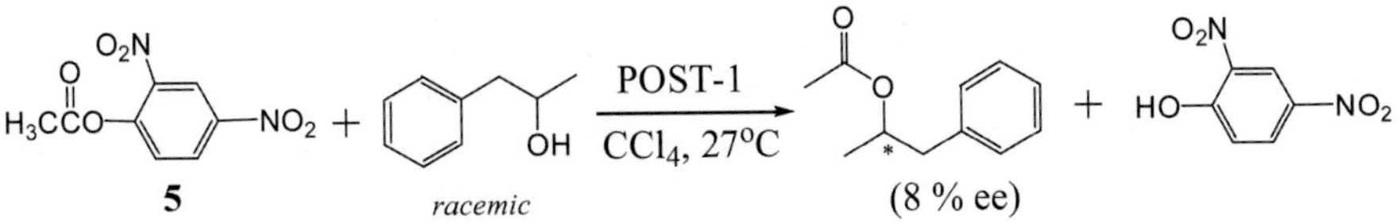

catalysis mainly occurs in the channels. **POST-1** was also used to catalyze kinetic resolution of *rac*-1-phenyl-2-propanol via transesterification of **5**. The reaction of **5** with a large excess of *rac*-1-phenyl-2-propanol in the presence of D-**POST-1** produces the corresponding esters with ~8% enantiomeric excess in favor of *S* enantiomer (Scheme VI). Although modest, this is the first observation of asymmetric induction by modular porous materials.

Lin et al. recently synthesized a family of homochiral porous lanthanide phosphonates and examined their applications in heterogeneous Lewis acid catalysis. Lanthanide bisphosphonates with the formula of [Ln(**L**-H$_2$)(**L**-H$_3$)(H$_2$O)$_4$]·x H$_2$O (Ln = La, Ce, Pr, Nd, Sm, Gd, Tb, x = 9 - 14, **6a-g**) were synthesized by slow evaporation of an acidic mixture of Ln(III) salts and 2,2'-diethoxy-1,1'-binaphthalene-6,6'-bisphosphonic acid (**L**-H$_4$) in methanol at room temperature (Scheme VII).

Scheme VII

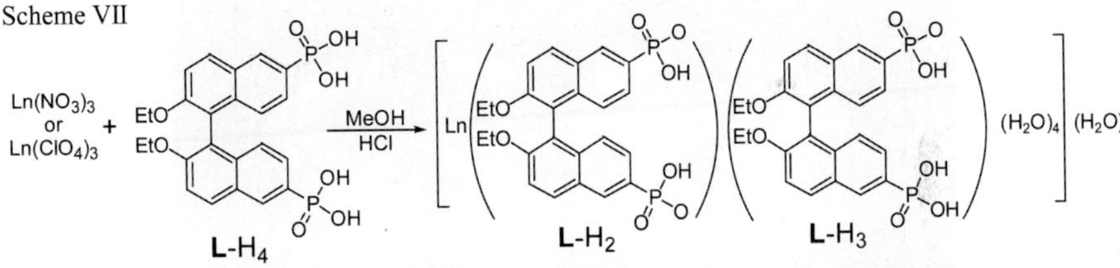

Ln = La, Ce, Pr, Nd, Sm, Gd, Tb; x = 9 - 14; **6a-g**

A single crystal X-ray diffraction study performed on [Gd(*R*-**L**-H$_2$)(*R*-**L**-H$_3$)(H$_2$O)$_4$]·12H$_2$O (*R*-**6f**) reveals a 2D lamellar structure consisting of 8-coordinate Gd centers and bridging binaphthylbisphosphonate groups. The Gd center adopts a square anti-prismatic geometry by coordinating to four water molecules and four phosphonate oxygen atoms of four different binaphthylbisphosphonate ligands. The skewed configuration of the binaphthyl subunits allows the formation of an elongated 2D rhombohedral grid lying in the *ac* plane (Figure 9). Such 2D grids stack along the *b* axis via interdigitation of the binaphthyl rings from the adjacent layers. There is still void space formed between the lamellae after such interdigitation of the binaphthyl rings, and a space-filling model of **6f** viewed along the *a* axis clearly indicates the presence of large asymmetric channels with a largest dimension of ~12 Å.

The framework stability of **6a-g** was demonstrated using PXRD. While the XRPD patterns of desolvated **6a-g** have broadened significantly, exposure of the desolvated solids to water vapor regenerates the XRPD patterns identical to those of the pristine samples. Such behavior is consistent with a distortion of the long-range lamellar-type structure with preservation of local coordination environments of **6a-g** during desolvation. The presence of both Lewis and Brönsted acid sites in **6a-g** has rendered them capable of catalyzing several organic transformations including cyanosilylation of aldehydes and ring opening of *meso*-carboxylic anhydrides. No enantioselectivity, however, was observed for these reactions. The lack of enantioselectivity is not surprising given highly symmetrical nature of catalytically active Ln(III) centers.

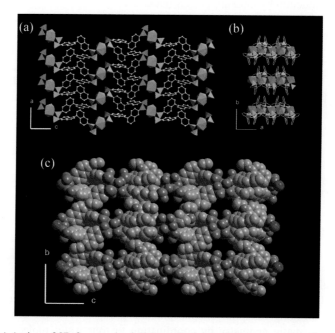

Figure 9. (a) A view of 2D framework of **6f** down the *b* axis. Ethoxy groups have been omitted for clarity. (b) A view of stacking of 2D framework of **6f** along the *c* axis showing interdigitation of binaphthyl rings from adjacent layers. The coordination environments of P atoms and Gd atoms are represented with blue and orange polyhedra, respectively. (c) A space-filling model of **6f** viewed down the *a* axis (Reprinted from Ref 20 with permission from American Chemical Society).

To enhance the enantioselectivity, Lin *et al.* developed metal phosphonates containing pendant chiral chelating bisphosphines. A chiral porous Zr phosphonate with approximate formula of $Zr[Ru(L_1)DMF)_2Cl_2]\cdot 2MeOH$ (Zr-Ru-L_1) was synthesized by refluxing $Zr(O^tBu)_4$ and 1 equiv of $Ru(L_1-H_4)(DMF)_2Cl_2$ in methanol [21]. $Zr[Ru(L_2)DMF)_2Cl_2]\cdot 2MeOH$ (Zr-Ru-L_2) was similarly prepared. N_2

Scheme VIII

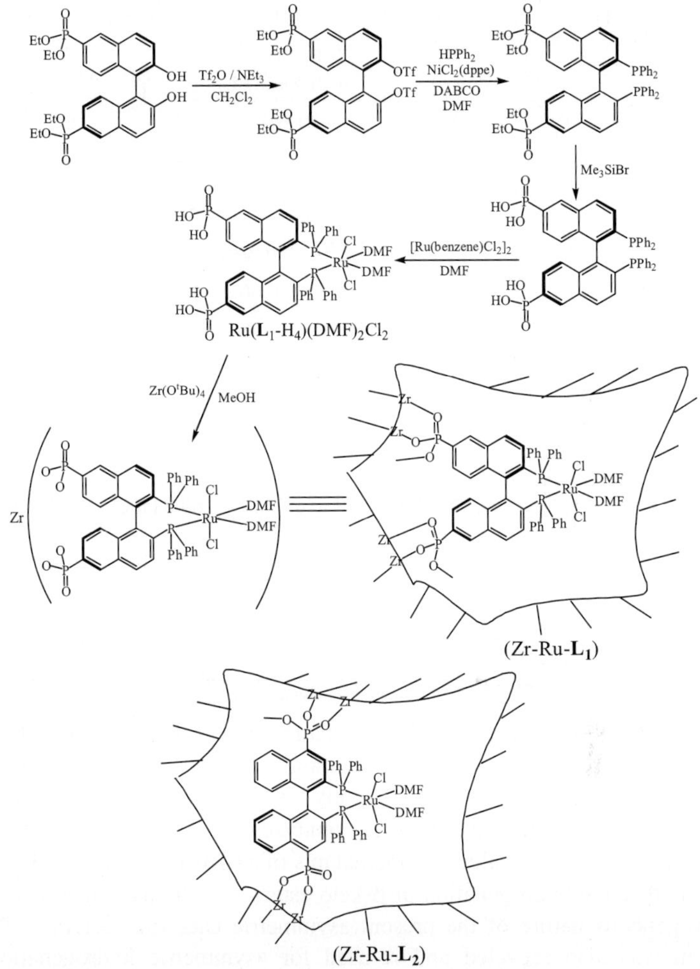

adsorption measurements indicate that both Zr-Ru-L_1 and Zr-Ru-L_2 are highly porous with rather wide pore size distributions. Zr-Ru-L_1 exhibits a total BET surface area of 475 m^2/g with a microporous surface area of 161 m^2/g and a pore volume of 1.02 cm^3/g. Zr-Ru-L_2 exhibits a total BET surface area of 387 m^2/g with a microporous surface area of 154 m^2/g and a pore volume of 0.53 cm^3/g. SEM images show that both solids are composed of sub-micrometer particles, while PXRD indicate that both solids are amorphous.

These chiral porous Zr phosphonates containing the BINAP-Ru moieties were used for heterogeneous asymmetric hydrogenation. As Table 1 shows, both Zr-Ru-L_1 and Zr-Ru-L_2 are highly active catalysts for asymmetric hydrogenation of β-keto

Table 1. Heterogeneous Asymmetric Hydrogenation of β-Keto Esters

$$R_1 \underset{O}{\overset{O}{\text{C}}} \underset{O}{\overset{O}{\text{C}}} O\text{-}R_2 + H_2 \xrightarrow[\text{CH}_3\text{OH}]{\text{Zr-Ru-}(R)\text{-}L_1 \text{ or Zr-Ru-}(R)\text{-}L_2} R_1 \underset{O}{\overset{OH}{\text{C}}} \underset{O}{\overset{O}{\text{C}}} O\text{-}R_2$$

Substrate	Catalyst Loading	Temp	H_2 Pressure	Zr-Ru-L_1 e.e. (yield)	Zr-Ru-L_2 e.e. (yield)
(methyl acetoacetate)	1%	60 °C	700 (psi)	94.0 (100)	
	1%	rt	1400	95.0 (100)	73.1 (90)
	0.1%	60 °C	700	93.3 (100)	
(ethyl β-keto propyl)	1%	rt	1400	92.0 (100)	65.0 (100)
(isopropyl β-keto ester)	1%	rt	1400	91.7 (100)	68.1 (85)
(phenyl β-keto ester)	1%	rt	1400	69.6 (100)	15.7 (50)
(t-butyl β-keto ester)	1%	rt	1400	93.1 (100)	64.0 (100)
(branched β-keto ester)	1%	rt	1400	93.3 (100)	78.8 (70)

esters. Zr-Ru-L_1 catalyzes the hydrogenation of a wide range of β-alkyl-substituted β-keto esters with complete conversions and e.e's ranging from 91.7 to 95.0%. This level of enantioselectivity is only slightly lower than that of their best homogeneous counterparts. In contrast, Zr-Ru-L_2 catalyzes the hydrogenation of β-keto esters with only modest e.e. values. Supernatants of Zr-Ru-L_1 and Zr-Ru-L_2 in MeOH did not catalyze the hydrogenation of β-keto esters, which unambiguously demonstrates heterogeneous nature of the present asymmetric catalytic systems. The Zr-Ru-L_1 system was also recycled and re-used for asymmetric hydrogenation of methyl acetoacetate without significant deterioration of enantioselectivity. The Zr-Ru-L_1 system was used for five cycles of hydrogenation with complete conversions and e.e. values of 93.5%, 94.2%, 94.0%, 92.4%, and 88.5%, respectively. Ready tunability of such a molecular building block approach will allow the optimization of the catalytic performance of these hybrid materials and lead to practically useful heterogeneous asymmetric catalysts.

3 MOCNs as Second-Order Optical Materials

Second-order NLO materials have the ability to double the frequency of incident photons and are key to the future photonics technology [22]. Bulk second-order NLO susceptibility (χ^2) is a third rank tensor and will vanish in a centrosymmetric environment [23]. Ideal NLO chromophores typically contain a good electron donor and acceptor connected through a conjugated hydrocarbon bridge. Efficient NLO chromophores are therefore electronically asymmetric and highly dipolar. Such highly dipolar molecules tend to adopt centrosymmetric arrangements due to the dominance of centric dipole-dipole repulsive interactions. Lin *et al.* has explored the rational design of polar solids based on MOCNs by

Scheme IX

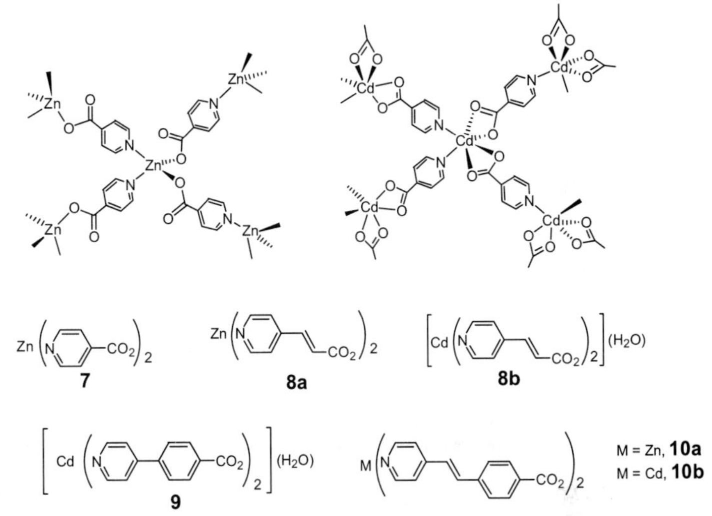

taking advantage of the strong and highly directional nature of metal-ligand coordination bonds. In this approach, metal-ligand coordination bonds were utilized to counteract unfavorable centric intermolecular interactions, thereby directing the assembly of noncentrosymmetric solids.

The most successful approach for the crystal engineering of polar solids is based on the diamondoid network (DN). DNs are not predisposed to pack in centrosymmetric space groups owing to the lack of inversion centers on the connecting points. An acentric DN will arise if unsymmetrical bridging ligands are used to connect tetrahedral metal centers. It is also well known that metal-organic DNs have a high propensity to interpenetrate in order to fill any void space. Despite the inherent acentricity of individual DNs, an even number-fold interpenetration could potentially lead to inversion centers relating pairs of mutually independent diamondoid nets. On the other hand, an odd number-fold interpenetrated DN synthesized from unsymmetrical bridging ligands will be necessarily acentric.

Metal centers (nodes) with tetrahedral extension were used to link unsymmetrical rigid linear *p*-pyridinecarboxylate ligands (spacers) to form a series of interpenetrated DNs (Scheme IX). The d^{10} metal ions Zn^{2+} and Cd^{2+} were used as connecting points in order to minimize optical losses from unwanted d→d transitions in the visible region. Unsymmetrical linking groups also introduce electronic asymmetry, while the rigidity of the bridging ligands is synergistic with good conjugation between the electron donors and acceptors.

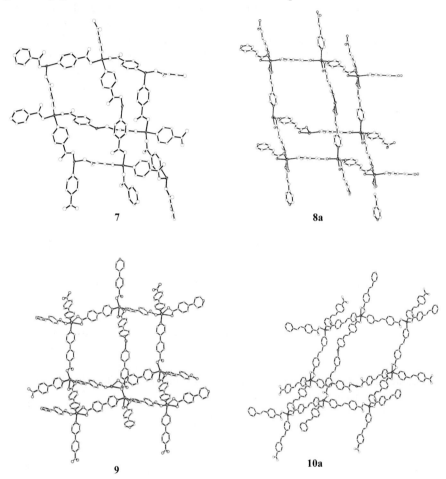

Figure 10. Diamondoid structures of **7**, **8a**, **9**, and **10a**.

Hydro(solvo)thermal reaction between appropriate metal ions and bridging ligands or their corresponding precursors afforded a series of interpenetrated DNs [24-27]. All these solids adopt DN structures (Figure 10), and they only differ in the degree of penetration which is defined as the number of independent DNs in each solid (Table 2). This work thus convincingly demonstrated that metal p-pyridinecarboxylate frameworks have a high propensity to form diamondoid solids. Moreover, the combination of unsymmetrical bridging ligands and metal centers with appropriate geometry has led to successful design of acentric solids. While an even number-fold interpenetration can (but may not) introduce unwanted inversion centers and lead to bulk centrosymmetry, an odd number-fold DN with unsymmetrical bridging ligands will necessarily crystallize in a polar space group. Furthermore, there is compelling evidence to suggest that the degree of interpenetration is entirely dependent upon the length of the bridging ligand (Figure 11). Therefore, with ligands of the appropriate length that induce the formation of odd number-fold interpenetrated DNs, polar solids can be readily crystal-engineered.

Table 2. NLO properties of Noncentrosymmetric Diamondoid Networks.

Compound	Degree of Interpenetration	Space Group	$I^{2\omega}/I^{2\omega}$(quartz)
7	3-fold	$P2_12_12_1$	1.5
8a	5-fold	Cc	126
8b	5-fold	Cc	18
9	7-fold	Ia	310
10a	8-fold	$C2$	400
10b	8-fold	$C2$	345
LiNbO$_3$			600

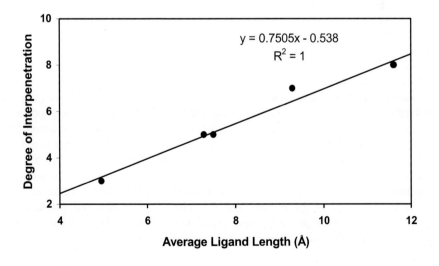

Figure 11. Dependence of degree of interpenetration of diamondoid networks on the length of bridging *p*-pyridinecarboxylate ligands (Reprinted from Ref 24 with permission from American Chemical Society).

The NLO properties of acentric diamondoid networks were examined using the Kurtz and Perry powder techniques (Table 2). Consistent with their acentric structures, all the compounds in Table 2 are SHG-active. In fact the second-order NLO properties of **9** and **10** approach that of technologically important lithium niobate ($I^{2\omega}$ of 600 versus α-quartz). These results are very encouraging considering the limited donor/acceptor ability of *p*-pyridinecarboxylates.

Nonlinear optically active coordination networks based on other structural motifs have also been designed. For example, noncentrosymmetric solids based on 2D grids have also been successfully synthesized by Lin *et al* [24, 28, 29]. However, the control of acentricity in these solids is significantly more difficult than in the 3D DN case. In the 2D grids, in addition to the choice of appropriate linking groups to avoid even number-fold interpenetration, efforts are also needed to ensure the lack of inversion centers between individual 2D grids.

4. Single-Crystalline Nanocomposites

The synthesis of nanocrystalline materials has undergone revolutionary growth over the past decade. Nanostructures of different morphologies, aspect ratios, and compositions are now available. Assembling such nanostructures into periodically ordered meso- or micro-structures is however significantly more challenging. Li *et al.* recently uncovered a novel approach towards covalently bonded hybrid composites of metal chalcogenides that not only possess a uniform and periodic structure, but also offer a significant variation of optical properties [30].

Two polymorphic compounds [α-ZnTe(en)$_{1/2}$], **11**, and [β-ZnTe(en)$_{1/2}$], **12**, were synthesized hydrothermally by treating Zn^{2+} and Te in ethylenediamine (en). X-ray structure determinations revealed that **11** and **12** are chalcogenide-based hybrid materials of which uniform structures are formed via direct, covalent bonds between the inorganic host (the II-VI semiconductor ZnTe) and the organic spacers. **11** adopts a 3D network structure containing 2D [ZnTe] slabs and bridging en molecules, as illustrated in Figure 12a. The [ZnTe] slabs stack along the *c*-axis and are interconnected by en molecules, each bridged to two Zn metal centers from the adjacent slabs. As shown in Figure 12b, the inorganic slab is a puckered 6^3 (honeycomb) net formed by alternating, three-coordinated Zn and Te. Such a 2D [ZnTe] slab may also be regarded as a "slice" from the zinc blende or wurtzite structure. **12** is a polymorph of **11**, with a similar 2D slab of [ZnTe]. **12** differ from **11** in the topology of the 6^3 nets and the relative orientation and connectivity between the inorganic slabs and organic pillars in the two structures. A third related phase with the formula of [ZnTe(pda)$_{1/2}$], **13**, was similarly obtained.

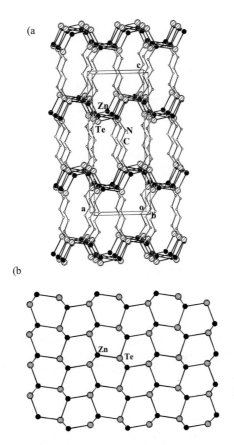

Figure 12. (a) A view of **11** down the *b*-axis. The large solid circles are Zn, shaded circles Te, and small open and singly shaded circles C and N, respectively. (b) The 2D [ZnTe] slab as viewed down the *c*-axis (Reprinted from Ref 30 with permission from American Chemical Society).

The optical absorption spectra measured by diffuse reflectance experiments gave absorption edges of 3.5, 3.3, and 3.4 eV for **11**, **12**, and **13**, respectively. These values have significantly blue-shifted from the absorption edge of 2.1 eV for bulk ZnTe (Figure 13). Calculations using density functional theory indicated that the blue shifts observed in **11-13** are a result of quantum confinement effect (QCE). It is interesting to note that such a large blue shift of optical absorption edge can be achieved only for very small, chemically grown, colloidal dots in which a uniform structure would be very difficult to achieve. In contrast, the single-crystalline uniform, periodically ordered nanocomposites such as **11-13** hybrid composites can be readily generated in a controllable manner.

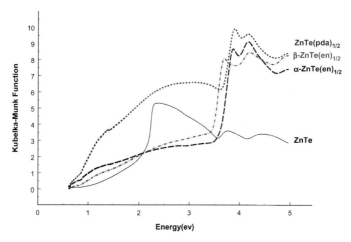

Figure 13. Optical absorption spectra for **11-13**, and bulk ZnTe (Reprinted from Ref 30 with permission from American Chemical Society).

Feng et al. more recently synthesized a covalent superlattice of chalcogenide nanoclusters based on a similar strategy [31]. A single-crystalline nanocomposite denoted as UCR-9 was obtained by heating a mixture of 4,4'-trimethylenedipyridine and a light yellow nanocluster intermediate with the composition $Cd_{10}S_4(SPh)_{12}$, which was prepared from $[N(CH_3)_4]_4[Cd_{10}S_4(SPh)_{16}]$, at 190°C for 3 days. X-ray crystal structure determinations showed that UCR-9 is a 3D superlattice formed by linking $[Cd_8(SPh)_{12}]^{4+}$ clusters with exo-tetradentate 1,2,4,5-tetra(4'-pyridyl)-benzene. The 1,2,4,5-tetra(4'-pyridyl)benzene molecule evidently results from oxidative coupling of two 4,4'-trimethylenedipyridine molecules under the hydrothermal condition.

UCR-9 can be considered as layers of $[Cd_8(SPh)_{12}]^{4+}$ clusters stacked along the c axis. Within each layer, $[Cd_8(SPh)_{12}]^{4+}$ clusters are joined into a square pattern encircling large square pores formed by four $[Cd_8(SPh)_{12}]^{4+}$ clusters. The linkage between adjacent layers is provided by benzene rings of 1,2,4,5-tetra(4'-pyridyl)benzene molecules (Figure 14).

The photoluminescent properties of UCR-9 reflect both inorganic and organic compositions and their synergetic effects. The emission peak at 580nm is likely due to the S^{2-} to Cd^{2+} charge transfer excitation similar to that observed for the bulk CdS, while the emission peaks at 415 and 440nm are probably due to the π-π* transitions in 1,2,4,5-tetra(4-pyridyl)benzene. The ability to combine the properties of both inorganic nanoclusters and organic bridging ligands in such a single-crystalline specimen presents an interesting opportunity to design new functional materials.

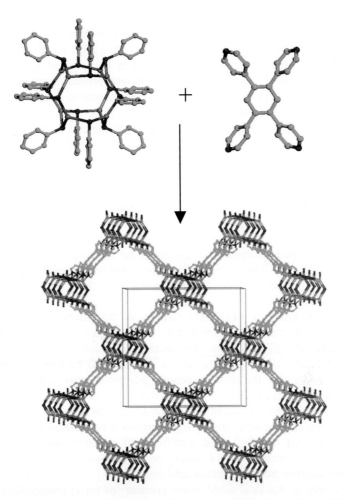

Figure 14. A schematic showing the formation of 3D network of UCR-9 from the the $[Cd_8(SPh)_{12}]^{4+}$ cubic unit and 1,2,4,5-tetra(4-pyridyl)benzene (Reprinted from Ref 31 with permission from American Chemical Society).

5. MOCNs with Interesting Magnetic Properties

There has been significant interest in the design of MOCNs which can not only be used as molecule-based magnets but also combine magnetism with other physical properties. Prussian blues and analogs are the most well studied system [32, 33]. The magnetic ordering of this family of materials can be rationalized based on the orbital orthogonality principle [34], which states that ferromagnetic coupling between cyano-bridged metal centers can be promoted if the participating d orbitals are orthogonal to each other. Adjacent metal centers with octahedral configurations should thus utilize t_{2g} and e_g orbitals, respectively. The most effective ferromagnetic coupling between the orthogonal orbitals can then be achieved when their energy difference if minimal. Guided with this simple orbital argument, polymeric cyanometallates of Prussian Blue structures with magnetic ordering temperatures higher than 100 °C have been designed [35].

Another family of coordination networks with interesting magnetic properties concerns the high spin—low spin equilibrium [36]. With properly designed nitrogen-based ligands, first-row transition metals, in particular Fe, can exhibit so called spin crossover phenomenon, i.e., the Fe(II) centers can adopt both high spin and low spin configurations depending on electromagnetic radiation and the ambient temperature and pressure. Such a cooperative transition between a low-spin and a high-spin state can be abrupt with a thermal hysteresis, and can thus be potentially utilized to construct molecular switches or memory devices.

A recent work by Kepert et al. combines the porosity with spin crossover in the same MOCN [37]. The nanoporous metal-organic framework $Fe_2(azpy)_4(NCS)_4\cdot$(ethanol), **14**·(ethanol), was synthesized by the slow diffusion of equimolar amounts of $Fe^{II}(NCS)_2$ and *trans*-4,4'-azopyridine (azpy) in ethanol. The crystal structure of **14**·(EtOH) consists of the double interpenetration of 2D rhombic grids that are constructed by the linkage of Fe(II) centers by azpy units (Figure 15). The compressed octahedral coordination around the two crystallographically distinct Fe(II) centers is completed by two axial thiocyanate ligands bound through the nitrogen donor. Interpenetration of the 2D grids also leads to 1-D channels parallel to the *c* axis that are occupied by ethanol molecules. Interestingly, the ethanol molecules in **14**·(EtOH) can be removed at 375 K with the retention of the 2D grid structure. A single crystal X-ray structure determination on **14** reveals several changes, most significant of which being an altered Fe(II) coordination environment in **14** owing to the removal of ethanol molecules hydrogen-bonded to the thiocyanato sulfur atoms.

This change in the local coordination geometry of Fe(II) centers has a profound effect on the magnetic property of **14**. Magnetic susceptibility and Mössbauer spectral measurements indicated that **14**·(EtOH) undergoes "half-spin" crossovers, i.e., only one of the two crystallographically distinct Fe(II) centers undergoes transition from high-spin state (S=2) to low-spin state (S=0) in the temperature range of 130-150 K (Figure 16). In contrast, the desolvated phase **14** does not exhibit spin crossover phenomenon. The host network of **14** thus displays reversible uptake and release of guest molecules and contains electronic switching centers that are sensitive to the nature of the sorbed guests. The generation of such a host lattice that interacts with exchangeable guest species in a switchable fashion has implications for the generation of previously undeveloped advanced materials with applications in areas such as molecular sensing.

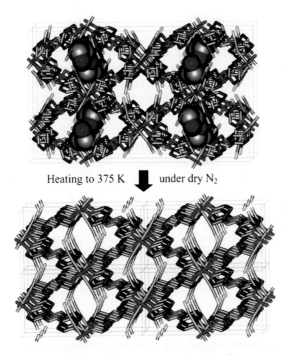

Figure 15. X-ray crystal structures of **14·(EtOH)** at 150 K and **14** at 375 K, viewed approximately down the 1-D channels (*c* axis) (Reprinted from Ref 37 with permission from the American Association for the Advancement of Science).

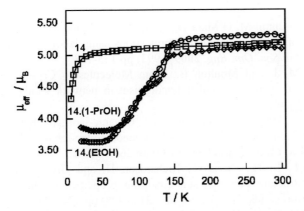

Figure 16. VT magnetic moment of **14·x(guest)**, recorded on a single sample at different stages of guest desorption and resorption, showing 50% spin crossover behavior between 50 and 150 K for the fully loaded phases and an absence of spin crossover for the fully desorbed phase **14** (Reprinted from Ref 37 with permission from the American Association for the Advancement of Science).

6. Conclusions and Outlook

This chapter highlights recent advances in the area of field of solid-state supramolecular chemistry. It is evident from the examples illustrated in this chapter that molecular networks of diverse topologies and functionalities can be designed via a modular approach. The facile tunability of such a molecular building blocks should allow precise engineering of a multitude of properties via systematic tuning of the molecular modules. Researchers in this area have just scratched the tip of an iceberg. Many interesting functional materials can be expected from this burgeoning field of solid-state supramolecular chemistry.

Acknowledgements

We thank Professors Ken Suslick, Kimoon Kim, Pingyun Feng, Jing Li, Omar Yaghi, and Susumu Kitagawa for providing figures. Funding support from NSF (DMR-9875544) makes the writing of this chapter possible.

References

1. Desiraju G. R. Crystal engineering: the design of organic solids. (Elsevier, New York, 1989).
2. Lehn J.-M. Supramolecular chemistry: concepts and perspectives, (VCH Publishers, New York, 1995).
3. Gavezzotti A., Are crystal structures predictable? *Acc. Chem. Res.* **27** (1994) pp.309-315.
4. Phillips W., Stanton M, O'Mara D., Li Y., Naday I. and Westbrook E., CCD-based detector for crystallographic applications using laboratory x-ray sources. *Proc. SPIE-Int. Soc. Opt. Eng.* **2009** (1993) pp.133-138.
5. Zaworotko M. J. and Moulton B., From Molecules to Crystal engineering: supramolecular isomerism and polymorphism in network solids. *Chem. Rev.* **101** (2001) pp.1629-1658.
6. Hagrman P. J., Hagrman D. and Zubieta J., Organic-inorganic hybrid materials: from "simple" coordination polymers to organodiamine-templated molybdenum oxides. *Angew. Chem. Int. Ed.* **38** (1999) pp.2638-2684.
7. Yaghi O. M., Li H., Davis C., Richardson D. and Groy T. L., Synthetic Strategies, structure patterns, and emerging properties in the chemistry of modular porous solids. *Acc. Chem. Res.* **31** (1998) pp.474-484.
8. Munakata M., Wu L. P. and Kuroda-Sowa T., Toward the construction of functional solid-state supramolecular metal complexes containing copper(I) and silver(I). *Adv. Inorg. Chem.* **46** (1999) pp.173-304.

9. Janiak C., Functional organic analogs of zeolites based on metal-organic coordination frameworks. *Angew. Chem., Int. Ed. Engl.* **36** (1997) pp.1431-1434.
10. Batten S. R. and Robson R., Interpenetrating nets: ordered, periodic entanglement. *Angew. Chem., Int. Ed. Eng.* **37** (1998) pp.1460-1494.
11. Breck D. W. Zeolite molecular sieves, structure, chemistry, and use. (John Wiley & Sons, New York, 1974).
12. Eddaoudi M., Moler D. B., Li H., Chen B., Reineke T. M., O'Keeffe M. and Yaghi O. M., Modular chemistry: secondary building units as a basis for the design of highly porous and robust metal-organic carboxylate frameworks. *Acc. Chem. Res.* **34** (2001) pp.319-330.
13. Li H., Eddaoudi M., O'Keeffe M. and Yaghi O. M., Design and synthesis of an exceptionally stable and highly porous metal-organic framework. *Nature* **402** (1999) pp.276-279.
14. Kosal M. E., Chou J-H., Wilson S. R. and Suslick K. S., A functional zeolite analogue assembled from metalloporphyrins. *Nature* **1** (2002) pp.118-121.
15. Noro S-i., Kitagawa S., Kondo M. and Seki K., A new, methane adsorbent, porous coordination polymer [{$CuSiF_6$(4,4'-bipyridine)$_2$}$_n$]. *Angew. Chem. Int. Ed.* **39** (2000) pp.2082-2084.
16. Noro S-i., Kitaura R., Kondo M., Kitagawa S., Ishii T., Matsuzaka H. and Yamashita M., Framework engineering by anions and porous functionalities of Cu(II)/4,4'-bpy coordination polymers. *J. Am. Chem. Soc.* **124** (2002) pp.2568-2583.
17. Mohamed E, Jaheon K, Nathaniel R, Vodak D, Wachter J, O'Keeffe M and Yaghi O. M., Systematic design of pore size and functionality in isoreticular MOFs and their application in methane storage. *Science* **295** (2002) pp.469-72.
18. Sawaki T., Dewa T. and Aoyama Y., Immobilization of soluble metal complexes with a hydrogen-bonded organic network as a supporter. A simple route to microporous solid Lewis acid catalysts. *J. Am. Chem. Soc.* **120** (1998) pp.8539-8540.
19. Seo J. S., Whang D., Lee H., Jun S I., Oh J., Jeon Y. J. and Kim K., A homochiral metal-organic porous material for enantioselective separation and catalysis. *Nature* **404** (2000) pp.982-985.
20. Evans O. R., Ngo H. L. and Lin W., Chiral porous solids based on lamellar lanthanide phosphonates, *J. Am. Chem. Soc.* **123** (2001) pp.10395-10396.
21. Hu A., Ngo H. L. and Lin W., Chiral porous hybrid solids for highly enantioselective heterogeneous asymmetric hydrogenation of β-keto esters, *Angew. Chem. Int. Ed.*, in press.
22. Bossi D. E. and Ade, R.W., Integrated-optic modulators benefit high-speed fiber links. *Laser Focus World*, **28** (1992) pp.135-142.
23. Zyss J., Molecular nonlinear optics: materials, physics, and devices. (Academic Press, New York, 1993).

24. Evans O. R. and Lin W., Crystal engineering of NLO materials based on metal-organic coordination networks *Acc. Chem. Res.* **35** (2002) pp.511-522.
25. Evans O. R., Xiong, R.-G., Wang, Z., Wong G. K. and Lin W., Crystal engineering of acentric diamondoid metal-organic coordination networks. *Angew. Chem. Int. Ed.* **38** (1999) pp.536-538.
26. Evans O. R. and Lin W. Crystal engineering of nonlinear optical materials based on interpenetrated diamondoid coordination networks. *Chem. Mater.* **13** (2001) pp.2705-2712.
27. Lin W., Ma L. and Evans O. R., NLO-active Zn(II) and cadmium(II) coordination networks with 8-fold diamondoid structures. *Chem. Commun.* (2000) pp.2263-2264.
28. Lin W., Evans O. R., Xiong R-G. and Wang Z., Supramolecular engineering of chiral and acentric 2D networks. Synthesis, structures, and second-order nonlinear optical properties of bis(nicotinato)zinc and bis{3-[2-(4-pyridyl)ethenyl]benzoato}cadmium. *J. Am. Chem. Soc.* **120** (1998) pp.13272-13273.
29. Evans O. R. and Lin W., Rational design of nonlinear optical materials based on 2D coordination networks. *Chem. Mater.* **13** (2001) pp.3009-3017.
30. Huang X. and Li J., The first covalent organic-inorganic networks of hybrid chalcogenides: structures that may lead to a new type of quantum wells. *J. Am. Chem. Soc.* **122** (2000) pp.8789-8790.
31. Zheng N., Bu N. and Feng P., Self-assembly of novel dye molecules and $[Cd_8(SPh)_{12}]^{4+}$ cubic clusters into three-dimensional photoluminescent superlattice. *J. Am. Chem. Soc.* **124** (2002) pp.9688-9689.
32. Entley W. R. and Girolami G. S., New high-temperature molecular magnets based on cyanovanadate building blocks: spontaneous magnetization at 230 K. *Science* **268** (1995) pp.397-400.
33. Mallah T., Thiebaut S., Verdaguer M. and Veillet P., High-T_c molecular magnets: ferrimagnetic mixed-valance chromium(III)-chromium(II) cyanides with T_c at 240 and 190 Kelvin. *Science* **262** (1993) pp.1554-1557.
34. Kollmar C. and Kahn O., Ferromagnetic spin alignment in molecular systems: an orbital approach. *Acc. Chem. Res.* **26** (1993) pp.259-265.
35. Holmes S. M. and Girolami, G. S., Sol-gel synthesis of $KV^{II}[Cr^{III}(CN)_6]\cdot 2H_2O$: a crystalline molecule-based magnet with a magnetic ordering temperature above 100 °C. *J. Am. Chem. Soc.* **121** (1999) pp.5593-5594.
36. Kahn O. and Martinez C. J., Spin-transition polymers: from molecular materials toward memory devices. *Science* **279** (1998) pp.44-48.
37. Halder G. J., Kepert C. J., Moubaraki B., Murray K. S. and Cashion J. D., Guest-dependent spin crossover in a nanoporous molecular framework material. *Science* **298** (2002) pp.1762-1765.

MOLECULAR CLUSTER MAGNETS

JEFFREY R. LONG

Department of Chemistry, University of California, Berkeley, CA 94720, USA

Molecular clusters with a high spin ground state and a large negative axial zero-field splitting possess an intrinsic energy barrier for spin reversal that results in slow relaxation of the magnetization. The characteristics of [$Mn_{12}O_{12}(MeCO_2)_{16}(H_2O)_4$]—the first example of such a single-molecule magnet—leading up to this behavior are described in detail. Other clusters known to exhibit an analogous behavior are enumerated, consisting primarily of oxo-bridged species containing Mn^{III}, Fe^{III}, Ni^{II}, V^{III}, or Co^{II} centers as a source of anisotropy. The progress to date in controlling the structures and magnetic properties of transition metal-cyanide clusters as a means of synthesizing new single-molecule magnets with higher spin-reversal barriers is summarized. In addition, the phenomenon of quantum tunneling of the magnetization, which has been unambiguously demonstrated with molecules of this type, is explained. Finally, potential applications involving high-density information storage, quantum computing, and magnetic refrigeration are briefly discussed.

1 Introduction

Over the course of the past decade, a rapidly increasing number of molecular clusters have been shown to exhibit magnetic bistability. These species, dubbed single-molecule magnets, possess a combination of high spin S and axial anisotropy D in the ground state that leads to an energy barrier U for reversing the direction of their magnetization. The ensuing slow magnetic relaxation observed at low temperatures is in many ways analogous to the behavior of a superparamagnetic nanoparticle below its blocking temperature [1].

With 2-30 transition metal centers and diameters in the 0.5-2 nm regime, established single-molecule magnets reside at the smaller end of the spectrum of nanostructured materials. As molecular compounds, they can generally be isolated in pure form and crystallized, permitting the precise determination of atomic structure via X-ray crystallography. Hence, with essentially no size dispersion, these species exhibit well-defined and remarkably reproducible physical properties. This, along with the massively parallel production scheme associated with solution-based molecular assembly reactions, provides impetus for the current optimism surrounding the development of spin-based molecular electronic devices. Other applications envisioned for such molecular magnets include high-density information storage, quantum computing, and magnetic refrigeration. Moreover, these clusters are positioned at the frontier between molecular and bulk magnetism, allowing study of new physical phenomena such as quantum tunneling of the magnetization.

Herein, the state of the nascent field of molecular cluster magnets is summarized, with an eye toward elucidating new and existing challenges. In view of the considerable effort devoted to this area of research of late, surprisingly few review articles on the subject are presently available [2-5].

2   A Mn$_{12}$ Cluster Magnet

In 1980, Lis reported the synthesis and structure of [Mn$_{12}$O$_{12}$(MeCO$_2$)$_{16}$(H$_2$O)$_4$]· 2MeCO$_2$H·4H$_2$O, a compound containing an unprecedented dodecanuclear cluster with the disc-shaped geometry depicted in Figure 1 [6]. Its structure features a central Mn$^{IV}_4$O$_4$ cubane unit surrounded by a ring of eight Mn^{III} centers connected through bridging oxo ligands. Bridging acetate and terminal water ligands passivate the surface, such that each Mn center possesses an approximate octahedral coordination environment. Variable temperature magnetic susceptibility data were also reported by Lis, along with the prescient observation that "such a complicated dodecameric unit should have interesting magnetic properties".

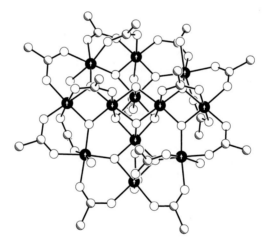

Figure 1. Structure of the disc-shaped cluster [Mn$_{12}$O$_{12}$(MeCO$_2$)$_{16}$(H$_2$O)$_4$] [6]. Black, shaded, and white spheres represent Mn, C, and O atoms, respectively; H atoms are omitted for clarity. White arrows indicate relative orientations of local spins in the ground state; note that the four central Mn^{IV} centers have a lower spin ($S = {}^3/_2$) than the eight outer Mn^{III} centers ($S = 2$).

More than a decade later, magnetization data for this compound collected at high magnetic field strengths and low temperatures were interpreted as indicating an $S = 10$ ground state with significant axial anisotropy [7,8]. Coordinated by weak-field oxo donor ligands, the Mn^{IV} and Mn^{III} centers possess the electron configurations $t_{2g}^3 e_g^0$ and $t_{2g}^3 e_g^1$, imparting local spins of $S = {}^3/_2$ and $S = 2$, respectively. The total spin of the ground state can then be understood as arising from a situation in which the spins of the four central Mn^{IV} centers are all aligned antiparallel to the spins of the eight outer Mn^{III} centers, to give $S = |(4 \times {}^3/_2) + (8 \times -2)| = 10$ (see Figure 1). Reduced magnetization curves for the compound were found to deviate significantly from a simple Brillouin function, suggesting that the $S = 10$ ground state is subject to a substantial zero-field splitting. Indeed, fits to the magnetization data indicate an axial zero-field splitting parameter with a magnitude of $|D| = 0.5$ cm^{-1}. High-field, high-frequency EPR spectra are consistent with this value, and further indicate the sign of D to be negative.

The negative axial zero-field splitting removes the degeneracy in the M_S levels of the ground state, placing the higher magnitude levels lower in energy, as depicted in Figure 2. Together with the selection rule of $\Delta M_S = \pm 1$ for allowed transitions, this results in an energy barrier U separating the two lowest energy levels of $M_S = +10$ and $M_S = -10$. In general, for an integral spin state, the energy barrier will be $U = S^2|D|$, while for a half-integral spin state it will be $U = (S^2 - {}^1/_4)|D|$. Thus, for the $S = 10$ ground state of the Mn$_{12}$ cluster, we have a spin-reversal energy barrier of $U = S^2|D| = 10^2|\cdot 0.5$ cm$^{-1}| = 50$ cm^{-1}. Note that a positive D value would result in the $M_S = 0$ level being lowest in energy, such that there is no energy cost for losing direction of the spin (i.e., in going, for example, from to $M_S = +10$ to $M_S = 0$).

As a consequence of the energy barrier U intrinsic to its ground state, the magnetization of the Mn$_{12}$ cluster can be pinned along one direction, and then

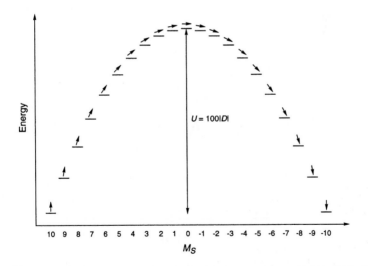

Figure 2. Energy level diagram for an $S = 10$ ground state with a negative axial zero-field splitting, D, in the absence of an applied magnetic field. Arrows represent the relative orientation of the spin with respect to the easy axis of the molecule. As indicated, the spin reversal barrier is given by $U = S^2|D| = 100|D|$. For [Mn$_{12}$O$_{12}$(MeCO$_2$)$_{16}$(H$_2$O)$_4$], $D = -0.5$ cm^{-1}, resulting in a barrier of $U = 50$ cm^{-1}.

relaxes only slowly at very low temperatures. This effect is readily probed through AC magnetic susceptibility measurements, which provide a direct means of gauging the relaxation rate. Here, the susceptibility of a sample is measured using a weak magnetic field (usually of ca. 1 G) that switches direction at a fixed frequency. As the switching frequency increases and starts to approach the relaxation rate for the magnetization within the molecules, the measured susceptibility—referred to as the in-phase or real component of the AC susceptibility and symbolized as χ'—begins to diminish. Accordingly, the portion of the susceptibility that cannot keep up with the switching field—referred to as the out-of-phase or imaginary component of the AC susceptibility and symbolized as χ''—increases. If just a single relaxation process is operational, then a plot of χ'' versus temperature will display a peak with a maximum at the temperature where the switching of the magnetic field matches the relaxation rate, $1/\tau$, for the magnetization of the molecules. Furthermore, since $1/\tau$ increases with temperature, this peak should shift to higher temperature when the switching frequency is increased. As shown in Figure 3, such behavior has indeed been observed for $[Mn_{12}O_{12}(MeCO_2)_{16}(H_2O)_4]\cdot 2MeCO_2H\cdot 4H_2O$ [8]. More precisely, the relaxation time for the magnetization in a single-molecule magnet can be expected to follow an Arrhenius relationship:

$$\tau = \tau_0 e^{(U_{eff}/k_B T)} \tag{1}$$

where the preexponential term τ_0 can be thought of as the relaxation attempt frequency. Thus, a plot of $\ln \tau$ versus $1/T$ should be linear, with the slope and intercept permitting evaluation of U_{eff} and τ_0. Analysis of data for the

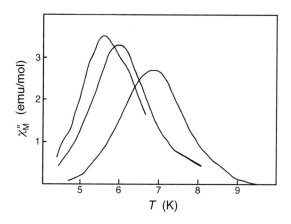

Figure 3. Schematic representation of the out-of-phase component of the molar AC magnetic susceptibility observed for a polycrystalline sample of $[Mn_{12}O_{12}(MeCO_2)_{16}(H_2O)_4]\cdot 2MeCO_2H\cdot 4H_2O$ in zero applied DC field (adapted from reference 8). From left to right, peaks correspond to data collected in an AC field oscillating at a frequency of 55, 100, and 500 Hz, respectively.

[Mn$_{12}$O$_{12}$(MeCO$_2$)$_{16}$(H$_2$O)$_4$] cluster in this manner gave $U_{eff} = 42$ cm^{-1} and $\tau_0 = 2.1 \times 10^{-7}$ s [9]. Note that, as is generally the case, the effective energy barrier U_{eff} obtained is slightly lower than the intrinsic spin reversal barrier U calculated from S and D, owing to the effects of quantum tunneling of the magnetization (see Section 5). Importantly, AC susceptibility measurements performed on the analogous [Mn$_{12}$O$_{12}$(EtCO$_2$)$_{16}$(H$_2$O)$_4$] cluster dissolved in polystyrene reveal the same behavior, indicating that the slow magnetic relaxation is indeed associated with individual clusters and is not a bulk phenomenon [10].

The slow relaxation of the magnetization in a single-molecule magnet also leads to magnetic hysteresis. Figure 4 depicts a hysteresis loop collected for a sample of [Mn$_{12}$O$_{12}$(MeCO$_2$)$_{16}$(H$_2$O)$_4$]·2MeCO$_2$H·4H$_2$O at 2.1 K [11]. This hysteresis has a substantially different origin from that in an ordered ferromagnet. Here, rather than inducing domain wall motion, increasing the applied magnetic field shifts the relative energies of the M_S levels (see Figure 2), decreasing the thermal activation barrier for reversing spin direction and thereby accelerating the relaxation process. As a consequence, the coercivity of the sample changes dramatically with temperature. For example, at 2.6 K, the hysteresis loop shown in Figure 4 narrows to one having a coercive field of less than 0.5 T [11]. As explained below in Section 5, the unusual steps apparent in the hysteresis curve are due to quantum tunneling of the magnetization.

The presence of an energy barrier for reversing spin orientation suggests the possibility of storing a bit of information as the direction of the spin in an individual molecule (in addition to the other applications discussed in Section 6). With $U_{eff} \approx 40$ cm^{-1}, the Mn$_{12}$ cluster exhibits a magnetization relaxation half-life of more than 2 months at 2 K; however, above 4 K this half-life is dramatically reduced and

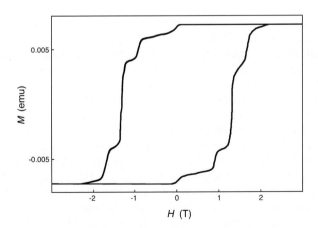

Figure 4. Schematic representation of a magnetic hysteresis loop observed for a single crystal of [Mn$_{12}$O$_{12}$(MeCO$_2$)$_{16}$(H$_2$O)$_4$]·2MeCO$_2$H·4H$_2$O at 2.1 K (adapted from reference 11).

magnetic hysteresis is no longer observed [9]. Thus, in order for these molecules to be capable of storing information at more practical temperatures there is a clear need for clusters possessing a larger spin reversal barrier U. To meet this challenge, one would like to identify or synthesize molecules bearing ground states with exceptionally high spin S and a large negative axial anisotropy D.

3 Other Oxo-Bridged Cluster Magnets

Since the discovery of the remarkable magnetic properties of the Mn_{12} cluster, much effort has been devoted to searching for other examples of molecules exhibiting such behavior. The fruits of that effort are enumerated in Table 1 [7-9,12-40], with the clusters arranged in order of decreasing U_{eff}. In most cases, U_{eff} was established using AC susceptibility measurements, while S and D were determined by fitting magnetization data and high-field EPR spectra [41], respectively. A variety of other physical techniques can also be of utility in probing the magnetic anisotropy of these systems, including high-field torque magnetometry and the use of micro-SQUID arrays [42,43].

A quick inspection of Table 1 reveals that by far the majority (30 out of 33) of the known single-molecule magnets are clusters in which the metal centers are bridged by oxygen donor atoms. Of these, most are manganese-containing species, with only a few clusters featuring iron or nickel and just one containing vanadium. Ground state spins range from $S = 3$ to $S = {}^{33}/_2$, while the D values measured vary between -0.037 and -1.5 cm^{-1}. Significantly, the highest values of U_{eff} are still held by $[Mn_{12}O_{12}(RCO_2)_{16}(H_2O)_4]$ clusters with an $S = 10$ ground state and the core structure depicted in Figure 1.

The all-important magnetic anisotropy of these clusters stems from anisotropy in the electronic structure of the individual metal centers, which in turn arises from spin-orbit coupling. Metal ions with orbital angular momentum and a strong tendency to undergo a Jahn-Teller distortion are particularly suitable for generating a large overall D value. For example, the manganese-oxo clusters listed in Table 1 all contain octahedral Mn^{III} centers with a $t_{2g}^3 e_g^1$ electron configuration. In the $[Mn_{12}O_{12}(MeCO_2)_{16}(H_2O)_4]$ cluster (see Figure 1) the Mn^{III} centers all display a distorted coordination environment consisting of a tetragonal elongation, which occurs roughly perpendicular to the disc of the molecule and coincident with its easy axis. Indeed, variants of the Mn_{12} cluster structure in which some of the tetragonal elongation axes are not so aligned show significantly reduced spin-reversal energy barriers [15,16]. The anisotropy in other metal-oxo single-molecule magnets results from pseudooctahedral V^{III} ($t_{2g}^2 e_g^0$), Fe^{III} ($t_{2g}^3 e_g^2$), or Ni^{II} ($t_{2g}^6 e_g^2$) centers. Very recently, $[Co_4(hmp)_4(MeOH)_4Cl_4]$ (hmpH = hydroxymethylpyridine), an $S = 6$ ground-state cluster containing pseudooctahedral Co^{II} ($t_{2g}^5 e_g^2$) centers, was reported to behave as a single-molecule magnet with an exceptionally large overall anisotropy estimated at $D = -3$ cm^{-1} [44]. It is worth noting that metal centers such

Table 1. Examples of Single-Molecule Magnets.

	S	D (cm^{-1})	U_{eff} (cm^{-1})	ref
[Mn$_{12}$O$_{12}$(CH$_2$BrCO$_2$)$_{16}$(H$_2$O)$_4$]	10		56^a	12
[Mn$_{12}$O$_{12}$(CHCl$_2$CO$_2$)$_8$(Bu^{t}CH$_2$CO$_2$)$_8$(H$_2$O)$_4$]	10	-0.45	50	13
[Mn$_{12}$O$_{12}$(CHCl$_2$CO$_2$)$_8$(EtCO$_2$)$_8$(H$_2$O)$_4$]	10	-0.42	49	13
[Mn$_{12}$O$_{12}$(MeCHCHCO$_2$)$_{16}$(H$_2$O)$_4$]	10	-0.44	45	14
[Mn$_{12}$O$_{12}$(p-PhC$_6$H$_4$CO$_2$)$_{16}$(H$_2$O)$_4$]	10	-0.44	45	14
[Mn$_{12}$O$_{12}$(p-MeC$_6$H$_4$CO$_2$)$_{16}$(H$_2$O)$_4$]^b	10		44	15
[Mn$_{12}$O$_{12}$(MeCO$_2$)$_{16}$(H$_2$O)$_4$]	10	-0.5	42	7-9
[Mn$_{12}$O$_{12}$(MeCO$_2$)$_8$(Ph$_2$PO$_2$)$_8$(H$_2$O)$_4$]	10	-0.41	42	16
[Mn$_{12}$O$_{12}$(PhCO$_2$)$_{16}$(H$_2$O)$_4$]$^{1-}$	$19/2$	-0.44	40	17
[Mn$_{30}$O$_{24}$(OH)$_8$(Bu^{t}CH$_2$CO$_2$)$_{32}$(H$_2$O)$_2$(MeNO$_2$)$_4$]	7	-0.79	39^c	18
[Mn$_{12}$O$_{12}$(CHCl$_2$CO$_2$)$_{16}$(H$_2$O)$_4$]$^{2-}$	10	-0.27	27^c	19
[Mn$_{12}$O$_{12}$(p-MeC$_6$H$_4$CO$_2$)$_{16}$(H$_2$O)$_4$]^d	10		26	15
[Mn$_{12}$O$_8$Cl$_4$(PhCO$_2$)$_8$(hmp)$_6$]^e	7	-0.6	21	20
[Mn$_9$O$_7$(MeCO$_2$)$_{11}$(thme)(py)$_3$(H$_2$O)$_2$]^f	$17/2$	-0.29	19	21
[Fe$_8$O$_2$(OH)$_{12}$(tacn)$_6$]$^{8+\,g}$	10	-0.19	15	22
[V$_4$O$_2$(EtCO$_2$)$_7$(bpy)$_2$]$^{1+\,h}$	3	-1.5	14^c	23
[Mn$_4$(MeCO$_2$)$_2$(pdmH)$_6$]$^{2+\,i}$	8	-0.24	12	24
[Mn$_4$O$_3$(p-MeC$_6$H$_4$CO$_2$)$_4$(dbm)$_3$]^j	$9/2$	-0.62	12^c	25
[Mn$_4$(hmp)$_6$Br$_2$(H$_2$O)$_2$]$^{2+\,e}$	9	-0.35	11	26
[(Me$_3$tacn)$_6$MnMo$_6$(CN)$_{18}$]$^{2+\,k}$	$13/2$	-0.33	10	27
[Fe$_{19}$O$_6$(OH)$_{14}$(metheidi)$_{10}$(H$_2$O)$_{12}$]$^{1+\,l}$	$33/2$	-0.035	9.5^c	28
[Mn$_4$O$_2$(MeO)$_3$(PhCO$_2$)$_2$L$_2$(MeOH)]$^{2+\,m}$	$7/2$	-0.77	9.2^c	29
[Mn$_4$O$_3$Br(MeCO$_2$)$_3$(dbm)$_3$]^j	$9/2$	-0.50	8.3	30
[Mn$_4$O$_3$Cl(MeCO$_2$)$_3$(dbm)$_3$]^j	$9/2$	-0.53	8.2	31
[Ni$_{12}$(chp)$_{12}$(MeCO$_2$)$_{12}$(H$_2$O)$_6$(THF)$_6$]^n	12	-0.047	7	32
[Mn$_{10}$O$_4$(biphen)$_4$Br$_{12}$]$^{4-\,o}$	12	-0.037	4.9	33
[(tetren)$_6$Ni$_6$Cr(CN)$_6$]$^{9+\,p}$	$15/2$		4.2	34
[Fe$_{10}$Na$_2$O$_6$(OH)$_4$(PhCO$_2$)$_{10}$(chp)$_6$(H$_2$O)$_2$(MeCO$_2$)$_2$]^n	11		3.7	35
[Ni$_4$(MeO)$_4$(sal)$_4$(MeOH)$_4$]^q	4		3.7	36
[Mn$_9$(O$_2$CEt)$_{12}$(pdm)$_2$(pdmH)$_2$(C$_{14}$H$_{16}$N$_2$O$_4$)$_2$]^i	$11/2$	-0.11	3.1	37
[Ni$_{21}$(OH)$_{10}$(cit)$_{12}$(H$_2$O)$_{10}$]$^{16-\,r}$	3		2.9	38
[Fe$_4$(MeO)$_6$(dpm)$_6$]^s	5	-0.2	2.4	39
[Fe$_2$F$_9$]$^{3-}$	5	-0.15	1.5	40

^{a}This sample also displays a second relaxation process with U_{eff} = 23 cm^{-1}. ^{b}Crystallized with three water solvate molecules. ^{c}Estimated value (U), based upon S and D; typically, the measured value (U_{eff}) is somewhat lower, owing to quantum tunneling of the magnetization. ^{d}Crystallized with one p-MeC$_6$H$_4$CO$_2$H solvate molecule. ^{e}hmpH = 2-hydroxymethylpyridine. ^{f}thmeH$_3$ = 1,1,1-tris(hydroxymethyl)ethane. ^{g}tacn = 1,4,7-triazacyclononane. ^{h}bpy = 2,2'-bipyridine. ^{i}pdmH$_2$ = pyridine-2,6-dimethanol. ^{j}dbmH = dibenzoylmethane. ^{k}Me$_3$tacn = N,N',N''-trimethyl-1,4,7-triazacyclononane. ^{l}metheidiH$_3$ = N-(1-hydroxynethylethyl)iminodiacetic acid. ^{m}L = 1,2-bis(2,2'-bipyridine-6-yl)ethane. ^{n}biphen = 2,2'-biphenoxide. ^{o}chp = 6-chloro-2-pyridonate. ^{p}tetren = tetraethylenepentamine. ^{q}salH = salicylaldehyde. ^{r}cit^{4-} = citrate. ^{s}dpmH = dipivaloylmethane.

as V^{III} and Co^{II}, which typically display an individual anisotropy where D is positive, can in fact give rise to clusters with a negative overall D value [23,44]. In fact, systems involving these two metal ions in particular would seem to hold considerable promise for the development of new single-molecule magnets with high spin-reversal barriers. Unfortunately, not much is yet understood about how to control or predict the overall magnetic anisotropy of a cluster, even knowing the nature of the anisotropy associated with its constituent metal centers [45].

The molecules listed in Table 1 represent only a rather low percentage of the metal-oxo clusters that have been prepared and investigated for their magnetic properties. It is not sufficient for a cluster simply to have a large number of interacting paramagnetic metal centers, since very frequently these will conspire to produce a ground state of low or zero net spin. For the most part, the clusters are synthesized in one-step self-assembly reactions, from which it is usually impossible to predict the structure of the product *a priori*. The difficulty lies in the enormous structural variability encountered in metal-oxo cluster systems, which, while fascinating from many perspectives, waylays most attempts at developing rational approaches to their synthesis. For a given cluster product considerable variation is possible in the nuclearity, the M-O-M angles (90-180°), and the coordination number of the bridging oxygen atoms (2-6). Moreover, the pairwise magnetic exchange interactions within a cluster are highly sensitive to geometry, making it all but impossible to predict the magnetic properties of a complex metal-oxo cluster. Thus, the discovery of new metal-oxo single-molecule magnets remains very much a serendipitous process.

4   Cyano-Bridged Clusters

As an alternative system where some control over structures and magnetic properties can be anticipated, a number of researchers have turned to metal-cyanide clusters. In a bridging coordination mode, cyanide binds only two metal centers and exhibits a distinct preference for a linear geometry. Thus, assembly reactions can be set up with the expectation that the product will feature linear M'-CN-M moieties. Moreover, given this bridging arrangement, it is possible to predict the nature of the magnetic exchange interactions between M' and M (see Figure 5) [46,47]. Assuming an octahedral coordination geometry for both metal centers, unpaired electrons in orbitals of compatible symmetry ($t_{2g} + t_{2g}$ or $e_g + e_g$) will couple antiferromagnetically, while those in orthogonal orbitals ($t_{2g} + e_g$) will couple ferromagnetically. The antiferromagnetic interaction is typically stronger than the ferromagnetic interaction, and will dominate the superexchange in a competitive situation. Furthermore, the strength of the exchange interaction depends critically upon the degree of overlap between the metal- and cyanide-based orbitals, and, consequently, is high when the radially-extended d orbitals of low-valent early transition metals are involved.

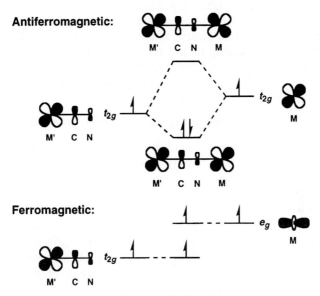

Figure 5. Orbital interactions across a bridging cyanide ligand giving rise to magnetic superexchange. Upper: Unpaired electrons in symmetry compatible t_{2g} orbitals interact through cyanide π^* orbitals, resulting in antiferromagnetic coupling (via the Pauli exclusion principle). In actuality, this is a bit of an oversimplification, as electronic structure calculations indicate that the π orbitals of cyanide are responsible to nearly the same extent [47]. Lower: Unpaired electrons from incompatible metal-based orbitals leak over into orthogonal cyanide-based orbitals, resulting in ferromagnetic coupling (via Hund's rules).

Much of the confidence in being able to control the structures and magnetic properties of metal-cyanide clusters is predicated by extensive investigations into magnetic Prussian blue type solids [46-54]. These compounds have proven to exhibit highly adjustable magnetic behavior, and are generally synthesized via aqueous assembly reactions of the following type.

$$x[M(H_2O)_6]^{y+} + y[M'(CN)_6]^{x-} \rightarrow M_x[M'(CN)_6]_y \cdot zH_2O \qquad (2)$$

The resulting structures are based on an extended cubic lattice of alternating M and M' centers connected through linear cyanide bridges. Two different metal sites are present in the framework, one in which the metal, M', is coordinated by the carbon end of cyanide and experiences a strong ligand field, and another in which the metal, M, is coordinated by the nitrogen end of cyanide and experiences a weak ligand field. Recognition of how the aforementioned factors influence magnetic superexchange through cyanide has enabled chemists to produce solids of this type with bulk magnetic ordering temperatures as high as 376 K [54].

A simple strategy for synthesizing molecular metal-cyanide clusters parallels that employed in reaction 2, but utilizes blocking ligands to hinder formation of an extended solid. The level of structural control possible is illustrated with the assembly of clusters consisting of just one of the fundamental cubic cage units comprising the Prussian blue structure type. Here, a tridentate ligand such as 1,4,7-triazacyclononane (tacn) is employed to block three *fac* sites in the octahedral coordination sphere of each precursor complex.

$$4[(tacn)M(H_2O)_3]^{x+} + 4[(tacn)M'(CN)_3]^{y-} \rightarrow [(tacn)_8M_4M'_4(CN)_{12}]^{4(x-y)+} \quad (3)$$

As in the reaction 2, the nitrogen end of the cyanide ligand displaces water to form linear M'-CN-M linkages. Now, however, the tacn ligands, which are not so readily displaced owing to the chelate effect, prevent growth of an extended Prussian blue framework, and direct formation of a discrete molecular cube. Successful implementation of this strategy has been demonstrated with the reaction between $[(tacn)Co(H_2O)_3]^{3+}$ and $[(tacn)Co(CN)_3]$ in boiling aqueous solution to form the cubic $[(tacn)_8Co_8(CN)_{12}]^{12+}$ cluster depicted in Figure 6 [55,56]. Analogous clusters capped by cyclopentadienyl, a mixture of cyclopentadienyl and carbonyl, or a mixture of 1,3,5-triaminocyclohexane and water ligands have also been reported [57-59].

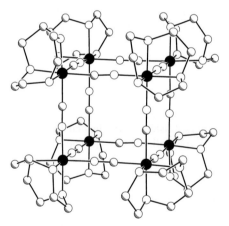

Figure 6. Structure of the cubic cluster $[(tacn)_8Co_8(CN)_{12}]^{12+}$, as crystallized in $[(tacn)_8Co_8(CN)_{12}]$-$(C_7H_7SO_3)\cdot 24H_2O$ [56]. Black, shaded, and white spheres represent Co, C, and N atoms, respectively; H atoms are omitted for clarity.

In distinct contrast to the situation with metal-oxo clusters, once a new metal-cyanide cluster has been discovered, one can be reasonably confident that it will be possible to substitute other transition metal ions having similar geometric proclivities into the structure. This provides a potent means for attempting to manipulate the strength of the magnetic exchange coupling, the overall spin of the ground state, and even the magnetic anisotropy within a cluster. The maximum spin ground state attainable for a cubic $M_4M'_4(CN)_{12}$ cluster, however, is $S = 10$, corresponding, for example, to the case where ferromagnetic coupling is expected to arise between M = Ni^{II} ($t_{2g}^6 e_g^2$) and M' = Cr^{III} ($t_{2g}^3 e_g^0$). Note that this is the same as the spin in the original Mn_{12} single-molecule magnet.

To produce the exceptionally large spin states ultimately sought in single-molecule magnets, it is necessary to develop methods for constructing higher nuclearity clusters in which an even greater number of metal centers are magnetically coupled. One simple idea for accomplishing this is to carry out assembly reactions with a blocking ligand on only one of the components in reaction 3. Accordingly, the following reaction performed in boiling aqueous solution was found to yield a fourteen-metal cluster ($Me_3tacn = N,N',N''$-trimethyl-1,4,7-triazacyclononane) [56].

$$8[(Me_3tacn)Cr(CN)_3] + 6[Ni(H_2O)_6]^{2+} \rightarrow [(Me_3tacn)_8Cr_8Ni_6(CN)_{24}]^{12+} \quad (4)$$

As depicted in Figure 7, the product exhibits a core structure consisting of a cube of eight Cr^{III} centers connected through cyanide bridges to six Ni^{II} centers positioned just above the faces of the cube. Note, however, that the carbon ends of the cyanide ligands are now bound to Ni^{II}, whereas they were initially bound to Cr^{III} in the reactants. Apparently, in the course of heating the reaction, sufficient thermal energy is available to induce isomerization of the cyanide ligand to give the

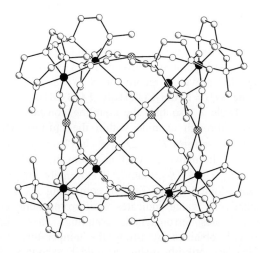

Figure 7. Structure of the face-centered cubic cluster $[(Me_3tacn)_8Cr_8Ni_6(CN)_{24}]^{12+}$, as crystallized in $[(Me_3tacn)_8Cr_8Ni_6(CN)_{24}](NO_3)\cdot 54H_2O$ [56]. Black, crosshatched, shaded, and white spheres represent Cr, Ni, C, and N atoms, respectively; H atoms are omitted for clarity.

thermodynamically-preferred Ni^{II}-CN-Cr^{III} orientation. Unfortunately, this has the added effect of driving the Ni^{II} centers toward a square planar coordination geometry with a low-spin diamagnetic electron configuration, thereby destroying any exchange coupling with the surrounding Cr^{III} centers. The isomerization of cyanide can be forestalled, however, by carrying out reaction 4 at -40 °C in methanol, resulting in a metastable cluster of nominal formula $[(Me_3tacn)_8(H_2O)_x(MeOH)_yNi_6Cr_8(CN)_{24}]^{12+}$. Magnetic susceptibility and magnetization data collected for samples containing this cluster are consistent with the expected ferromagnetic coupling between Cr^{III} and high-spin Ni^{II}, giving rise to an $S = 18$ ground state.

Similar approaches have led to numerous other molecular metal-cyanide clusters [52,60]. The highest nuclearity geometries yet uncovered occur in the tetracapped edge-bridged cubic $[(Me_3tacn)_{12}Cr_{12}Ni_{12}(CN)_{48}]^{12+}$ and double face-centered cubic $[(Me_3tacn)_{14}Cr_{14}Ni_{13}(CN)_{48}]^{20+}$ clusters recently reported [61]. Although these both contain diamagnetic Ni^{II} centers as a consequence of isomerization of the cyanide ligands, isolation of the high-spin form of the latter species would presumably result in a cluster with an $S = 32$ ground state. While a spin state of this magnitude has not yet been realized, many other high-spin metal-cyanide clusters have now been characterized. Table 2 presents these in order of decreasing S. Topping the list are body-centered, face-capped cubic clusters produced through reactions between Mn^{2+} ions and $[M(CN)_8]^{3-}$ (M = Mo, W) complexes in methanol or ethanol. Magnetic susceptibility and magnetization data for $[(EtOH)_{24}Mn_9W_6(CN)_{48}]$ clearly indicate antiferromagnetic coupling between the Mn^{II} and W^V centers to give an $S = {}^{39}/_2$ ground state [63]. Surprisingly, the analogous $[(MeOH)_{24}Mn_9Mo_6(CN)_{48}]$ cluster instead appears to exhibit ferromagnetic coupling and an $S = {}^{51}/_2$ ground state, although there may still be some question as to whether desolvation has complicated the measurements [62]. Regardless, these are the highest spin ground states yet observed for a molecular cluster, surpassing the previous record of $S = {}^{33}/_2$ held by $[Fe_{19}O_6(OH)_{14}(heidi)_{10}(H_2O)_{12}]^{1+}$ (heidiH_3 = $N(CH_2COOH)_2(CH_2CH_2OH))$ [81].

In most cases, Table 2 also lists the coupling constant J characterizing the exchange interactions through cyanide within the clusters. Here, an exchange Hamiltonian of the pairwise form $\hat{H} = -2J\hat{S}_1 \cdot \hat{S}_2$ has been employed. Occasionally, researchers will use an alternative convention where the exchange Hamiltonian is instead of the form $\hat{H} = -J\hat{S}_1 \cdot \hat{S}_2$, making it extremely important to cite which convention has been adopted for the sake of comparing J values. Note that with either form, a negative J value indicates antiferromagnetic coupling, while a positive J value indicates ferromagnetic coupling. These magnetic coupling constants are normally obtained by fitting the temperature dependence of the measured magnetic susceptibility using a model exchange Hamiltonian that incorporates all of the pairwise exchange parameters for the pertinent cluster geometry. With metal-cyanide clusters, it is usually sufficient to include only the exchange between metal centers directly connected to each other through a cyanide bridge. For the clusters in Table 2, measured J values range in magnitude from 0.8 to 17.6 cm^{-1}. Overall, the coupling tends to be a bit weaker than observed for oxo-bridged clusters. Given an appropriate choice of metal ions, however, the coupling through a cyanide bridge can be much stronger, and the highest J value yet observed is -113 cm^{-1}, occurring in the dinuclear molybdenum(III) complex $[Mo_2(CN)_{11}]^{5-}$ [82].

Table 2. Examples of High-Spin Metal-Cyanide Clusters.

	S	J (cm^{-1})^a	ref
$[(MeOH)_{24}Mn^{II}{}_9Mo^V{}_6(CN)_{48}]$	51/2	+	62
$[(EtOH)_{24}Mn^{II}{}_9W^V{}_6(CN)_{48}]$	39/2	-	63
$[(Me_3tacn)_8(H_2O)_x(MeOH)_yNi^{II}{}_6Cr^{III}{}_8(CN)_{24}]^{12+\ b}$	18	+	56
$[(TrispicMeen)_6Mn^{II}{}_6Cr^{III}(CN)_6]^{9+\ c}$	27/2	-4.0	64
$[(dmptacn)_6Mn^{II}{}_6Cr^{III}(CN)_6]^{9+\ d}$	27/2	-5	65
$[(MeOH)_{24}Ni^{II}{}_9M^V{}_6(CN)_{48}]$ (M = Mo, W)	12	ca. +16	66
$[(IM2-py)_6Ni^{II}{}_3Cr^{III}{}_2(CN)_{12}]^e$	9	+5	67
$[(tetren)_6Ni^{II}{}_6Cr^{III}(CN)_6]^{9+\ f}$	15/2	+8.4	68
$[(IM2-py)_6Ni^{II}{}_3Fe^{III}{}_2(CN)_{12}]^e$	7	+3.4	69
$[(Me_3tacn)_6Mn^{II}Cr^{III}{}_6(CN)_{18}]^{2+\ b}$	13/2	-3.1	70
$[(Me_3tacn)_6Mn^{II}Mo^{III}{}_6(CN)_{18}]^{2+\ b}$	13/2	-6.7	27
$[(bpy)_6(H_2O)_2Mn^{II}{}_3W^V{}_2(CN)_{16}]^g$	13/2	-6.0	71
$[(HIM2-py)_6Ni^{II}{}_3Cr^{III}{}_2(CN)_{12}]^e$	6	+6.8	72
$[(tach)_4(H_2O)_{12}Ni^{II}{}_4Fe^{III}{}_4(CN)_{12}]^{8+\ h}$	6	+6.1	59
$[(Me_3tacn)_2(cyclam)_3(H_2O)_2Ni^{II}{}_3Mo^{III}{}_2(CN)_6]^{6+\ bi}$	6	+8.5, +4.0	73
$[(5-Brsalen)_2Mn^{III}{}_2Fe^{III}(CN)_6]^{1-\ j}$	9/2	+2.3	74
$[(Tp)_3(H_2O)_3Fe^{III}{}_4(CN)_9]^k$	4	+	75
$[(Me_3tacn)_2(cyclam)Ni^{II}Cr^{III}{}_2(CN)_6]^{2+\ bi}$	4	+10.9	56
$[(bpm)_6Ni^{II}{}_3Fe^{III}{}_2(CN)_{12}]^l$	4	+5.3, -1.7	76
$[(bpy)_6Ni^{II}{}_3Fe^{III}{}_2(CN)_{12}]^g$	4	+3.9	77
$[(Me_3tacn)_2(cyclam)Ni^{II}Mo^{III}{}_2(CN)_6]^{2+\ bi}$	4	+17.6	73
$[(H_2L)_2Ni^{II}{}_2Fe^{III}{}_3(CN)_{18}]^{1-\ m}$	7/2	+2.1	78
$[(tach)(H_2O)_{15}Ni^{II}{}_3Fe^{III}(CN)_3]^{6+\ h}$	7/2	+0.8	59
$[(dmbpy)_4(IM2-py)_2Cu^{II}{}_2Fe^{III}{}_2(CN)_4]^{6+\ n}$	3	+4.9	79
$[(edma)_3Cu^{II}{}_3Cr^{III}(CN)_6]^o$	3	+9.2	80

^{a}For clusters where it has not been explicitly determined, only the sign of J is given. ^{b}Me$_3$tacn = N,N',N''-trimethyl-1,4,7-triazacyclononane. ^{c}TrispicMeen = N,N',N''-(tris(2-pyridylmethyl)-N'-methylethan)1,2-diamine. ^{d}dmptacn = 1,4-bis(2-methylpyridyl)-1,4,7-triazacyclononane. ^{e}IM2-py = 2-(2-pyridyl)-4,4,5,5-tetramethyl-4,5-dihydro-1H-imidazolyl-1-oxy. ^{f}tetren = tetraethylenepentamine. ^{g}bpy = 2,2'-bipyridine. ^{h}tach = 1,3,5-triaminocyclohexane. ^{i}cyclam = 1,4,8,11-tetraazacyclotetradecane. ^{j}5-Brsalen = N,N'-ethylenebis(5-bromosalicylideneiminato) dianion. ^{k}Tp = hydrotris(1-pyrazolyl)borate. ^{l}bpm = bis(1-pyazolyl)methane. ^{m}L = 3,10-bis(2-aminoethyl)-1,3,6,8,10,12-hexaazacyclotetradecane. ^{n}dmbpy = 4,4'-dimethyl-2,2'-bipyridine. ^{o}edma = ethylenediaminemonoacetate.

Ultimately, the strength of the magnetic exchange coupling is quite important if single-molecule magnets are to be produced that retain their unusual properties at higher temperatures. This is because J dictates how high excited spin states are above the ground state, and if these are close in energy then the spin reversal barrier may be compromised by their thermal population. Thus, identification of a linear bridging ligand that could be utilized in place of cyanide and deliver stronger magnetic exchange coupling would be of considerable value. A little-explored means of potentially achieving molecules with well-isolated, high-spin ground states involves use of electron delocalization to generate strong magnetic exchange coupling via a double-exchange mechanism [83].

Although high-spin, the clusters occupying the top six entries of Table 2 all approximate O_h symmetry. While this does not necessarily preclude development of magnetic anisotropy (since a very slight structural distortion could accompany magnetic polarization), it certainly might act against it. Indeed, none of these very high-spin molecules have been shown to exhibit single-molecule magnet behavior. Consequently, the development of strategies for synthesizing high-nuclearity metal-cyanide clusters with a more anisotropic overall geometry presents an important goal. An example of a high-spin species with a lower-symmetry geometry is the trigonal prismatic cluster [(Me$_3$tacn)$_6$MnCr$_6$(CN)$_{18}$]$^{2+}$ depicted in Figure 8 [70]. This molecule has an $S = {}^{13}/_2$ ground state, and was obtained serendipitously from a reaction between Mn^{2+} and [(Me$_3$tacn)Cr(CN)$_3$] in aqueous solution. Although it displays a unique molecular axis, the cluster still does not behave as a single-molecule magnet, owing to the negligible single-ion anisotropy associated with its Mn^{II} ($t_{2g}^3 e_g^2$) and Cr^{III} ($t_{2g}^3 e_g^0$) centers.

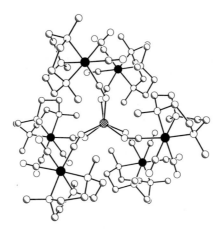

Figure 8. Structure of the trigonal prismatic cluster [(Me$_3$tacn)$_6$MnCr$_6$(CN)$_{18}$]$^{2+}$, as crystallized in K[(Me$_3$tacn)$_6$MnCr$_6$(CN)$_{18}$](ClO$_4$)$_3$ [70]. Black, crosshatched, shaded, and white spheres represent Mo, Mn, C, and N atoms, respectively; H atoms are omitted for clarity.

In fact, very few of the clusters listed in Table 2 contain metal ions likely to contribute to a large overall magnetic anisotropy. As stated previously, however, it is frequently possible to substitute other metal ions into a known metal-cyanide geometry, thereby potentially altering the ground state anisotropy. Accordingly, a straightforward approach to producing single-molecule magnets might be to replace the metal centers of the lower-symmetry clusters in Table 2 with ions known to deliver the anisotropy in metal-oxo single-molecule magnets, particularly Mn^{III}, V^{III}, and Co^{II}. Another idea is simply to move down the column in the periodic table and utilize second- or third-row transition metal ions. Since spin-orbit coupling is a relativistic phenomenon, this will generally enhance significantly any single-ion anisotropy. For example, while [Cr(acac)$_3$] (acac = acetylacetonate) has an axial zero-field splitting of $|D|$ = 0.59 cm^{-1} [84], [Mo(acac)$_3$] exhibits a large negative D value of -6.3 cm^{-1} [85]. Hence, replacing the Cr^{III} centers in a cluster with Mo^{III} might be expected to impart magnetic anisotropy while preserving the spin of the ground state. As an added advantage, the strength of the magnetic exchange coupling in the cluster should also increase, owing to the more diffuse valence d orbitals of Mo^{III}. Synthesis of the octahedral complex [(Me$_3$tacn)Mo(CN)$_3$] [73] enabled a demonstration of precisely these effects. Simply employing it in place of [(Me$_3$tacn)Cr(CN)$_3$] in the preparation established for the trigonal prismatic cluster [(Me$_3$tacn)$_6$MnCr$_6$(CN)$_{18}$]$^{2+}$ (see Figure 8), resulted in an isostructural product containing [(Me$_3$tacn)$_6$MnMo$_6$(CN)$_{18}$]$^{2+}$ [27]. Incorporating Mo^{III}, this cluster still has a ground state of $S = {}^{13}/_2$, but with an axial zero-field splitting of D = -0.33 cm^{-1} and an exchange coupling constant that has increased in magnitude from -3.1 cm^{-1} to -6.7 cm^{-1}. AC susceptibility measurements indeed show it to behave as a single-molecule magnet with U_{eff} = 10 cm^{-1}.

Ultimately, it is anticipated that similar substitutions in higher-spin metal-cyanide clusters may lead to new examples of single-molecule magnets with significantly enhanced spin-reversal energy barriers.

5 Quantum Tunneling of the Magnetization

Owing to the extremely high level of interest from physicists, a substantial body of literature already exists on quantum tunneling of the magnetization in single-molecule magnets. Rather than attempting a complete overview of the subject here, we will give only a very basic description of the phenomenon.

In 1996, two groups of researchers independently proposed an explanation for the unusual steplike features apparent in the magnetic hysteresis loops obtained from samples of [Mn$_{12}$O$_{12}$(MeCO$_2$)$_{16}$(H$_2$O)$_4$]·2MeCO$_2$H·4H$_2$O [11,86,87]. Inspection of Figure 4 reveals, for example, four steps on each side of the hysteresis loop collected at 2.1 K. These steps originate from a loss of spin polarization in the molecules due to tunneling of the magnetization through the energy barrier U rather than simple thermal activation. This tunneling only occurs with the resonant alignment of two (or more) M_S levels on the left and right sides of the energy diagram displayed in Figure 2. As shown, in zero applied magnetic field each M_S = +N level is aligned with the corresponding M_S = -N level, fulfilling the resonance condition. Indeed, the first step in the hysteresis loop shown in Figure 4 arises at H = 0. The tunneling does not necessarily transpire between the pair of levels lowest in energy, and in fact the probability of tunneling increases as one progresses upward toward M_S = 0. Thus, much of the loss of magnetization is a result of thermally-assisted tunneling between higher energy levels. As the strength of the applied magnetic field is increased, the M_S levels shift in energy, with the lower levels going up on one side and down on the other until eventually the spin reversal barrier disappears. Between these two extremes lie a number of field strengths at which a resonance occurs that again permits tunneling of the magnetization. Accordingly, the positions of the steps in the hysteresis loops can be used to map out the M_S energy levels of the ground state.

Certain differences in the rate of tunneling are observed between the various single-molecule magnets. The distinction can be particularly evident at extremely low temperatures, where the thermal energy no longer assists the tunneling process. For example, this temperature-independent regime is much more readily attained in the cluster $[Fe_8O_2(OH)_{12}(tacn)_6]^{8+}$ (which also has an S = 10 ground state) than in the Mn_{12} cluster [88]. The discrepancy has been attributed to the greater transverse component to the anisotropy of the former molecule. That is, the Fe_8 cluster shows a significantly larger value for the rhombic zero-field splitting parameter E (as well as fourth order terms), resulting in a greater degree of mixing between the M_S = ±N levels, which, in turn, facilitates tunneling [89].

6 Potential Applications

Several applications for single-molecule magnets have been envisioned.

Perhaps the most obvious of these is their potential future use as magnetic data storage media. Here, the idea is that an individual molecule would be capable of storing a single bit of binary information as the direction of its magnetization—i.e, with spin up representing, say, a 0 and spin down representing a 1. With a diameter of just 1-2 nm, this could lead to surface storage densities as high as 200,000 Gbits/in^2, approximately three orders of magnitude greater than can be achieved with current magnetic alloy film technology. Data storage density is of particular import in computer hard drives, where the distance between bits of information places constraints upon the speed and efficiency of the computer. Thus, the use of self-assembled monolayers of single-molecule magnets as a storage media could one day lead to extremely fast computer hard drives. Clearly, a significant challenge that must be met in order for this idea to come to fruition is the development of methods for reading and writing such miniscule magnetic moments. Another challenge, and one that is more amenable to chemists, lies in the synthesis of molecules having larger spin reversal barriers U, which would then permit storage of information at more accessible temperatures. As discussed previously, this entails constructing molecules possessing a well-isolated ground state with a very high spin S and a large negative axial zero-field splitting D. In addition, since quantum tunneling of the magnetization could contribute to the loss of information, one would ideally like these molecules to exhibit no transverse anisotropy.

Recently, it has been shown theoretically that single-molecule magnets could be used as the memory components in quantum computing [90-92]. A single crystal of the molecules could potentially serve as the storage unit of a dynamic random access memory device in which fast electron spin resonance pulses are used to read and write information [90]. This application would take advantage of the quantized transitions between the numerous M_S levels in the ground state of a single-molecule magnet, but must be implemented at very low temperature (below ca. 1 K) to avoid transitions due to spin-phonon interactions. Such a device would have an estimated clock speed of 10 GHz and could store any number between 0 and 2^{2S-2} ($= 2.6 \times 10^5$ for $S = 10$). Thus, to maximize the capacity of the memory, one would like to have a single-molecule magnet with as large a spin S as feasible. In addition, to ensure that resolution of the level structure within the ground state manifold is maintained, it is important for the magnitude of D also to be large.

Finally, it has been suggested that single-molecule magnets cluster might be of utility as low-temperature refrigerants via the magnetocaloric effect [92,93]. This application would take advantage of the large entropy change that occurs upon application of a magnetic field to a sample of randomized spins. Since each cluster magnet is identical, the change occurs only over a very narrow temperature window centered at its blocking temperature (3 K for the Mn_{12} cluster). In order to extend the range of accessible refrigeration temperatures, single-molecule magnets with higher blocking temperatures (i.e., with larger spin-reversal energy barriers U) would be of value.

Acknowledgments

This work was funded by NSF Grant No. CHE-0072691. I thank Ms. J. J. Sokol for helpful discussions.

References

1. Gatteschi D. and Sessoli R., Assembling magnetic blocks or how long does it take to reach infinity? In *Magnetism: A Supramolecular Function*, ed. Kahn O. (Kluwer, Dordrecht, 1996) pp. 411-430.

2. Eppley H. J., Aubin S. M. J., Wemple M. W., Adams D. M., Tsai H.-L., Grillo V. A., Castro S. L., Sun Z., Folting K., Huffman J. C., Hendrickson D. N. and Christou G., Single-molecule magnets: characterization of complexes exhibiting out-of-phase ac susceptibility signals, *Mol. Cryst. Liq. Cryst.* **305** (1997) pp. 167-179.
3. Gatteschi D., Sessoli R. and Cornia A., Single-molecule magnets based on iron(III) oxo clusters, *Chem. Commun.* (2000) pp. 725-732.
4. Gatteschi D., Single molecule magnets: a new class of magnetic materials, *J. Alloys Compd.* **317-318** (2001) pp. 8-12.
5. Hendrickson D. N., Christou G., Ishimoto H., Yoo J., Brechin E. K., Yamaguchi A., Rumberger E. M., Aubin S. M. J., Sun Z. and Aromi G., Molecular nanomagnets, *Mol. Cryst. Liq. Cryst.* **376** (2002) pp. 301-313.
6. Lis T., Preparation, structure, and magnetic properties of a dodecanuclear mixed-valence manganese carboxylate, *Acta Crystallogr. B* **36** (1980) pp. 2042-2046.
7. Caneschi A., Gatteschi D., Sessoli R., Barra A. L., Brunel L. C. and Guillot M., Alternating current susceptibility, high field magnetization, and millimeter band EPR evidence for a ground $S = 10$ state in $[Mn_{12}O_{12}(CH_3COO)_{16}(H_2O)_4]\cdot 2CH_3COOH\cdot 4H_2O$, *J. Am. Chem. Soc.* **113** (1991) pp. 5873-5874.
8. Sessoli R., Tsai H.-L., Schake A. R., Wang S., Vincent J. B., Folting K., Gatteschi D., Christou G. and Hendrickson D. N., High-spin molecules: $[Mn_{12}O_{12}(CH_3COO)_{16}(H_2O)_4]$, *J. Am. Chem. Soc.* **115** (1993) pp. 1804–1816.
9. Sessoli R., Gatteschi D., Caneschi A. and Novak M. A., Magnetic bistability in a metal-ion cluster, *Nature* **365** (1993) pp. 141-143.
10. Eppley H. J., Tsai H.-L., deVries N., Folting K., Christou G. and Hendrickson D. N., High-spin molecules: unusual magnetic susceptibility relaxation effects in $[Mn_{12}O_{12}(O_2CEt)_{16}(H_2O)_3]$ ($S = 9$) and the one-electron reduction product $(PPh_4)[Mn_{12}O_{12}(O_2CEt)_{16}(H_2O)_4]$ ($S = {}^{19}/_2$), *J. Am. Chem. Soc.* **117** (1995) pp. 301-317.
11. Thomas L., Lionti F., Ballou R., Gatteschi D., Sessoli R. and Barbara B., Macroscopic quantum tunnelling of magnetization in a single crystal of nanomagnets, *Nature* **383** (1996) pp. 145-147.
12. Tsai H.-L., Chen D.-M., Yang C.-I, Jwo T.-Y., Wur C.-S., Lee G.-H. and Wang Y., A single-molecular magnet: $[Mn_{12}O_{12}(O_2CCH_2Br)_{16}(H_2O)_4]$, *Inorg. Chem. Commun.* **4** (2001) pp. 511-514.
13. Soler M., Artus P., Folting K., Huffman J. C., Hendrickson D. N. and Christou G., Single-molecule magnets: preparation and properties of mixed-carboxylate complexes $[Mn_{12}O_{12}(O_2CR)_8(O_2CR')_8(H_2O)_4]$, *Inorg. Chem.* **40** (2001) pp. 4902-4912.
14. Ruiz-Molina D., Gerbier P., Rumberger E., Amabilino D. B., Guzei I. A., Folting K., Huffman J. C., Rheingold A., Christou G., Veciana J. and Hendrickson D. N., Characterization of nanoscopic $[Mn_{12}O_{12}(O_2CR)_{16}(H_2O)_4]$

single-molecule magnets: physicochemical properties and LDI- and MALDI-TOF mass spectrometry, *J. Mater. Chem.* **12** (2002) pp. 1152-1161.
15. Aubin S. M. J., Sun Z., Eppley H. J., Rumberger E. M., Guzei I. A., Folting K., Gantzel P. K., Rheingold A. L., Christou G. and Hendrickson D. N., Single-molecule magnets: Jahn-Teller isomerism and the origin of two magnetization relaxation processes in Mn_{12} complexes, *Inorg. Chem.* **40** (2001) pp. 2127-2146.
16. Boskovic C., Pink M., Huffman J. C., Hendrickson D. N. and Christou G., Single-molecule magnets: ligand-induced core distortion and multiple Jahn-Teller isomerism in $[Mn_{12}O_{12}(O_2CMe)_8(O_2PPh_2)_8(H_2O)_4]$, *J. Am. Chem. Soc.* **123** (2001) pp. 9914-9915.
17. Aubin S. M. J., Sun Z., Pardi L., Krzystek J., Folting K., Brunel L.-C., Rheingold A. L., Christou G. and Hendrickson D. N., Reduced anionic Mn_{12} molecules with half-integer ground states as single-molecule magnets, *Inorg. Chem.* **38** (1999) pp. 5329-5340.
18. Soler M., Rumberger E., Folting K., Hendrickson D. N. and Christou G., Synthesis, characterization and magnetic properties of $[Mn_{30}O_{24}(OH)_8(O_2CCH_2C(CH_3)_3)_{32}(H_2O)_2(CH_3NO_2)_4]$: the largest manganese carboxylate cluster, *Polyhedron* **20** (2001) pp. 1365-1369.
19. Soler M., Chandra S. K., Ruiz D., Davidson E. R., Hendrickson D. N. and Christou G., A third isolated oxidation state for the Mn_{12} family of single-molecule magnets, *Chem. Commun.* (2000) pp. 2417-2418.
20. Boskovic C., Brechin E. K., Streib W. E., Folting K., Bollinger J. C., Hendrickson D. N. and Christou G., Single-molecule magnets: a new family of Mn_{12} clusters of formula $[Mn_{12}O_8X_4(O_2CPh)_8L_6]$, *J. Am. Chem. Soc.* **124** (2002) pp. 3725-3736.
21. Brechin E. K., Soler M., Davidson J., Hendrickson D. N., Parsons S., Christou G., A new class of single-molecule magnet: $[Mn_9O_7(OAc)_{11}(thme)(py)_3(H_2O)_2]$ with an $S = 17/2$ ground state, *Chem. Commun.* (2002) pp. 2252-2253.
22. Barra A.-L., Debrunner P., Gatteschi D., Schulz C. E. and Sessoli R., Superparamagnetic-like behavior in an octanuclear iron cluster, *Europhys. Lett.* **35** (1996) pp. 133-138.
23. Castro S. L., Sun Z., Grant C. M., Bollinger J. C., Hendrickson D. N. and Christou G., Single-molecule magnets: tetranuclear vanadium(III) complexes with a butterfly structure and an $S = 3$ ground state, *J. Am. Chem. Soc.* **120** (1998) pp. 2365-2375.
24. Yoo J., Brechin E. K., Yamaguchi A., Makano M., Huffman J. C., Maniero A. L., Brunel L.-C., Awage K., Ishimoto H., Christou G. and Hendrickson D. N., Single-molecule magnets: a new class of tetranuclear manganese magnets, *Inorg. Chem.* **39** (2000) pp. 3615-3623.
25. Aliaga N., Folting K., Hendrickson D. N. and Christou G., Preparation and magnetic properties of low symmetry $[Mn_4O_3]$ complexes with $S = 9/2$, *Polyhedron* **20** (2001) pp. 1273-1277.

26. Yoo J. Y., Yamaguchi A., Nakano M., Krzystek J., Streib W. E., Brunel L.-C., Ishimoto H., Christou G. and Hendrickson D. N., Mixed-valence tetranuclear manganese single-molecule magnets, *Inorg. Chem.* **40** (2001) pp. 4604-4616.
27. Sokol J. J., Hee A. G. and Long J. R., A cyano-bridged single-molecule magnet: slow magnetic relaxation in a trigonal prismatic MnMo$_6$(CN)$_{18}$ cluster, *J. Am. Chem. Soc.* **124** (2002) pp. 7656–7657.
28. Goodwin J. C., Sessoli R., Gatteschi D., Wernsdorfer W., Powell A. K. and Heath S. L., Towards nanostructured arrays of single molecule magnets: new Fe oxyhydroxide clusters displaying high ground state spins and hysteresis, *J. Chem. Soc., Dalton Trans.* (2000) pp. 1835-1840.
29. Sañudo E. C., Grillo V. A., Knapp M. J., Bollinger J. C., Huffman J. C., Hendrickson D. N. and Christou G., Tetranuclear manganese complexes with dimer-of-dimer and ladder structures from the use of a bis-bipyridyl ligand, *Inorg. Chem.* **41** (2002) pp. 2441-2450.
30. Andres H., Basler R., Güdel H.-U., Aromí G., Christou G., Büttner H. and Rufflé B., Inelastic neutron scattering and magnetic susceptibilities of the single-molecule magnets [Mn$_4$O$_3$X(OAc)$_3$(dbm)$_3$] (X = Br, Cl, OAc, and F): variation of the anisotropy along the series, *J. Am. Chem. Soc.* **122** (2000) pp. 12469–12477.
31. Aubin S. M. J., Dilley N. R., Pardi L., Krzystek J., Wemple M. W., Brunel L.-C., Maple M. B., Christou G. and Hendrickson D. N., Resonant magnetization tunneling in the trigonal pyramidal Mn^{IV}Mn$^{III}_3$ complex [Mn$_4$O$_3$Cl(O$_2$CCH$_3$)$_3$(dbm)$_3$], *J. Am. Chem. Soc.* **120** (1998) pp. 4991-5004.
32. Cadiou C., Murrie M., Paulsen C., Villar V., Wernsdorfer W. and Winpenny R. E. P., Studies of a nickel-based single molecule magnet: resonant quantum tunelling in an S = 12 molecule, *Chem. Commun.* (2001) pp. 2666-2667.
33. Barra A. L., Caneschi A., Goldberg D. P. and Sessoli R., Slow magnetic relaxation of [Et$_3$NH][Mn(CH$_3$CN)(H$_2$O)$_2$][Mn$_{10}$O$_4$(biphen)$_4$Br$_{12}$] (biphen = 2,2′-biphenoxide) at very low temperature, *J. Solid State Chem.* **145** (1999) pp. 484-487.
34. Vernier N., Bellesa G., Mallah T. and Verdaguer M., Nonlinear magnetic susceptibility of molecular nanomagnets: tunneling of high-spin molecules, *Phys. Rev. B* **56** (1997) pp. 75-78.
35. Benelli C., Cano J., Journaux Y., Sessoli R., Solan G. A. and Winpenny R. E. P., A Decanuclear iron(III) single molecule magnet: use of Monte Carlo methodology to model the magnetic properties, *Inorg. Chem.* **40** (2001) pp. 188-189.
36. Nakano M., Matsubayashi G.-E., Muramatsu T., Kobayashi T. C., Amaya K., Yoo J., Christou G. and Hendrickson D. N., Slow magnetization reversal in [Ni$_4$(OMe)$_4$(sal)$_4$(MeOH)$_4$], *Mol. Cryst. Liq. Cryst.* **376** (2002) pp. 405-410.
37. Boskovic C., Wernsdorfer W., Folting K., Huffman J. C., Hendrickson D. N. and Christou G., Single-molecule magnets: novel Mn$_8$ and Mn$_9$ carboxylate

clusters containing an unusual pentadentate ligand derived from pyridine-2,6-dimethanol, *Inorg. Chem.* **41** (2002) pp. 5107-5118.
38. Ochsenbein S. T., Murrie M., Rusanov E., Stoeckli-Evans H., Sekine C., Güdel H. U., Synthesis, Structure, and Magnetic Properties of the Single-Molecule Magnet $[Ni_{21}(cit)_{12}(OH)_{10}(H_2O)_{10}]^{16-}$, *Inorg. Chem.* **41** (2002) pp. 5133-5140.
39. Barra A. L., Caneschi A., Cornia A., Fabrizi de Biani F., Gatteschi D. Sangregorio C., Sessoli R. and Sorace L., Single-molecule magnet behavior of a tetranuclear iron(III) complex. The origin of slow magnetic relaxation in iron(III) clusters, *J. Am. Chem. Soc.* **121** (1999) pp. 5302-5310.
40. Schenker R., Leuenberger M. N., Chaboussant G., Güdel H. U. and Loss D., Butterfly hysteresis and slow relaxation of the magnetization in $(Et_4N)_3Fe_2F_9$: manifestations of a single-molecule magnet, *Chem. Phys. Lett.* **358** (2002) pp. 413-418.
41. Barra, A.-L., Brunel L.-C., Gatteschi D., Pardi L. and Sessoli R., High-frequency EPR spectroscopy of large metal ion clusters: from zero field splitting to quantum tunneling of the magnetization, *Acc. Chem. Res.* **31** (1998) pp. 460-466.
42. Cornia A., Gatteschi D. and Sessoli R., New experimental techniques for magnetic anisotropy in molecular materials, *Coord. Chem. Rev.* **219-221** (2001) pp. 573-604.
43. Wernsdorfer W., Classical and quantum magnetization reversal studied in nanometer-sized particles and clusters, *Adv. Chem. Phys.* **118** (2001) pp. 99-190.
44. Yang E.-C., Hendrickson D. N., Wernsdorfer W., Nakano M., Zakharov L. N., Sommer R. D., Rheingold A. L., Ledezma-Gairaud M. and Christou G., Cobalt single-molecule magnet, *J. Appl. Phys.* **91** (2002) pp. 7382-7384.
45. Gatteschi D. and Sorace L., Hints for the control of magnetic anisotropy in molecular materials, *J. Solid State Chem.* **159** (2001) pp. 253-261.
46. Entley W. R., Trentway C. R. and Girolami G. S., Molecular magnets constructed from cyanometalate building blocks, *Mol. Cryst. Liq. Cryst.* **273** (1995) pp. 153-166.
47. Weihe H. and Güdel H. U., Magnetic exchange across the cyanide bridge, *Comments Inorg. Chem.* **22** (2000) pp. 75-103.
48. Gadet V., Mallah T., Castro I. and Verdaguer M., High-T_C molecular-based magnets: a ferromagnetic bimetallic chromium(III)-nickel(II) cyanide with T_C = 90 K, *J. Am. Chem. Soc.* **114** (1992) pp. 9213-9214.
49. Mallah T., Thiébaut S., Verdaguer M. and Veillet P., High-T_C molecular-based magnets: ferrimagnetic mixed-valence chromium(III)-chromium(II) cyanides with T_C at 140 and 190 Kelvin, *Science* **262** (1993) pp. 1554-1557.
50. Entley W. R. and Girolami G. S., High-temperature molecular magnets based on cyanovanadate building blocks: spontaneous magnetization at 230 K, *Science* **268** (1995) pp. 397-400.

51. Ferlay S., Mallah T., Ouahès R., Veillet P. and Verdaguer M., A room-temperature organometallic magnet based on Prussian blue, *Nature* **378** (1995) pp. 701-703.
52. Dunbar K. R. and Heintz R. A., Chemistry of transition metal cyanide compounds: modern perspectives, *Prog. Inorg. Chem.* **45** (1997) pp. 283-391 and references therein.
53. Hatlevik Ø., Buschmann W. E., Zhang J., Manson J. L. and Miller J. S., Enhancement of the magnetic ordering temperature and air stability of a mixed valent vanadium hexacyanochromate (III) magnet to 99 °C (372 K), *Adv. Mater.* **11** (1999) pp. 914-918.
54. Holmes S. M. and Girolami G. S., Sol-gel synthesis of $KV^{II}[Cr^{III}(CN)_6] \cdot 2H_2O$: a crystalline molecule-based magnet with a magnetic ordering temperature above 100 °C, *J. Am. Chem. Soc.* **121** (1999) pp. 5593-5594.
55. Heinrich J. L., Berseth P. A. and Long J. R., Molecular Prussian blue analogues: synthesis and structure of cubic $Cr_4Co_4(CN)_{12}$ and $Co_8(CN)_{12}$ clusters, *Chem. Commun.* (1998) pp. 1231-1232.
56. Berseth P. A., Sokol J. J., Shores M. P., Heinrich J. L. and Long J. R., High-nuclearity metal-cyanide clusters: assembly of a $Cr_8Ni_6(CN)_{24}$ cage with a face-centered cubic geometry, *J. Am. Chem. Soc.* **122** (2000) pp. 9655-9662.
57. Klausmeyer K. K., Rauchfuss T. B. and Wilson S. R., Stepwise assembly of $[(C_5H_5)_4(C_5Me_5)_4Co_4Rh_4(CN)_{12}]^{4+}$, an "organometallic box", *Angew. Chem., Int. Ed. Engl.* **37** (1998) pp. 1694-1696.
58. Klausmeyer K. K., Wilson S. R. and Rauchfuss T. B., Alkali metal-templated assembly of cyanometalate "boxes" $(NEt)\{M[Cp*Rh(CN)_3]_4[Mo(CO)_3]_4\}$ (M = K, Cs). Selective binding of Cs^+, *J. Am. Chem. Soc.* **121** (1999) pp. 2705-2711.
59. Yang J. Y., Shores M. P., Sokol J. J. and Long J. R., High-nuclearity metal-cyanide clusters: synthesis, magnetic properties, and inclusion behavior of open cage species incorporating $[(tach)M(CN)_3]$ (M = Cr, Fe, Co) complexes, *Inorg. Chem.* **42** (2003) pp. 1403–1419.
60. Fehlhammer W. P. and Fritz M., Emergence of a CNH and cyano complex based organometallic chemistry, *Chem. Rev.* **93** (1993) pp. 1243-1280.
61. Sokol J. J., Shores M. P. and Long J. R., Giant metal-cyanide coordination clusters: tetracapped edge-bridged cubic $Cr_{12}Ni_{12}(CN)_{48}$ and double face-centered cubic $Cr_{14}Ni_{13}(CN)_{48}$ species, *Inorg. Chem.* **41** (2002) pp. 3052-3054.
62. Larionova J., Gross M., Pilkington M., Andres H., Stoeckli-Evans H., Güdel H. U. and Decurtins S., High-spin molecules: a novel cyano-bridged $Mn^{II}_9Mo^V_6$ molecular cluster with a $S = {}^{51}/_2$ ground state and ferromagnetic intercluster ordering at low temperatures, *Angew. Chem., Int. Ed.* **39** (2000) pp. 1605-1609.
63. Zhong Z. J., Seino H., Mizobe Y., Hidai M., Fujishima A., Ohkoshi S. and Hashimoto K., A high-spin cyanide-bridged Mn_9W_6 cluster ($S = {}^{39}/_2$) with a full-capped cubane structure, *J. Am. Chem. Soc.* **122** (2000) pp. 2952-2953.
64. Scuiller A., Mallah T., Verdaguer M., Nivorozhin A., Tholence J. L. and Veillet P., A rational route to high-spin molecules via hexacyanometalates: a

new μ-cyano $Cr^{III}Mn^{II}_6$ heptanuclear complex with a low-lying $S = {}^{27}/_2$ ground state, *New J. Chem.* **20** (1996) pp. 1-3.
65. Parker R. J., Spiccia L., Berry K. J., Fallon G. D., Moubaraki B. and Murray K. S., Structure and magnetic properties of a high-spin $Mn^{II}_6Cr^{III}$ cluster containing cyano bridges and Mn centres capped by pentadentate ligands, *Chem. Commun.* (2001) pp. 333-334.
66. Bonadio F., Gross M., Stoeckli-Evans H. and Decurtins S., High-spin molecules: synthesis, X-ray characterization, and magnetic behavior of two new cyano-bridged $Ni^{II}_9Mo^V_6$ and $Ni^{II}_9W^V_6$ clusters with a $S = 12$ ground state, *Inorg. Chem.* **41** (2002) pp. 5891-5896.
67. Marvilliers A., Pei Y., Cano Boquera J., Vostrikova K. E., Paulsen C., Rivière E., Audière J.-P. and Mallah T., Metal-radical approach to high spin molecules: a pentanuclear μ-cyano $Cr^{III}Ni^{II}(radical)_2$ complex with a low-lying $S = 9$ ground state, *Chem. Commun.* (1999) pp. 1951-1952.
68. Mallah T., Auberger C., Verdaguer M. and Vaillet P., A heptanuclear $Cr^{III}Ni^{II}_6$ complex with a low-lying $S = {}^{15}/_2$ ground state, *J. Chem. Soc., Chem. Commun.* (1995) pp. 61-62.
69. Vostrikova K. E., Luneau D., Wernsdorfer W., Rey P. and Verdaguer M., A $S = 7$ ground spin-state cluster built from three shells of different spin carriers ferromagnetically coupled, transition-metal ions and nitroxide free radicals, *J. Am. Chem. Soc.* **122** (2000) pp. 718-719.
70. Heinrich J. L., Sokol J. J., Hee A. G. and Long J. R., Manganese-chromium-cyanide clusters: molecular $MnCr_6(CN)_{18}$ and $Mn_3Cr_6(CN)_{18}$ species and a related $MnCr_3(CN)_9$ chain compound, *J. Solid State Chem.* **159** (2001) pp. 293-301.
71. Podgajny R., Desplanches C., Sieklucka B., Sessoli R., Villar V., Paulsen C., Wernsdorfer W., Dromzée Y., and Verdaguer M., Pentanuclear octacyanotungstate(V)-based molecule with a high-spin ground state $S = {}^{13}/_2$, *Inorg. Chem.* **41** (2002) pp. 1323-1327.
72. Marvilliers A., Hortholary C., Rogez G., Audière J.-P., Rivière E., Cano Boquera J., Paulsen C., Villar V. and Mallah T., Pentanuclear cyanide-bridged complexes with high spin ground states $S = 6$ and 9: characterization and magnetic properties, *J. Solid State Chem.* **159** (2001) pp. 302-307.
73. Shores M. P., Sokol J. J. and Long J. R., Nickel(II)-molybdenum(III)-cyanide clusters: synthesis and magnetic behavior of species incorporating [(Me₃tacn)Mo(CN)₃], *J. Am. Chem. Soc.* **124** (2002) pp. 2279-2292.
74. Miyasaka H., Matsumoto N., Okawa H., Re N., Gallo E. and Floriani C., Complexes derived from the reaction of manganese(II) Schiff base complexes and hexacyanoferrate(II): syntheses, multidimensional network structures, and magnetic properties, *J. Am. Chem. Soc.* **118** (1996) pp. 981-994.
75. Lescouëzec R., Vaissermann J., Lloret F., Julve M. and Verdaguer M., Ferromagnetic coupling between low- and high-spin iron(III) ions in the tetranuclear complex fac-{[Fe^{III}{HB(pz)₃}(CN)₂(μ-CN)]₃Fe^{III}(H₂O)₃}·6H₂O

$(HB(pz)_2)^-$ = hydrotris(1-pyrazolyl)borate, *Inorg. Chem.* **41** (2002) pp. 5943-5945.
76. Van Langenberg K., Batten S. R., Berry K. J., Hockless D. C. R., Moubaraki B. and Murray K. S., Structure and magnetism of a bimetallic pentanuclear cluster $[(Ni(bpm)_2)_3(Fe(CN)_6)_2] \cdot 7H_2O$ (bpm = bis(1-pyrazolyl)methane). The role of the hydrogen-bonded $7H_2O$ "cluster" in long-range magnetic ordering, *Inorg. Chem.* **36** (1997) pp. 5006-5015.
77. Van Langenberg K., Hockless D. C. R., Moubaraki B. and Murray K. S., Long-range magnetic order displayed by a bimetallic pentanuclear cluster complex $[(Ni(2,2'-bipy)_2)_3(Fe(CN)_6)_2] \cdot 13H_2O$ and a Cu(II) analog, *Synth. Met.* **122** (2001) pp. 573-580.
78. Kou H.-Z., Zhou B. C., Liao D.-Z., Wang R-J. and Li Y., From one-dimensional chain to pentanuclear molecule. Magnetism of cyano-bridged Fe(III)-Ni(II) complexes, *Inorg. Chem.* **41** (2002) pp. 6887-6891.
79. Oshio H., Yamamoto M., Ito T., Cyanide-bridged molecular squares with ferromagnetically coupled $d\pi$, $d\sigma$, and $p\pi$ spin system, *Inorg. Chem.* **41** (2002) pp. 5817-5820.
80. Fu D. G., Chen J, Tan X. S., Jiang L. J., Zhang S. W., Zheng P. J. and Tang W. X., Crystal structure and magnetic properties of an infinite chainlike and a tetranuclear bimetallic copper(II)-chromium(III) complex with bridging cyanide ions, *Inorg. Chem.* **36** (1997) pp. 220-225.
81. Powell A. K., Heath S. L., Gatteschi D., Pardi L., Sessoli R., Spina G., Del Giallo F. and Pieralli F., Synthesis, structures, and magnetic properties of Fe_2, Fe_{17}, and Fe_{19} oxo-bridged iron clusters: the stabilization of high ground state spins by cluster aggregates, *J. Am. Chem. Soc.* **117** (1995) pp. 2491-2502.
82. Beauvais L. G. and Long J. R., Cyanide-limited complexation of molybdenum(III): synthesis of octahedral $[Mo(CN)_6]^{3-}$ and cyano-bridged $[Mo_2(CN)_{11}]^{5-}$, *J. Am. Chem. Soc.* **124** (2002) pp. 2110-2112.
83. Shores M. P. and Long J. R., Tetracyanide-bridged divanadium complexes: redox switching between strong antiferromagnetic and strong ferromagnetic coupling, *J. Am. Chem. Soc.* **124** (2002) pp. 3512-3513.
84. Elbers G., Remme S. and Lehmann G., EPR of chromium(3+) in tris(acetylacetonato)gallium(III) single crystals, *Inorg. Chem.* **25** (1986) pp. 896-987.
85. Averill B. A. and Orme-Johnson W. H., Electron paramagnetic resonance spectra of molybdenum(III) complexes: direct observation of molybdenum-95 hyperfine interaction and implications for molybdoenzymes, *Inorg. Chem.* **19** (1980) pp. 1702-1705.
86. Friedman J. R., Sarachik M. P., Tejada J., Maciejewski J. and Ziolo R., Steps in the hysteresis loops of a high-spin molecule, *J. Appl. Phys.* **79** (1996) pp. 6031-6033.

87. Friedman J. R., Sarachik M. P., Tejada J. and Ziolo R., Macroscopic measurement of resonant magnetization tunneling in high-spin molecules, *Phys. Rev. Lett.* **76** (1996) pp. 3830-3833.
88. Sangregorio C., Ohm T., Paulsen C., Sessoli R. and Gatteschi D., Quantum tunelling of the magnetization in an iron cluster nanomagnet, *Phys. Rev. Lett.* **78** (1997) pp. 4645-4648.
89. Barra A. L., Gatteschi D. and Sessoli R., High-frequency EPR spectra of $[Fe_8O_2(OH)_{12}(tacn)_6]Br$: a critical appraisal of the barrier for the reorientation of the magnetization in single-molecule magnets, *Chem. Eur. J.* **6** (2000) pp. 1608-1614.
90. Leuenberger M. N. and Loss D., Quantum computing in molecular magnets, *Nature* **410** (2001) pp. 789-793.
91. Tejada J., Chudnovsky E. M., del Barco E., Hernandez J. M. and Spiller T. P., Magnetic qubits as hardware for quantum computers, *Nanotechnology* **12** (2001) pp. 181-186.
92. Tejada J., Quantum behavior of molecule-based magnets: basic aspects (quantum tunneling and quantum coherence) and applications (hardware for quantum computers and magnetic refrigeration. A tutorial, *Polyhedron* **20** (2001) pp. 1751-1756.
93. Torres F., Hernández J. M., Bohigas X. and Tejada J., Giant and time-dependent magnetocaloric effect in high-spin molecular magnets, *Appl. Phys. Lett.* **77** (2000) pp. 3248-3250.

BLOCK COPOLYMERS IN NANOTECHNOLOGY

NITASH P. BALSARA AND HYEOK HAHN
Department of Chemical Engineering, and Materials Sciences Division, Lawrence Berkeley National Laboratory, University of California, Berkeley, California 94720, USA

Block copolymers provide a versatile platform for creating a wide variety of nanostructures. Individual structures such as nanocapsules as well as arrays of structures such as nanowires can be produced by self-assembly. Directed-assembly can be used to eliminate defects and create macroscopically aligned arrays. A-B diblock copolymers lead to the formation of simple structures such as spheres and cylinders, while A-B-C triblock copolymers lead to the formation of more complex structures such as cylinders with rings on the surface and knitted structures. Recent advances in polymer synthesis and templating enable the creation of nanostructures that are responsive to stimuli such as light, electricity, and pH.

1 Introduction

There is an increasing reliance on polymers to fulfill a diverse array of material needs. Current applications rely mainly on macroscopic mechanical properties of polymeric materials such as hardness, softness, adhesive strength, tensile modulus, etc [1]. Homopolymers like polystyrene (PS) are hard and glassy at room temperature whereas polymers like polyisoprene (PI) are soft and rubbery. Conventional polymer materials, regardless of their chemical identity, are composed of interpenetrating chain-like molecules that obey random walk statistics. Figure 1a shows a schematic of a typical polymer chain. The size (radius of gyration, R_g) of a typical synthetic polymer chain (e.g. PS, PI, etc.) with molecular weight of 10^5 g/mol is about 10 nm. The exact size depends on molecular weight and the chemical identity of the monomers. R_g is proportional to the square root of the molecular weight. We note in passing that biological polymers such as DNA in solution also obey random walk statistics. The effective step length of the random walk for synthetic polymers is about 1 nm while that for DNA chains is 50 nm [4]. An important feature of polymer chains is *flexibility* due to the dominance of entropy. Relatively small forces can lead to biased chain configurations. Collections of polymer chains can thus be packed into different kinds of nanometer-scale structures (spheres, cylinders, plates, etc.) with relative ease.

The size of synthetic polymer molecules makes them ideally suited for nanotechnology. The flexibility of polymer chains enables their use as a versatile platform for creating a wide array of nanostructures. In order to do so, however, the chains must be organized into molecular-scale assemblies.

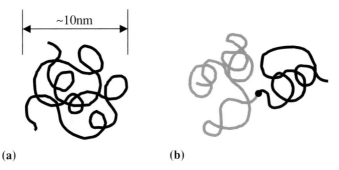

Figure 1. Schematic representation of polymers chain conformation in (a) homopolymers and (b) A-B diblock copolymers.

A simple method for constructing molecular polymeric assemblies is by synthesizing block copolymers. Block copolymers are composed of covalently bonded, chemically dissimilar chains such as polystyrene and polyisoprene. A simple example is an A-B diblock copolymer, shown schematically in Figure 1b. Most chemically dissimilar polymers are immiscible in each other and thus block copolymers are usually in a phase separated state [5,6]. The covalent bond restricts the length scale of the phase separation to molecular dimensions. The phase separation is thus usually referred to as microphase (or nanophase) separation. The shape and arrangement of the microphases is determined mainly by the composition of the copolymer [7,8]. Alternating lamellae of A and B are obtained if ϕ_A, the volume fraction of one of the blocks, say A, is in the vicinity of 0.5. Three other morphologies, a cubic gyroid phase, cylinders arranged on a hexagonal lattice, and spheres arranged on a body centered cubic lattice, are obtained in the ranges $0.36 \geq \phi_A \geq 0.32$, $0.32 \geq \phi_A \geq 0.15$, $0.15 \geq \phi_A > 0$, respectively. The structures obtained in diblock copolymer melts are shown in Figure 2.

Microphase separation (Figure 2) leads to the formation of a large number of internal interfaces per unit volume. Large microphases result is low interfacial area per unit volume, i.e. low interfacial energy, and are thus preferred from the energetic point of view. However the polymer chains of the block copolymer must stretch across length scales comparable to the size of the microphases. This stretching,

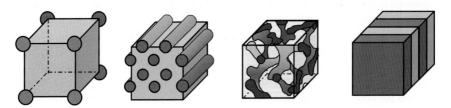

Figure 2. Morphology of A-B diblock copolymers. From left to right: spheres arranged on a body centered cubic lattice, hexagonally packed cylinders, gyroid, and lamellae.

which can be viewed as a distortion of the natural chain configuration (Figure 1a), leads to a reduction of entropy and this reduction increases with increasing microphase size. The size of the microphases at equilibrium is given by the traditional balance between energy and entropy. The entropic elasticity of polymer chains is thus a crucial component of microphase separation. The phases formed by diblock copolymers (Figure 2) are analogous to those formed by self-assembly of other amphiphilic molecules such as detergents and phospholipids.

Recently researchers have begun to study A-B-C triblock copolymers [9-15]. Some of the structures that have been identified thus far, shown in Figure 3, include

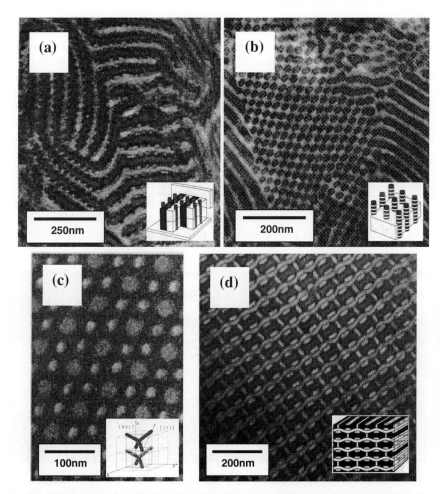

Figure 3. Transmission electron microscopy image and schematic of morphologies formed by A-B-C triblock copolymers: (a) cylinders at walls [ref.9] (b) rings on cylinders [ref.9] (c) tricontinuous double-diamond structure [ref.11] (d) knitting pattern [ref.12].

cylinders between lamellae, rings around cylinders, tricontinuous diamond networks, and a morphology that is referred to as the "knitting pattern". The number of equilibrium phases that these copolymers can form is, however, extremely large (perhaps 10^4). This is due to the presence of numerous kinds of internal interfaces (A/B, B/C, A/C), and the entropic elasticity of polymer chains. Many of the phases formed in ABC triblock copolymers are unique, i.e. they have not been identified in other self-assembling systems.

2 Synthesis and Applications

The block copolymer structures shown above can be used for nanolithography. In ref. 16, a thin film of a sphere-forming polystyrene-polybutadiene (PS-PB) block copolymer was spun cast on a silicon nitride (SiN) substrate. Annealing the film led to an array of 5 nm polybutadiene spheres in a polystyrene matrix, arranged on a locally coherent, hexagonal lattice (Figure 4a). The polymer film was then used as a mask to either obtain 5 nm holes or dots on the SiN substrate. To obtain holes, the film and substrate assembly were put in contact with ozone, which preferentially etches the polybutadiene spheres. Reactive ion etching was then used to transfer the pattern of holes into the SiN substrate (Figure 4b). This example illustrates how organic block copolymers can be used to template inorganic nanostructures.

Other kinds of structures like nanowire arrays can also be produced using block copolymers. In ref. 17 a polystyrene-polymethylmethacrylate (PS-PMMA) block copolymer with cylindrical morphology was spun cast on an electrically conducting

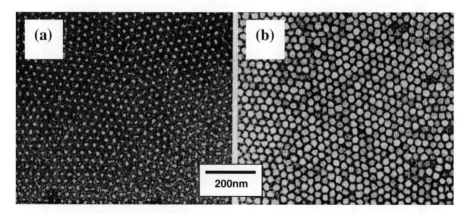

Figure 4. Transmission electron microscopy image and schematic of (a) A thin film of densely packed polybutadiene-spheres in polystyrene matrix before reactive ion etching (RIE) process (b) The transferred pattern on the silicon nitride wafer after RIE [ref.16].

gold substrate. The film was capped with Al. Annealing the film under applied electric fields of 30 V/μm led to the formation of hexagonally arranged PMMA cylinders in a PS matrix. The cylinders were then etched by exposure to UV radiation. The gold substrate and polymer film assembly were then used as the working electrode in a cobalt-electroplating bath. This led to the migration of Co^{2+} ions through the polymer film to the gold substrate where they were reduced to Co. The result was an array of Co nanowires with 14 nm diameter on a hexagonal lattice with a lattice constant of 24 nm. Such arrays may enable the production of high density magnetic storage media.

There are many mechanisms for controlling the size and spacing (d) of the periodic structures formed by block copolymers. For example, increasing the chain length of the block copolymers (N) leads to an increase in d ($d\sim N^{\nu}$ where ν lies between 0.5 and 0.7, depending on the interactions between the blocks). One could also "swell" the domains of an A-B block copolymer by the addition of A and B homopolymers of suitable molecular weight. These strategies have been used to obtain structures with micron-sized periodicity [18] for possible use in photonic applications due to the presence of an optical bandgap.

The similarity of block copolymers and phospholipids has been exploited to make bio-inspired nanostructures [19]. Polyethyleneoxide-polyethylethylene (PEO-PEE) block copolymer chains form a 20 nm sheath that closes on itself to form a closed capsule (vesicle) in excess water (Fig. 5). The hydrophobic PEE chains are in the middle of the sheath while the hydrophilic PEO chains extend into the aqueous phase. Similar capsule-like structures formed by natural phospholipids are currently used for drug delivery. The PEO-PEE based vesicles are much less permeable to water and have significantly higher toughness, relative to their natural

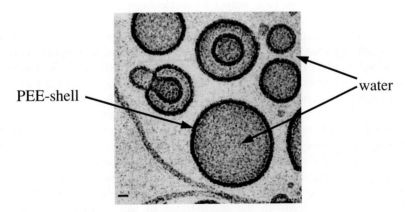

Figure 5. A cryogenic transmission electron microscopy image of the vesicles formed by polyethyleneoxide-*block*-polyethylethylene copolymer in water [ref.19]. The scale bar at lower left is 20 nm.

analogs. These block copolymers may thus enable drug delivery in regimes that are outside the scope of phospholipid-based vesicles.

An important aspect of block copolymer nanostructures is that they form by spontaneous *self-assembly*, i.e. no special processing steps are necessary The production of these structures can readily be commercialized because the raw materials required for synthesis are abundant and cheap.

Block copolymers are synthesized under "living" conditions where in all of the chain growth is initiated at the same time. The required monomers (e.g. A, B, C) are added sequentially to the reactor after the previous monomer (monomers) is (are) completely incorporated into the chains. The term "living" implies that the chains grow in a termination-free reaction and this requires stringent control over reaction conditions and extremely pure reactants [20-26]. The current workhorse for block copolymer synthesis is anionic polymerization [20-22]. The palate of monomers that can be polymerized by this technique is limited to simple organic monomers such as PS, PB, PI, and PMMA. The nanostructures formed by these kinds of block copolymers are responsive to classical chemical stimuli such as temperature and pressure. For example, increasing the temperature of cylindrical block copolymers with $\phi_A=0.32$ leads to a transformation to a lamellar phase [27]. In other words, the boundaries between the different phases shown in Figure 2 are weakly dependent on temperature. This transformation from one kind of nanostructure to another can be accomplished repeatedly and reversibly because both phases are at equilibrium. Block copolymers can thus be used to make responsive nanostructures.

Many applications would benefit if the nanocomponents were responsive to optical, electronic, magnetic, or biological stimuli. An obvious way to accomplish this would be by synthesizing copolymers with blocks that are stimuli-responsive. This may appear relatively easy because there are several examples of stimuli-responsive homopolymers {e.g. poly(1,4-phenylenevinylene) (PPV) is a polymer that emits light when stimulated by an electric field [28]}. Incorporating chains such as PPV in block copolymers is not straightforward because of the stringent requirements that must be met in order to perform living polymerization. In recent years, however, many new methods for synthesizing well-defined block copolymers such as genetic engineering [29,30], living radical polymerization [23-26] and ring-opening metatheisis polymerization (ROMP) [31-35] have been discovered. These methods hold promise for synthesizing a new class of stimuli-responsive block copolymers. For example, ROMP can be used to synthesize redox-active ferrocene-containing block copolymers [35].

A strategy for producing metal-containing nanostructures from block copolymers is templating. Certain blocks such as PB and polyvinylpyridine (PVP) chains have an affinity for salts. Thus, exposing ordered phases of PS-PVP block copolymers to salts leads to the sequestering of salt molecules in the PVP nanostructures [36]. Similarly, exposing PS-PB block copolymers to vapors such as OsO_4 leads to the preferential chemisorption of OsO_4 in the PB domains [37].

Subsequent chemical treatment such as reduction of metal cations enables the creation of metallic nanoscale patterns. Ceramic and semiconducting nanostructures can also be templated using block copolymers [38-43].

The applications for block copolymer nanostructures fall into three broad categories: (1) where well-ordered arrays of nanostructures are needed (e.g. addressable high density magnetic storage using nanowires), (2) where arrays of nanostructures are needed but long-range order is unnecessary (e.g. filters based on nanoporous films), and (3) where individual nanostructures are needed (e.g. vesicles).

Like all ordered phases formed by symmetry-breaking phase transitions, the array of nanostructures are only ordered on a local scale. Both two- and three-dimensional systems are made up of randomly oriented grains. Thus applications in categories (2) and (3) can be readily addressed by block copolymers. Addressing applications in category (1), which require single crystals of ordered nanostructures, require elimination of defects. These applications cannot be addressed by spontaneously self-assembled nanostructures; additional processing is necessary for defect-free single crystals.

There are several strategies for reducing the defect-density in block copolymer structures. It has been noted that the application of flow leads to macroscopic alignment of nanostructrues [44]. Electric fields [45] have the same effect. The structure in block copolymer thin films can be organized into defect-free arrays by graphotaxy: the block copolymer film is placed between lithographically created trenches with parallel walls that are commensurate with the block copolymer structure [46]. Applications in category (1) can thus be addressed by these kinds of *directed-assembly* processes. These processes may be viewed as analogs of zone-refining that enabled the practical use of Si in solid-state devices.

It is clear that block copolymers can serve as a platform for creating a variety of individual nanostructures and arrays of nanostructures. Concrete examples of technological applications for these structures are just beginning to emerge [47].

Acknowledgements

We thank the National Science Foundation (Grant Nos. DMR-213508 CTS-0196066) and the Department of Energy (Basic Energy Sciences) for supporting our efforts in this field.

References

1. Joseph C. Salamone (ed.) *Polymeric Materials Encyclopedia* (CRC press, New York, 1996).
2. Flory, P.J. *Principles of Polymer Chemistry* (Cornell University Press, 1953).

3. Larson, R.G., *The Structure and Rheology of Complex Fluids*. (Oxford University Press, New York, 1999).
4. Fetters, L.J., Lohse, D.J., and Colby, R.H., Chain Dimensions and Entanglement Spacings. In *Physical Properties of Polymers Handbook*. Ed. by James E. Mark (AIP press, New York, 1997)
5. Helfand, E., Block Copolymer Theory. III. Statistical Mechanics of the Microdomain Structure. *Macromolecules* **8** (1975) pp. 552-556.
6. Leibler, L. Theory of Microphase Separations in Block Copolymers. *Macromolecules* **13** (1980) pp. 1602-1617.
7. Bates, F.S., Schulz, M.F., Khandpur, A.K. Forster, S., Rosedale, J.H., Almdal, K., and Mortensen, K., Fluctuations, Conformational Asymmetry and Block Copolymer Phase Behavior. *Farad. Disc.* **98** (1994) pp. 7-18.
8. Hamley I.W. *The Physics of Block Copolymers*. (Oxford University Press, New York, 1998).
9. Stadler, R., Auschra, C., Beckmann, J., Krappe, U., Voigt-Martin, I., and Leibler, L., Morphology and Thermodynamics of Symmetric Poly(A-*block*-B-*block*-C) Triblock Copolymers. *Macromolecules* **28** (1995) pp. 3080-3097. pp.55–63.
10. Mogi, Y., Mori, K., Matsushita, Y., and Noda, I., Tricontinuous Morphology of Triblock Copolymers of the ABC type. *Macromolecules* **25** (1992) pp. 5412-5415.
11. Matsushita, Y., Tamura, M., and Noda, I., Tricontinuous Double-Diamond Structure Formed by a Stryrene-Isoprene-2-Vinylpyridine Triblock Copolymer. *Macromolecules* **27** (1994) pp. 3680-3682.
12. Breiner, U., Krappe, U., Thomas, E.L., and Stadler, R., Structural Characterization of the "Knitting Pattern" in Polystyrene-*block*-poly(ethylene-*co*-butylene)-*block*-poly(methyl methacrylate) Triblock Copolymers. *Macromolecules* **31** (1998) pp. 135-141.
13. Nakazawa, H., Ohta, T., Microphase Separation of ABC-Type Triblock Copolymers. *Macromolecules* **26** (1993) pp. 5503-5511.
14. Zheng, W. and Wang, Z.G., Morphology of ABC Triblock Copolymers. *Macromolecules* **28** (1995) pp. 7215-7223.
15. Phan, S. and Fredrickson, G.H., Morphology of Symmetric ABC Triblock Copolymers in the Strong Segregation Limit. *Macromolecules* **31** (1998) pp. 59-63.
16. Park, M., Harrison, C., Chaikin, P.M., Register, R.A., and Adamson, D.H., Block Copolymer Lithography: Periodic Arrays of $\sim 10^{11}$ Holes in 1 Square Centimeter. *Science* **276** (1997) pp. 1401-1404.
17. Urbas, A., Sharp, R., Fink, Y., Thomas, E.L., Xenidou, M., and Fetters, L.J., Tunable Block Copolymer/Homopolymer Photonic Crystals. *Adv. Mater.* **12** (2000) pp. 812-814.

18. Thurn-Albrecht, T., Schotter, J., Kastle, G.A., Emley, N., Shibauchi, T., Krusin-Elbaum, L., Guarini, K., Black, C.T., Touminen, M.T., and Russell, T.P., Ultrahigh-Density Nanowire Arrays Grown in Self-Assembled Diblock Copolymer Templates. *Science* **290** (2000) pp. 2126-2129.
19. Discher, B.M., Won, Y., Ege, D.S., Lee, J.C., Bates, F.S., Discher, D.E., and Hammer, D.A., Polymersomes: Tough Vesicles Made from Diblock Copolymers. *Science* **284** (1999) pp. 1143-1146.
20. Hadjichristidis, N., Pitsikalis, M., Pispas, S., and Iatrou, H., Polymers with Complex Architecture by Living Anionic Polymerization. *Chem.Rev.* **101** (2001) pp. 3747-3792.
21. Beylen, M.V., Bywater, S., Smets, G., Szwarc, M., and Worsfold, D.K., Developments in Anionic Polymerization- A Critical Review. *Adv. Pol. Sci.* **86** (1988) pp. 87-143.
22. Rempp, P., Franta, E., Herz, J.E., Macromolecular Engineering by Anionic Methods. *Adv. Pol. Sci.* **86** (1988) pp. 145-173.
23. Davis, K.A., Matyjaszewski, K., Statistical, Gradient, Block and Graft Copolymers by Controlled/Living Radical Polymerizations. *Adv. Pol. Sci.* **159** (2002) pp. 1-157.
24. Fischer, H., The Persistent Radical Effect: A Principle for Selective Radical Reactions and Living Radical Polymerizations. *Chem.Rev.* **101** (2001) pp. 3581-3610.
25. Kamigaito, M., Ando, T., and Sawamoto, M., Metal-Catalyzed Living Radical Polymerization. *Chem. Rev.* **101** (2001) pp. 3689-3745.
26. Hawker, C.J., Bosman, A.W., and Harth, E., New Polymer Synthesis by Nitroxide Mediated Living Radical Polymerizations. *Chem. Rev.* **101** (2001) pp. 3661-3688.
27. Hajduk, D.A., Gruner, S.M., Rangarajan, P., Register, R.A., Fetters, L.J., Honeker, C., Albalak, R.J., and Thomas, E.L., Observation of a Reversible Thermotropic Order-Order Transition in a Diblock Copolymer. *Macromolecules* **27** (1994) pp. 490-501.
28. Shim, H. and Jin, J., Light-Emitting Characteristics of Conjugated Polymers. *Adv. Pol. Sci.* **158** (2002) pp. 193-243.
29. McGrath, K.P., Fournier, M.J., Mason, T.L., and Tirrell, D.A., Genetically directed syntheses of new polymeric materials. Expression of artificial genes encoding proteins with repeating -(AlaGly)3ProGluGly- elements. *J. Am. Chem. Soc.* **114** (1992) pp. 727-733.
30. Kobayashi, S., Uyama, H., and Kimura, S., Enzymatic Polymerization. *Chem.Rev.* **101** (2001) pp. 3793-3818.
31. Ivin, K.J. and Mol, J.C., Olefin Metathesis and Metathesis Polymerization (Academic Press, San Diego, 1997).
32. Bielawski, C.W., Benitez, D., and Grubbs, R.H., An "Endless" Route to Cyclic Polymers. *Science* **297** (2002) pp. 2041-2044.

33. Trnka, T.M. and Grubbs, R.H., The Development of $L_2X_2Ru=CHR$ Olefin Metathesis Catalysts: An Organometallic Success Story. *Acc.Chem.Res.* **34** (2001) pp. 18-29.
34. Schrock, R.R., Living Ring-Opening Metathesis Polymerization Catalyzed by Well-Characterized Transition-Metal Alkylidene Complexes. *Acc.Chem.Res.* **23** (1990) pp. 158-165.
35. Buretea, M.A. and Tilley, T.D., Poly(ferrocenylenevinylene) from Ring-Opening Metathesis Polymerization of *ansa*-(Vinylene)ferrocene. *Organometallics* **16** (1997) pp. 1507-1510.
36. Spatz J.P., Eibeck P., Mossmer S., Moller M., Herzog T., and Ziemann P., Ultrathin diblock copolymer/titanium laminates - A tool for nanolithography. *Adv. Mater.* **10** (1998) pp. 849-853.
37. Chan, Y.N.C., Schrock, R.R., and Cohen, R.E., Synthesis of Single Silver Nanoclusters within Spherical Microdomains in Block Copolymer Films. *J. Am. Chem. Soc.* **114** (1992) pp. 7295-7296.
38. Sankaran, V., Yue, J., Cohen, R.E., Schrock, R.R., and Silbey, R.J., Synthesis of Zinc Sulfide Clusters and Zinc Particles within Microphase-Separated Domains of Organometallic Block Copolymers. *Chem. Mat.* **5** (1993) pp. 1133-1142.
39. Zubarev, E.R., Pralle, M.U., Li, L., and Stupp, S.I., Conversion of Supramolecular Clusters to Macromolecular Objects. *Science* **283** (1999) pp. 523-526.
40. Yang, P., Zhao, D., Margolese, D.I., Chmelka, B.F., and Stucky, G.D., Block Copolymer Templating Syntheses of Mesoporous Metal Oxides with Large Ordering Lengths and Semicrystalline Framework. *Chem. Mat.* **11** (1999) pp. 2813-2826.
41. Yu, C., Tian, B., Fan, J., Stucky, G.D., and Zhao, D., Nonionic Block Copolymer Synthesis of Large-Pore Cubic Mesoporous Single Crystals by Use of Inorganic Salts *J. Am. Chem. Soc.* **124** (2002) pp. 4556-4557.
42. Feng, P., Bu, X., Stucky, G.D., and Pine, D.J., Monolithic Mesoporous Silica Templated by Microemulsion Liquid Crystals. *J. Am. Chem. Soc.* **122** (2000) pp. 994-995.
43. Chan, V.Z.H., Hoffman, J., Lee, V.Y., Iatrou, H., Avgeropoulos, A., Hadjichrisidis, N., Miller, R.D., and Thomas, E.L., Ordered Bicontinuous Nanoporous and Nanorelief Ceramic Films from Self Assembling Polymer Precursors. *Science* **286** (1999) pp. 1716-1719.
44. Folkes, M.J. and Keller, A., The birefringence and mechanical properties of a 'single crystal' from a three-block copolymer. *Polymer* **12** (1971) pp. 222-236.
45. Amundson, K., Helfand, E., Davis, D.D., Quan, X., and Patel, S.S., Effect of an Electric Field on Block Copolymer Microstructure. *Macromolecules* **24** (1991) pp. 6546-6548.

46. Segalman, R.A., Yokoyama, H., and Kramer, E.J., Graphotaxy of Spherical Domain Block Copolymer Films. *Adv. Mater.* **13** (2001) pp.1152-1155.
47. Asakawa, K., Hiraoka, T., Hieda, H., Sakurai, M., Kamata, Y., Naito, K., Nano-Patterning for Patterned Media using Block-Copolymer. *J. Photopolymer Sci. Tech.* **15** (2002) pp. 465-470.

THE EXPANDING WORLD OF NANOPARTICLE AND NANOPOROUS CATALYSTS

ROBERT RAJA

Department of Chemistry, University of Cambridge, Lensfield Road, Cambridge CB2 1EW, United Kingdom

JOHN MEURIG THOMAS

The Royal Institution of Great Britain, Davy Faraday Research Laboratory, 21 Albemarle Street, London, W1S 4BS United Kingdom.
Department of Materials Science, University of Cambridge, Pembroke Street, Cambridge, CB2 3QZ, United Kingdom.

> Extremely good catalytic performance – large turnover frequencies and high molecular selectivities – is exhibited by two new and growing categories of heterogeneous catalysts described in this article. The first category consists of discrete bimetallic nanoparticles of diameter 1 to 1.5 nm, distributed more or less uniformly along the inner walls of so-called mesoporous (*ca* 3 to 10 nm diameter) silica supports. Such bimetallic entities prepared by us invariably contain ruthenium. The second consists of extended, crystallographically ordered inorganic solids possessing nanopores (apertures and channels), the diameters of which fall in the range of *ca* 0.3 to 1.0 nm. Onto the inner walls of mesoporous silica one may also anchor organometallic moieties which, acting in a spatially constrained environment, display exceptional enantioselectivity in the hydrogenation of prochiral species. A wide range of hydrogenations, of solid acid-catalyzed dehydration of alkanols (of direct relevance to the hydrogen economy), as well as an eclectic range of selective oxidation of alkanes (in air) – usually under solvent-free conditions – may be carried out with these two categories of nanocatalysts.

1 Introduction

First we define our terms: by nanoparticle catalysts we mean bimetallic entities that fall in the size range of 1 to 1.5 nm diameter and are supported on high-area (generally silica) supports; and by nanoporous catalysts we mean those that have sharply defined pore diameters that may be as little as *ca* 0.3 nm or as large as 1.0 nm diameter, depending upon the active sites that are "implanted" (by means described below) on to their inner walls.

Before we proceed to summarize and illustrate the salient (and often unique) performance of these novel catalysts, it is relevant to recall that up until twenty or so years ago two noteworthy features of what were then generally called microporous materials stood out. First, the vast majority of such porous materials then in widespread use were of poor crystallinity, in the sense that the framework structures (i.e. the walls of the pores) were more-or-less atomically disordered. Typical examples were silica, silica-alumina and porous carbon in any of its numerous variants. Second, the very small minority of porous solids that did exhibit high-degrees of crystallinity, namely the zeolites (naturally occurring aluminosilicates) and a small number of mineral metalophosphates, all possessed pore diameters that ranged from a low of 0.3 to a high of *ca* 0.8 nm.

In the intervening years, as the chapters in this monograph and many recently published reviews testify [1-11], there has been an enormous increase in the number of nanoporous materials (as we defined them in our opening statement) that may now be fashioned into desirable catalysts that effect important (and often hitherto non-feasible) chemical conversions.

It is also noteworthy to recall that a very large number of metal oxides, non-metal oxides, including ternary, quaternary and higher variants thereof, as well as chalcogenides, oxynitrides, an extraordinary range of alumino–, vanado–, molybdeno– and a variety of transition-metal phosphates — in which isomorphous replacements (such as Si^{IV} for either Al^{III} and/or P^{V}) abound — have all been synthesized and, in general, well-characterized during the course of the last two decades [12-15]. Synthetic methods, involving adroit choice of organic templates (structure-directing agents) are incorporated into the nutrient sol or gel from which the (templated) novel nanoporous product is extracted and ultimately de-templated (usually by relatively gentle thermal treatment), are, by now, many and varied and well-documented in a number of reviews [3,6-10].

One of the most comprehensive source book enumerating the topology and other salient features of the framework of well-crystallized (aluminosilicate, aluminophosphate as well as gallo– and germano– analogues) of nanoporous solids of diameter *ca* 0.3 nm upwards and potentially capable of exhibiting catalytic activity is the regularly updated compilation of so-called zeolitic structures published by the International Zeolite Association [16]. Regular synoptic reviews on new mesoporous solids of diameter ca 1 nm upwards appear in *Current Opinion in Solid State Materials Science,* (see, for example, refs [4,11]). A recent review [17] on porous materials, with emphasis on their use other than as catalysts (e.g. slow-release drugs, optical devices and as novel adsorbents) contains some useful pointers to new applications in which catalytic performance may be unusual.

Techniques of characterization have also become very powerful and versatile. Whereas in the "classical period of porous solids," typified by the well-known text by Gregg and Sing, [18], pore diameters and pore-size distributions were derived from adsorption isotherms or (for larger pore material) by such techniques as mercury porosimetry, these days a suite of definitive (both direct and indirect) methods of characterizing nanoporous and nanoparticle materials are at our disposal. These methods include high-resolution electron microscopy, electron diffraction, X-ray absorption spectroscopy and, occasionally, direct X-ray crystallographic structure determinations, and the manner in which such techniques guide us is illustrated in the examples that we cite below [19-21].

2 Bimetallic Nanoparticle Catalysts

We have established that nanoparticles such as Ru_6Pd_6, $Ru_{10}Pt_2$, Ru_5Pt, Ru_6Sn, $Ru_{12}Cu_4$ and $Ru_{12}Ag_4$) anchored within silica nanopores [22-26] display high activities and often high selectivities (depending upon the composition of the nanocatalysts) in a number of single-step (and sometimes solvent-free [22,26]) hydrogenations at low temperatures. (The need to develop powerful new catalysts of this kind, operating with molecular hydrogen under mild conditions, is rather pressing in view of the looming importance of the hydrogen economy).

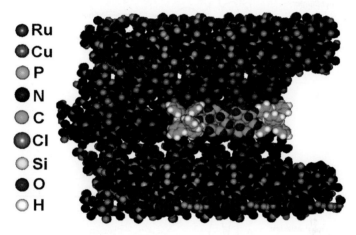

Figure 1. Illustration of the manner in which large molecular ions (such as $[PPN]^+[Ru_{12}Cu_4C_2(CO)_{32}Cl_2]^{2-}[PPN]^+$) are incorporated and closely packed inside a mesopore. (PPN stands for bis-(triphenylphosphane) iminium).

In the definition used by Sinfelt [27,28], a pioneer in this field, "bimetallic clusters refer to bimetallic entities which are highly dispersed on the surface carrier." All our bimetallic catalysts conform to Sinfelt's definition, but with the crucial difference that, in his examples, surface atoms (of the cluster) constitute only a relatively small fraction (a few percent) of the total number of metal atoms in the catalysts, whereas in ours a very high fraction (more than 90 percent) does. The reason for this resides primarily in the manner in which we prepare and introduce the clusters into the high-area silica supports. We start with a mixed-metal carbonylate anion (and its associated cation). Then, by an appropriate choice of solvent and procedure [29], we allow the molecular ion-pairs to occupy the nanopores of the silica, a process that is facilitated by the hydrogen bonding that occurs between the carbonyls of the mixed-metal complex and the pendant silanol groups on the inner walls of the silica nanopores (–M–C–O· · · · H–O–Si–). Thermolysis, which is monitored (*in situ*) by XAFS analysis and FTIR [23], and

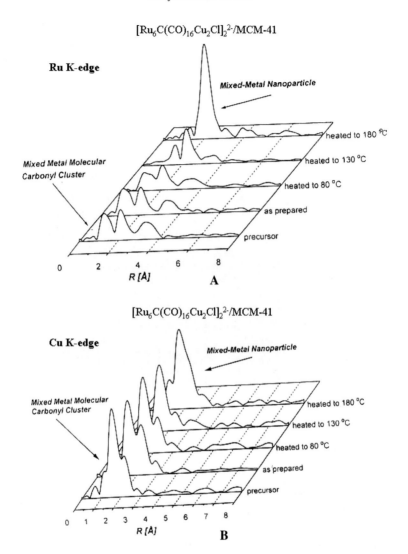

Figure 2. Fourier transform of (**A**) the Ru K-edge and of (**B**) the Cu K-edge EXAFS data recorded during the thermolysis of the $Ru_{12}Cu_4C_2$-silica catalyst.

which causes the gradual loss of the CO groups (and, in certain instances, depending upon the composition of the mixed-metal carbonyl), and of the traces of phosgene, then yields the denuded (i.e. bare) bimetallic clusters anchored as discrete entities in side the nanopores of the silica. The whole course of preparation and of their ultimate characterization is depicted in Figures 1-3.

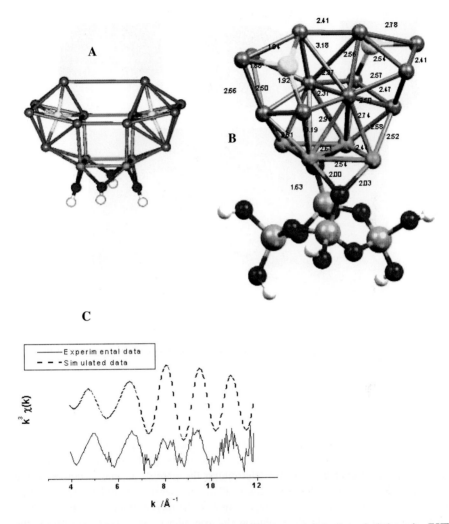

Figure 3. (**A**) The model derived from a simple forcefield, energy-minimization calculation using ESFF interatomic potentials starting from the bond distances and coordination numbers extracted from the EXAFS analysis of the raw data. (**B**) DFT-optimized structure of the silica-bound $Ru_{12}Cu_4C_2$ cluster. The distances (in Å) include all independent bonds. (**C**) Comparison of the Cu K-edge EXAFS data with the spectrum calculated taking the structural details obtained from the DFT calculation using this structure.

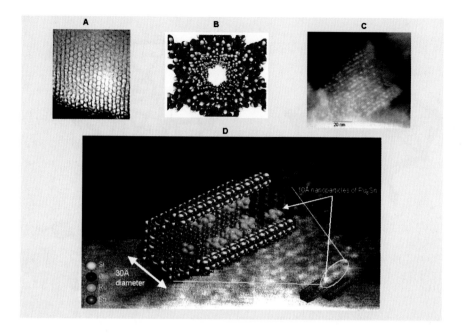

Figure 4. (**A**) High-resolution electron micrograph of a typical hexagonal array of nanopores in silica (diameter 10 nm). (**B**) Computer graphic representation of the interior of a single pore of the silica showing the pendant silanol groups. (**C**) High-angle-annular-dark field (scanning-electron-transmission) micrograph showing the distribution of anchored Ru_6Sn nanoparticles within the nanopores of the siliceous host. (**D**) Computer-graphic illustration of the Ru_6Sn nanoparticles superimposed on an enlargement of the electron micrograph shown in (**C**).

The preparation and characterization of other bimetallic clusters anchored within nanopores of silica, follows similar procedures [29]. It is to be noted that XAFS plays an important role in determining the precise atomic structure of the cluster: seldom, if ever, does the denuded mixed-metal carbonylate adopt, when anchored, a structure that is equivalent to that of the mixed-metal core in the parent carbonylate [19,20,23]. High-resolution electron microscopy (HREM) [30] in the scanning transmission mode, when conducted using Rutherford-scattered electrons (i.e. those that are scattered through high angles, so as to yield an "atomic-number-sensitive" signal) is also a powerful technique of characterization [31,32].

Figures 4 and 5 illustrate how effectively HREM, used in this fashion, pinpoints the precise location of the anchored bimetallic entities. Moreover, using electron tomography (under so-called dark-field conditions), as described fully elsewhere [32], the topography of the individual bimetallic clusters may be directly imaged, as shown in Figure 6, where the Ru_6Pd_6 cluster has been color-coded.

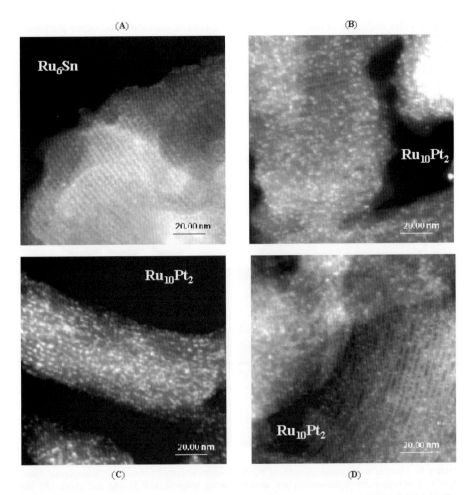

Figure 5. Electron micrographs (HAADF images) of Ru_6Sn/MCM-41 (**A**) and of $Ru_{10}Pt_2$/MCM-41 (**B-D**) after catalysis [ref].

3 The Catalytic Performance of Bimetallic Clusters

The activity and selectivity of the bimetallic catalysts are exceptionally good and may be gauged from Figures 7 and 8 as well as from Table 1.

Figure 7 reveals an instructive example of molecularly selective bimetallic nanocatalysts is seen in the solvent-free conversion of polyenes by Ru_6Sn and some other Ru-containing bimetallic analogues. The polyenes in question are: 1,5,9-cyclododecatriene; 1,5-cyclooctadiene; and 2,5-norbornadiene [26]. The monoenes of all these polyenes are used extensively as intermediates in the synthesis of bicarboxylic aliphatic acids, ketones, cyclic alcohols, lactones and other purposes. Moreover, the selective hydrogenation of 1,5,9-cyclododecatriene to cyclododecene and cyclododecane is industrially important in the synthesis of valuable organic and polymeric intermediates, such as 12-laurolactam and dodecanedioic acid, which are important monomers for Nylon 12, Nylon 612, copolyamides, polyesters, and applications in coatings [26].

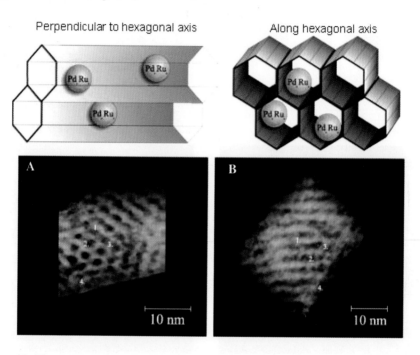

Figure 6. Electron tomographic pictures of bimetallic nanocatalysts Pd_6Ru_6 inside mesoporous silica. Views of nanoparticles (color-coded) are shown looking perpendicular to (**A**) and down (**B**) the nanopores of the silica.

A wide variety of homogeneous and heterogeneous hydrogenation catalysts such as Raney nickel, palladium, platinum, cobalt and mixed transition-metal complexes have been previously used for the above-mentioned hydrogenations. But all these reactions entailed the use of organic solvents (such as n-heptane and benzonitrile) and some required utilization of hydrogen donors (such as 9,10-dihydroanthracene) — often at temperatures in excess of 300 °C — to achieve the desired selectivities.

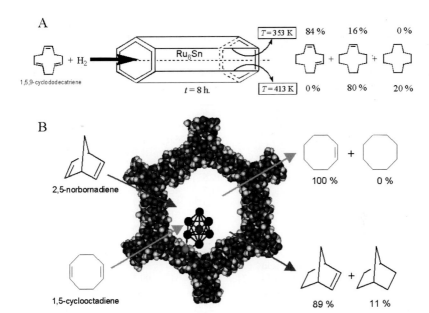

Figure 7. (A) The selectivity of Ru$_6$Sn bimetallic nanocatalysts anchored inside mesoporous silica in the hydrogenation of 1,5,9-cyclododecatriene is a sensitive function of temperature [ref]. (B) 2,5-norbornadiene and 1,5-cyclooctadiene are also selectively hydrogenised in H$_2$ by Ru$_6$Sn bimetallic nanocatalysts [26].

Anchored Ru$_6$Sn nanocatalysts, such as that depicted in Figure 4 (**D**), as well as three other bimetallic nanocatalysts (see Table 1) are very active — note the high turnover frequencies (TOF) enumerated in Table 1 — for the solvent-free hydrogenation of 1,5,9-cyclododecatriene. Importantly, the selectivities may be fine-tuned by adjusting the temperature {Figure 7 (**A**)} and, to some degree, the contact times (Table 1). The solvent-free hydrogenation of 1,5-cyclooactadiene (Figure 7 **B**) is also efficiently catalyzed by Ru$_6$Sn as well as by other Ru-containing nanocatalysts. For the hydrogenation of 2,5-norbornadiene (Figure 7 **B**), the TOF for both Ru$_6$Sn and Ru$_6$Pd$_6$ are exceptionally large, but in the former (at 333 K) the ratio of norbornene to norbornane is 8:1, whereas for the latter it is 1:3.

Preliminary experiments also show that low-temperature, solvent-free selective hydrogenation of benzene to cyclohexene (an important intermediate in a new industrial method for producing adipic acid via cyclohexanol [33]) may be effected with some of our bimetallic nanocatalysts (Table 1).

We turn next to *low-temperature, single-step hydrogenations* of some key organic compounds, and, in particular, we highlight the promising performance of two, related new bimetallic nanocatalysts Ru$_{10}$Pt$_2$ and Ru$_5$Pt, for the hydrogenation of (i) dimethyl terephthalate (DMT) to 1,4-cyclohexanedimethanol (CHDM); and (ii) benzoic acid to cyclohexane carboxylic acid [25].

Table 1. Single-step hydrogenations of cyclic polyenes and aromatics – comparison of catalysts (see Figure 8 for description of products).

Catalyst anchored in mesoporous SiO_2	Substrate	t h	Conv mol %	TOF h^{-1}	Product distribution, mol %			
					A	B	C	D
Pd_6Ru_6	1,5,9-CDT	8	64.9	5350	–	11.7	88.5	–
Ru_6Sn	1,5,9-CDT	8	17.2	1940	17.2	82.4	–	–
Ru_5Pt_1	benzene	6	36.2	2625	8.7	91.2	–	–
$Ru_{10}Pt_2$	benzene	6	27.6	1790	24.9	75.2	–	–
Pd_6Ru_6	benzene	6	58.7	3216	–	100	–	–
Ru_6Sn	benzene	6	17.5	953	11.7	88.1	–	–
Cu_4Ru_{12}	benzene	6	11.7	480	–	100	–	–
Ru_5Pt_1	DMT	4	7.5	155	58.7	33.5	–	6.9
$Ru_{10}Pt_2$	DMT	4	23.3	714	42.6	52.3	–	–
Pd_6Ru_6	DMT	8	15.5	125	22.5	4.2	–	74.2
Ru_6Sn	DMT	8	5.3	54	77.2	–	22.6	–
Cu_4Ru_{12}	DMT	24	14.2	45	25.3	63.2	–	11.3
Ru_5Pt_1	benzoic acid	24	61.2	167	86.5	13.3	–	–
$Ru_{10}Pt_2$	benzoic acid	24	78.5	317	99.5	–	–	–
Pd_6Ru_6	benzoic acid	24	44.5	126	61.5	39.2	–	–
Ru_6Sn	benzoic acid	24	15.9	24	9.0	42.5	48.3	–
Cu_4Ru_{12}	benzoic acid	24	21.8	48	–	79.6	21.2	–

Reaction conditions: Catalyst = 20 mg; H_2 pressure = 20 bar; TOF = $[(mol_{substr})(mol_{cluster})^{-1}h^{-1}]$; *Note*: CDT = cyclododecatriene; DMT = dimethyl terephthalate; temp = 373 K (benzene = 353 K); mesoporous SiO_2 used here is of the MCM-41 type.

It is noteworthy that CHDM is nowadays preferred over ethylene glycol as a stepping-stone in the production of polyester fibers (for extensive use in photography and in other applications involving polycarbonates and polyurethanes). It is also relevant to note that, to date, the only reported feasible catalytic route to CHDM is a two-step process, via dimethyl hexahydroterephthalate, DMHT, using a rhodium complex tethered on silica-supported palladium using Angelici's special catalyst [34].

Since DMT is a solid at the low temperatures at which our catalytic hydrogenations are conducted, an environmentally benign solvent (ethanol) was used to facilitate liquid-solid catalysis. Table 1 shows that more CHDM is produced than DMHT (the only other product) when $Ru_{10}Pt_2$ nanocatalysts are used. With $Ru_{12}Cu_4$, tested for comparison, there is an even higher ratio of CHDM to DMHT than with $Ru_{10}Pt_2$, but at substantially lower conversion and turnover frequency (Table 1).

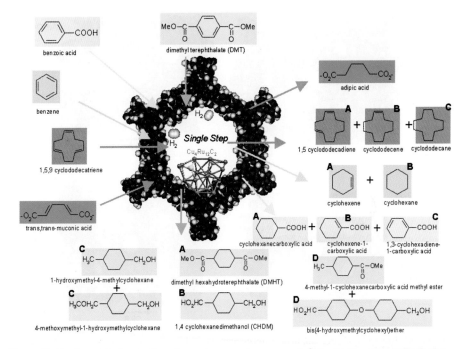

Figure 8. Single-step hydrogenation of some key organic compounds using highly active and selective anchored, bimetallic nanoparticle catalysts ($Cu_4Ru_{12}C_2$, in this case). Note: **A**, **B**, **C** and **D** correspond to the products as listed in Table 1 (under product distribution).

The industrial interest in cyclohexane carboxylic acid stems from the fact that, upon treatment with $NOHSO_4$ (in the presence of strong acids), it yields caprolactam, a stepping-stone in one of the routes in the manufacture of Nylon [25]. The bimetallic nanocatalyst, $Ru_{10}Pt_2$, is exceptionally efficient both in selectivity (99+ per cent towards the desired product) and in extent of conversion and TOF (activity) compared with other analogous nanocatalysts studied by us (see Table 1).

4 Shape-Selective and Regio-Selective Nanoporous Catalysts

Here we focus on two distinct kind of chemical conversion: (a) solid acid catalysis, in which the Brönsted acidity of the solid facilitates processes such as dehydration of the alkanols to yield olefins or the oligomerization of small organic molecules; and (b) selective oxidation catalysts, in which the oxyfunctionalization of alkanes in air to yield desirable products (as building block or stepping stones to other materials). Although many distinct structural types of nanoporous solids (with well-defined pore diameters in the range 0.35 to 1.0 nm) exist, we restrict our discussions

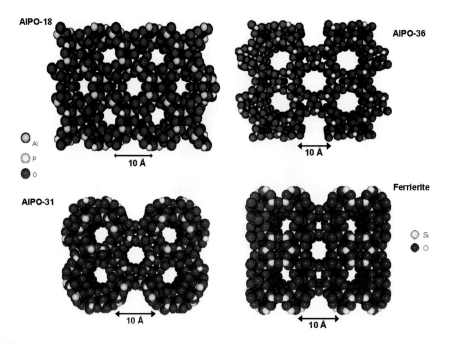

Figure 9. Examples of three aluminophosphate and one aluminosilicate nanoporous catalysts used for conversions described herein: AlPO-18 (pore diameter 3.8 Å), AlPO-36 (pore diameter 6.5 x 7.5 Å), AlPO-31 (pore diameter 5.4 Å) and ferrierite (pore diameter 4.2 x 5.4 Å). Transition metal ions M, typically Co and Mn, occupy a fraction of the Al^{III} framework sites in MAlPOs. The ferrierite structure is represented in its idealized pure silica form.

to just four (Figure 9): three of these are derived from open-structure aluminophosphates, and one is a synthetic form of the aluminosilicate zeolites, ferrierite.

4.1 Solid-acid nanoporous catalysts

The analogy between the mode of operation of enzymes on the one hand and molecular sieve nanopore catalysts on the other has frequently been drawn [35]. In each case, cavities in the catalysts impose shape-selectivity that governs the "choice" of reactant species which is to be transformed (or released), and the molecular complementarity of the microenvironment at the active site facilitates ensuing chemical conversion. In each case also, spatially isolated single sites are the loci of catalytic action. In addition, the aim is to achieve high selectivity and high activity, as well as to function as close as possible to ambient temperatures.

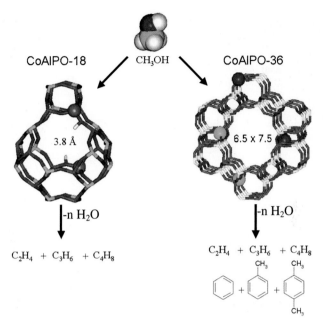

Figure 10. Two examples of acidic nanoporous catalysts for the conversion (by dehydration) of methanol to light olefins [36].

Earlier work by one of us (JMT) had shown [37] that members of the family of MAlPO-18 (where M ≡ Zn^{II}, Mg^{II}, Co^{II}, Mn^{II} which substitute for Al^{III} ions isomorphously in the framework, thereby requiring a proton to be loosely attached to a framework oxygen adjacent to the site of the M^{II} ion) are powerful solid acid catalysts that shape-selectively convert methanol to light olefins. On the other hand, MAlPO-36 [38], in which there are larger pores and cavities, allows the proton-catalyzed dehydration of methanol to generate larger molecules including benzene, toluene and xylene (Figure 10). These are classic examples of shape-selective catalysis effected by nanoporous solid acids, a phenomenon that was first clearly identified by Weisz et al [35]. Numerous other examples have since been reported [refs]. Here we select one interesting example: the cyclodimerization of 3-hydroxy-3-methylbutan-2-one (HMB) to yield the cyclic dimer [39].

In acidic solutions HMB readily undergoes reaction to generate a variety of products. In an acid molecular sieve, such as synthetic ferrierite (idealized formula $Na_2Mg_2(Al_6Si_{30}O_{72}) \cdot 18H_2O$), only one product – the cyclic dimer – is observed. The mechanistic details are as follows:

(It is to be noted that, by prior treatment, most of the exchangeable cations of the synthetic ferrierite are protons).

4.2 Nanoporous catalysts for the regio-selective and shape-selective oxyfunctionalization of alkanes in air [40]

Saturated hydrocarbons are among the most abundant of all naturally occurring organic molecules, and although they are readily oxidized to completion (burnt) at elevated temperature, they are also among the most difficult to oxyfunctionalize in a controlled manner at lower temperature. The incentives in doing this are pressing, if only because terminally oxidized (linear) alkanes, such as n-alkanols and n-alkanoic acids are extremely desirable potential feedstocks for the chemical industry. We have demonstrated elsewhere [40-43] that nanoporous aluminophosphates (AlPOs) have great merit as catalysts for selective oxyfunctionalization. Not least among these advantages are their large (internal) surfaces, which are accessible to reactant molecules of certain size and shape, and from which those products of appropriate dimension could diffuse out. Other advantages include:

- the ability for small quantities of isolated transition-metal ions (M), such as Co^{III}, Mn^{III} and Fe^{III}, to be incorporated into the AlPO frameworks, thereby conferring "*redox*" catalytic activity at the sites of their incorporation into the resulting MAlPO, microporous solid;
- relative ease of preparation, using appropriately chosen structure-directing organic templates, of a considerable range of MAlPO structures differing in their micropore and cage characteristics (e.g. pore diameters, extent of pore intersection); and
- good thermal stability.

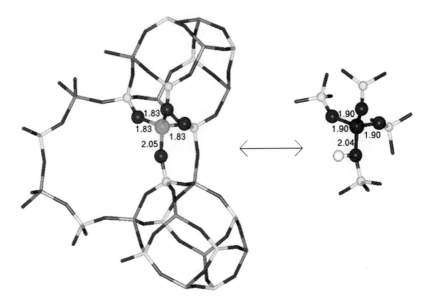

Figure 11. When $Co^{II}AlPO$-18 is converted, by calcination in oxygen, to $Co^{III}AlPO$-18, there is a decrease in the Co-O bond length.

An added bonus is that the precise structure and electronic state [40,41] of the catalytically active sites (Co^{III} or Mn^{III} ions in the framework) have been determined in our laboratories (by XAFS, FTIR and computational analysis). In CoAlPO-18, for example, the transition metal ion occupies a framework site (in place of Al^{III}) as shown in Figure 11. Note the small changes in the lengths of the Co-O bonds as between the tetrahedrally sited Co^{II} and Co^{III} forms. A similar situation exists for the corresponding Mn^{II}– and Mn^{III}– framework-substituted AlPO-18 structures.

To achieve (with our MAlPO catalysts) single, terminal methyl group attack, one must harness the fact that the M^{III} ion (redox) active site must, during reaction, be readily accessible in a preferential manner to the terminal methyl group of the alkane. One must also ensure that the alkane can gain entry into the interior of the MAlPO nanopore catalyst only by an "*end-on*" approach (Figure 12), otherwise oxyfunctionalization of the CH_2 groups at C_2 and C_3 will also occur. (It is to be recalled that during reaction the catalyst and alkanes are immersed in air, as the O_2 (and N_2) are small enough to reach the M^{III} and other sites within the nanopores.

Co^{III} (Mn^{III}) AlPO-18 nanopore catalysts are very effective in the regioselective oxidation of *n*-alkanes. Apart from the qualitative arguments given above for facilitating terminal attack, the quantitative results of energy-minimization (Figure 12 C) provide a deeper understanding of the regioselectivity. The computation, which combines Monte Carlo, molecular dynamics and docking procedures, reveals that the terminal methyl group is significantly closer to a tetrahedral framework site than either C_2 or C_3 carbons in the alkane chain. Moreover, the end of the alkane becomes slightly bent, and all this favors oxidation of the terminal (mainly) and penultimate carbon atoms, just as is seen in the experimental results (Figure 13).

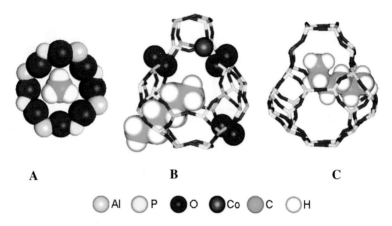

Figure 12. (**A**) & (**B**) Only end-on approach of alkane chains can secure access to active sites situated inside the channels of MAlPO-18 catalysts. (**C**) Energy-minimized configuration adopted by *n*-hexane inside the cages of the AlPO-18 framework [41].

Two other considerations prompted us to harness the selective, oxidative power of Mn^{III} (Co^{III}) framework-substituted nanoporous catalysts, and both these arose more because of auto-suggestion than prior fact. First, Mn^{III} (Co^{III}) ions accommodated substitutionally in place of Al^{III} ions possess close to tetrahedral coordination [42], and just like 4-coordinated Ti^{IV} ions in powerful selective oxidation catalysts, or the 5-coordinated Mn ions in manganese superoxide dismutase, they are coordinatively unsaturated, a feature that facilitates redox catalytic action. Moreover, the occurrence of isolated Mn^{III} (Co^{III}) ions in the framework of the MAlPO would, in the presence of O_2 and those alkanes whose size and shape permit access to these high-oxidation-state ions, favor the production of free radicals, which are known to be implicated in the selective oxidation of saturated hydrocarbons with oxygen [41].

Using the correctly designed nanoporous catalyst containing M^{III} ions strategically placed at its interior surface, we have succeeded in converting *n*-alkanes (C_nH_{2n+2}, $n = 6$ to 8, 10, 12) to the corresponding *n*-alkanols and *n*-alkanoic acids in air at moderate temperatures (100 to 130°C). By judicious modification of the siting of single-site active centers within the right nanocavity (of, for example, Co^{III}AlPO-18) it also because possible to achieve double terminal oxyfunction-alization, thereby enabling *n*-hexane to be converted in air to adipic acid [43].

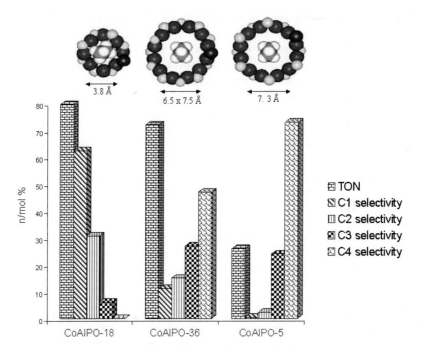

Figure 13. Bar chart showing the regio-selective oxyfunctionalization of n-octane when exposed to air and a $Co^{III}AlPO$-18 catalyst. 93 per cent of the products are oxyfunctionalized at the terminal methyl (C_1) and at the penultimate methylene (C_2) carbons, in marked contrast to what occurs under identical conditions (373K, 1.5MPa, 24h) with CoAlPO-36 or CoAlPO-5.

4.3 A nanoporous catalyst for the shape-selective oxidation of cyclohexane to cyclohexanol, cyclohexanone and adipic acid

Selectively oxyfunctionalizing alicyclic hydrocarbons such as cyclohexane and adamantane as tasks just as demanding as terminally oxyfunctionalizing n-alkanes (see, for example, ref. [44]). Here again, practical considerations loom large, since selective oxidation of cyclohexane into a mixture of cyclohexanol and cyclohexanone (called K-A oil) is of considerable importance in the industrial production of Nylon-6 and Nylon-6,6. Adipic acid (AA) is commercially important in that it is also centrally implicated in the manufacture of Nylon-6,6.

In designing appropriate molecular sieve (MAlPO) catalysts for these conversions, it is necessary to optimize the dimensions of the pores (with respect to the size and shape of the reactants and desirable products) and also to maximize the fraction of the M cation (Co, Mn, Fe) in the III oxidation state. Clearly CoAlPO-36 is superior to CoAlPO-18 even though a smaller fraction (0.45 to be exact [40,41]) of the Co ions in framework sites are Co^{III} largely because the former, with its larger pore dimension, readily permits entry of the cyclohexane into the interior of the catalyst [40]. Even though all the Co ions in CoAlPO-18 are in their Co^{III} state, cyclohexane cannot gain access to the active sites, which are situated at the interior surfaces of the catalyst. X-ray absorption spectroscopic results, indicate [45] that in FeAlPO-5, all the Fe substituted in the framework is in the Fe^{III} state, whereas with CoAlPO-5, only some 20 percent of the framework Co is there as Co^{III}. These facts are reflected in the catalytic performances summarized in Figure 14.

If, however, our objective is to achieve greater catalytic production of adipic acid, we argued [46] that it is necessary to assemble an Fe^{III}-substituted MAlPO structure in which the pore aperture is significantly smaller than the 7.3 Å diameter of FeAlPO-5, as it is necessary to create a constrained environment for the cyclohexane oxidation, so as to modify the selectivity of the reaction. Again we capitalize upon *shape-selectivity* using a carefully designed MAlPO catalyst. This time it is *product shape selectivity* we must attain: only those products with appropriate molecular dimensions may diffuse easily out of the pores, whereas larger ones formed during the course of the reaction are trapped inside, their diffusion being significantly retarded.

We selected FeAlPO-31, which could readily be prepared in a phase-pure fashion, with proof (from X-ray absorption spectroscopy [46]) that the Fe^{III}-centered active site is indistinguishable in this catalyst from what it is in FeAlPO-5 and Fe-ZSM-5 [46].

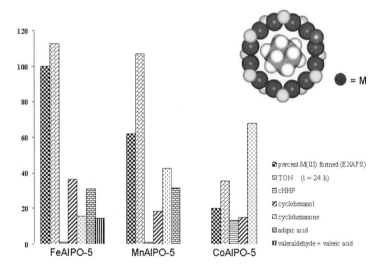

Figure 14. Comparison of the trend in catalytic performance (TON) of Fe-, Mn- and Co-based AlPO-5 catalysts and fraction of the M^{III} ions present (estimated from EXAFS) for the oxidation of cyclohexane. Individual product selectivities are also shown.

The selectivity pattern for FeAlPO-31 is completely different from that of FeAlPO-5 (Figure 15), even though the active sites are identical [45,46]. This is entirely rationalized in terms of the differences in pore dimensions (Figure 15); and one sees that adipic acid, as a sinewy molecule, would diffuse out more readily than any intermediates formed in the free-radical oxidation reaction. Owing to the shape constraint, the diffusion of cyclohexane and the cyclic intermediates in the oxidation reaction within the FeAlPO-31 channel system is severely limited, and therefore further oxidation of the cyclic intermediates to linear products such as adipic acid is facilitated.

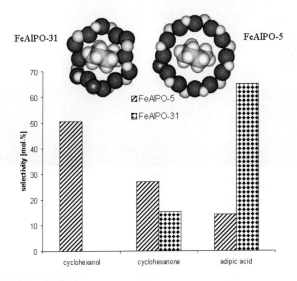

Figure 15. Bar chart comparing the product selectivities (at similar levels of conversion) in the aerial oxidation of cyclohexane, after 24 h, at 373 K, with FeAlPO-31 and FeAlPO-5. Graphical representation of the cyclohexane molecule in the differently sized pore openings of FeAlPO-31 (5.4 Å) and FeAlPO-5 (7.3 Å) is also shown. The more puckered inner walls of the former introduce a constrained environment for the cyclohexane molecule as compared to the latter.

5 Enantioselective Hydrogenations Using Tethered, Complexed, Noble-Metal Catalysts Inside Nanoporous Silica

By extending a principle first conceived [47] a few years ago (and summarized pictorially in Figures 16 and 17), whereby a designed active centre attached to a chiral ligand is tethered at the inner walls of mesoporous silica (pore diameter ca 3 nm), it is possible to produce an efficient enantioselective catalyst for the direct hydrogenation of a number of unsaturated organic molecules that are of considerable commercial importance.

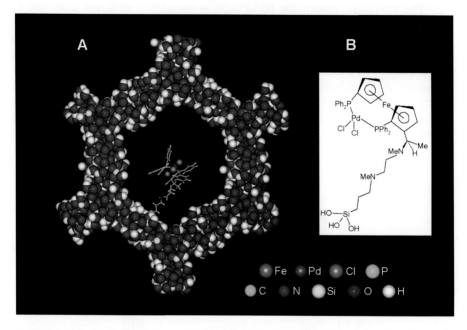

Figure 16. Schematic drawing illustrating the nature of a tethered Pd (diphenyl phosphino-ferrecenyl-diamine) complex, chirally constrained within a siliceous nanopore.

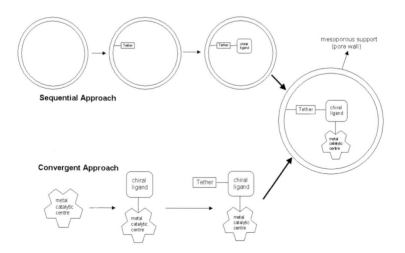

Figure 17. Schematic of the two approaches used to anchor the chiral catalyst onto the mesoporous silica surface.

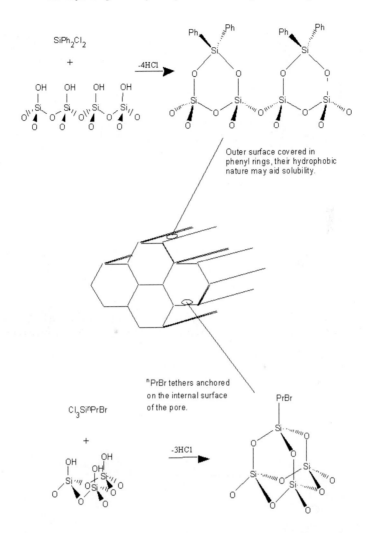

Figure 18. Sequence of steps outlining the synthesis of the activated MCM-41: the mesoporous framework was first treated with Ph$_2$SiCl$_2$ under non-diffusive conditions to deactivate the exterior walls of the material followed by the derivatization of the interior walls with 3-bromopropyltrichlorosilane to yield the activated MCM-41.

The conceptual methodology entails identifying a proven homogeneous catalytic system and tethering it within the mesopores of the silica. The exterior surface is first deactivated (as described earlier — see Figure 18); and the inner walls of the pores are derivatized so as to facilitate the subsequent act of tethering (Figure 19). The purpose in tethering a chirally based homogeneous catalyst inside

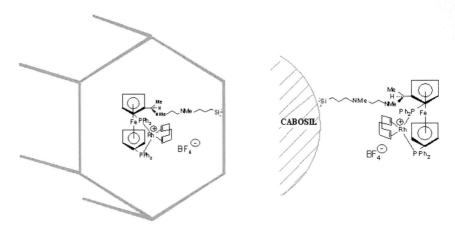

Figure 19. Depiction of the catalytically active dppf–ferrocenyl– Rh–COD–catalyst, bound in a constrained manner, to the inner walls of mesoporous silica (left) and attached to a non-porous silica (right).

a mesopores is to capitalize on the spatial restrictions imposed upon the reactant by the pore walls so as to boost the enantioselectivity of the reaction to be catalyzed [47,48].

Proof that such enantioselective enhancement may be generated in this way has been recently demonstrated in the allylic amination of cinnamyl acetate using a chiral catalyst from the ligand 1,1'-bis(diphenylphosphino)ferrocene (dppf) tethered to the inner walls of mesoporous silica [48].

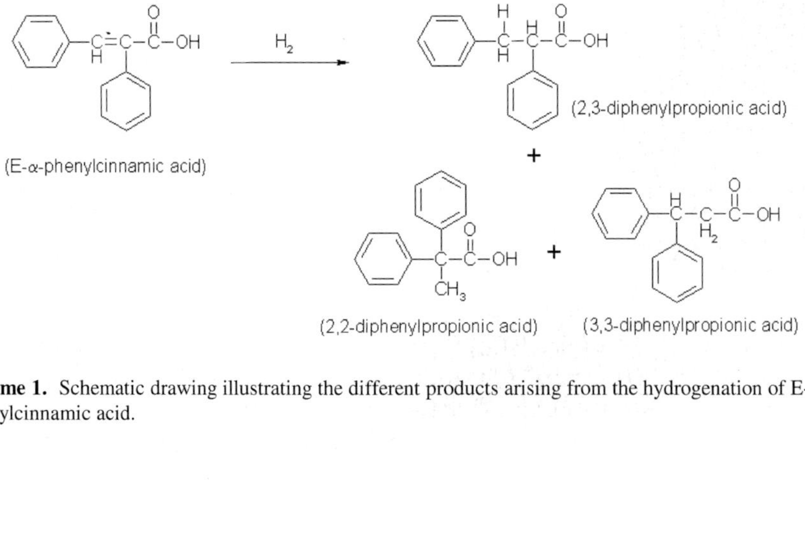

Scheme 1. Schematic drawing illustrating the different products arising from the hydrogenation of E-α-phenylcinnamic acid.

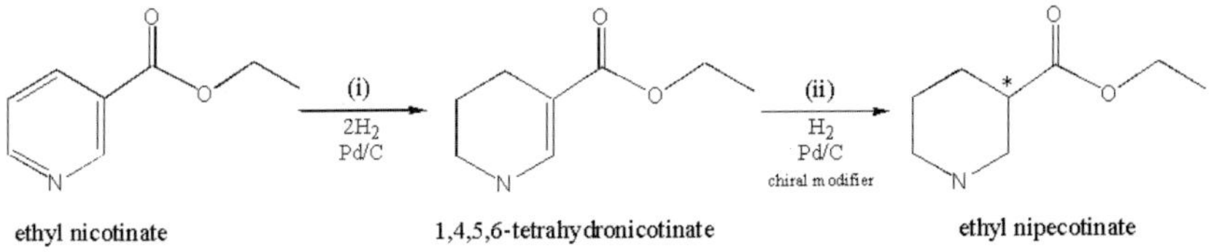

Scheme 2. The two-step hydrogenation of ethyl nicotinate to ethyl nipecotinate via 1,4,5,6-tetrahydronicotinate.

Insofar as enantioselective hydrogenations are concerned, we first demonstrated the viability of this kind of constrained, chiral catalyst in the hydrogenation of E-α-phenylcinnamic acid (Scheme 1). Next, we attacked the enantioselective catalytic hydrogenation of ethyl nicotinate to ethyl nipecotinate [49] in a single step, so as to replace current practice (Scheme 2) that uses a two-step process. Using a Pd (I) dppf-based tethered catalyst, a one-step conversion with an enantioselectivity higher than previously reported was achieved (Tables 2-4). And with another constrained, chiral catalyst, this time involving Rh (I), [50,51] an even better enantioselectivity was achieved (Figure 20 and Table 5).

Table 2. Enantioselective hydrogenation of E-α-phenyl cinnamic acid (Sequential approach).

dppf-ferrocenyl-diamine-Pd-Catalyst	t (h)	Conv (%)	TON	Product Distribution		ee (%)
				A	B	
homogeneous	48	11.5	69	32.5	67.3	35.0
	72	17.2	90	26.0	73.7	24.7
tethered to non porous silica	48	16.2	215	20.5	79.4	-
	72	19.5	238	13.0	86.7	-
confined within mesoporous silica (dia. 30 Å)	48	35.2	416	31.0	68.8	51.2
	72	39.7	447	26.9	73.0	46.3

Reaction conditions: substrate = 5 g; catalyst = 250 mg; H_2 pressure = 20 bar; solvent ≡ methanol : THF = 1: 9; temp = 40 °C;
Products: A = 2,3- diphenyl propionic acid; B = 2,2-/3,3- diphenyl propionic acid (see Scheme 1).

Scope clearly exists here to improve further the performance of these novel, enantioselective heterogeneous catalysts (for hydrogenations and other reactions). Their prime advantage (over homogeneous analogs) is that they greatly facilitate separation of product from reactant, and are also readily amenable to recycling.

Table 3. Enantioselective hydrogenation of E-α-phenyl cinnamic acid (Convergent approach).

dppf-ferrocenyl-diamine-Pd-Catalyst	t (h)	Conv (%)	TON	Product Distribution		ee (%)
				A	B	
homogeneous	48	23.6	379	83.5	16.8	-
	72	38.3	513	76.2	23.1	-
tethered to non porous silica	48	19.8	253	44.1	55.3	36.1
	72	26.0	320	38.4	62.0	27.2
confined within mesoporous silica (dia. 30 Å)	48	62.0	526	100	-	90.7
	72	90.5	715	100	-	96.5

Reaction conditions: substrate = 5 g; catalyst = 250 mg; H_2 pressure = 20 bar; solvent ≡ methanol : THF = 1: 9; temp = 40 °C;
Products: A = 2,3- diphenyl propionic acid; B = 2,2-/3,3- diphenyl propionic acid (see Scheme 1).

Table 4. Enantioselective hydrogenation of ethyl nicotinate.

dppf-ferrocenyl-diamine-Pd-Catalyst	t (h)	Conv (%)	TON	Product Distribution		ee (%)
				A	B	
homogeneous	72	15.9	379	27.3	72.5	-
	120	27.2	513	4.8	95.0	-
tethered to non porous silica	72	12.6	253	16.9	83.0	2
	120	19.2	320	-	100	2
confined within mesoporous silica (dia. 30 Å)	48	35.5	526	5.2	94.5	17
	72	53.7	715	-	100	20

Reaction conditions: substrate = 5 g; catalyst = 250 mg; H_2 pressure = 20 bar; solvent ≡ methanol : THF = 1: 9; temp = 40 °C;
Products: A = 1,4,5,6-tetrahydronicotinate; B = ethyl nipecotinate (see Scheme 2).

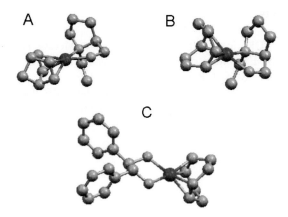

Figure 20. The crystal structures (**A**-**C**) of [Rh(1,5-cyclooctadiene)(S)-(-)-2-aminomethyl-1-ethylpyrrolidine]BF$_4$, [Rh(norbornadiene)(S)-(-)-2-amino methyl-1-ethylpyrrolidine]BF$_4$ and [Rh(1,5-cyclooctadiene)(1R,2R)-(+)-1,2-diphenyl-ethylenediamine]BF$_4$. *N.B.* the BF$_4^-$ has been omitted for clarity; color code: grey represents carbon atoms, blue nitrogen and purple rhodium.

Table 5. Hydrogenation of E-α-phenylcinnamic acid using the rhodium amine complexes.

Amine[a]	H / E	T (h)	Conv	Sel.	ee
(S)-(-)-2-aminomethyl-1-ethyl-pyrrolidine	H[b]	24	74	73	93
(S)-(-)-2-aminomethyl-1-ethyl-pyrrolidine	E[c]	24	70	77	96
(S)-(-)-2-aminomethyl-1-ethyl-pyrrolidine	H[b]	24	88	76	64
(S)-(-)-2-aminomethyl-1-ethy-lpyrrolidine	E[c]	24	99	66	91
(1R,2R)-(+)-1,2-diphenylethyl-enediamine	H[b]	24	57	84	81
(1R,2R)-(+)-1,2-diphenylethyl-enediamine	E[c]	24	98	80	93

Reaction conditions and notes. The reactions were carried out at 20 bar of H$_2$, temp = 40 °C, using 10 mgs of the homogeneous catalyst and 50 mgs of heterogeneous, 500 mgs of substrate and using methanol as the solvent. [a]the amines are shown in the crystal structures (Figure 20), [b]cyclooctadiene, [c]norbornadiene; H = homogeneous catalyst; E = heterogeneous catalyst.

In this section we have concentrated exclusively on nanoporous silica as the support for anchoring high-performance nanoparticle catalysts. But the general principle may be extended to other solids that have recently been prepared where there are sharply defined nanopores in the range 3 to 30 nm. Indeed, Joo *et al* [52] have recently prepared ordered nanoporous arrays of carbon that support high dispersions of platinum nanoparticles (dia. less than 3 nm) that display promising electro-catalytic activity for oxygen reduction. These nanocatalysts are likely to prove important for fuel cell technologies, based on the proven behavior of other (e.g. Pd-Pt and Pt-Co-Cr) similar systems.

Acknowledgements

We thank EPSRC (UK) for a rolling grant to Sir John Meurig Thomas, The Royal Commission for the Exhibition of 1851 and Bayer AG, Leverkusen, Germany for their support to Dr. Robert Raja, and our colleagues, especially Prof. B.F.G. Johnson, Drs Sophie Hermans, Tetyana Khimyak, Sang-OK Lee, Stuart Raynor, Gopinathan Sankar, Robert Bell, Dewi Lewis, Paul Midgley, Thomas Maschmeyer, and Mr. Matthew Jones, for their valuable stimuli.

References

1. Thomas, J.M. and Thomas, W.J., in *'Heterogeneous Catalysis: Principles and Practices'* (Wiley-VCH, Weinheim: 1997).
2. Thomas, J.M. and Zamaraev K.I., (eds.), in *"Perspectives in Catalysis"* (IUPAC, Blackwell Science Publishers, Oxford, 1992).
3. Thomas, J.M., Turning points in catalysis. *Angew. Chem. Int. Ed. Engl.* **33** (1994) pp. 913-937.
4. Zhao, D.Y., Yang, P.D., Huo, Q.S., Chmelka, B.F. and Stucky, G.D., Topological construction of mesoporous materials. *Curr. Opin. Solid State Mat. Sci.* **3** (1998) pp. 111-121.
5. Corma, A., Inorganic solid acids and their use in acid-catalyzed hydrocarbon reactions. *Chem. Rev.* **95** (1995) pp. 559-614.
6. van Bekkum, H., Flanigen, E.M. and Jansen, J.C. (eds.), in *"Introduction to Zeolite Science and Practice"* (Elsevier, Amsterdam, 1991).
7. Thomas, J.M. and Raja, R., Nanoporous and nanoparticle catalysts. *The Chemical Record* **1** (2001) pp. 448-466.
8. Thomas, J.M. and Raja, R., Catalytically active centers in porous oxides: design and performance of highly selective new catalysts. *Chem. Commun.* (2001) pp. 675-687.
9. Thomas, J.M., Design, synthesis and in situ characterization of new solid catalysts. *Angew. Chem. Int. Ed. Engl.* **38** (1999) pp. 3588-3628.
10. Thomas, J.M. and Raja, R., The materials chemistry of inorganic catalysts. *Aust. J. Chem.* **54** (2001) pp. 551-560.
11. Maschmeyer, T., Derivatized mesoporous solids. *Curr. Opin. Solid State Mat. Sci.* **3** (1998) pp. 71-78.
12. Corbin, D.R., Whitney, J.F., Fultz, W.C., Stucky, G.D., Eddy, M.M. and Cheetham, A.K., Synthesis of open-framework transition metal phosphates using organometallic precursors in acidic media – Preparation and structural characterization of $Fe_5P_4O_{20}H_{10}$ and $NaFe_3P_3O_{121}$. *Inorg. Chem.* **25** (1986) pp. 2279-2280.

13. Sun, T. and Ying, J.Y., Synthesis of microporous transition metal oxide molecular sieves by supramolecular templating mechanism. *Nature* **389** (1997) pp. 704-706.
14. Tanaka, K., Optical material – Light-induced anisotropy in amorphous chalcogenides. *Science* **277** (1997) pp. 1786-1787 and references therein.
15. Cheetham, A.K., Ferey, G. and Loiseau, T., Open-framework inorganic materials. *Angew. Chem. Int. Ed. Engl.* **38** (1999) pp. 3269-3292.
16. Meier, W.M. and Olson, D.H., *Atlas of Zeolite Structure Types*; (Butterworth-Heinemann: 3rd Revised Edition, 1992).
17. Davis, M.E., Ordered porous materials for emerging applications. *Nature* **417** (2002) pp. 813-821.
18. Gregg, S.J. and Sing, K.S.W., in *'Adsorption, Surface Area and Porosity'* (Academic Press, London: 1967).
19. Sankar, G. and Thomas, J.M., *In situ* combined X-ray absorption spectroscopic and X-ray diffractometeric studies of solid catalysts. *Top. Catal.* **8** (1999) pp. 1-21.
20. Thomas, J.M. and Sankar, G., The role of XAFS in the *in situ* elucidation of active sites in designed solid catalysts. *J. Synchrotron Rad.* **8** (2001) pp. 55-64.
21. Thomas, J.M., Terasaki, O., Gai, P.L., Zhou, W.Z. and Gonzalez-Calbet, J., Structural elucidation of microporous and mesoporous catalysts and molecular sieves by high-resolution electron microscopy. *Acc. Chem. Res.* **34** (2001) pp. 583-594.
22. Thomas, J.M., Raja, R., Sankar, G., Johnson, B.F.G. and Lewis, D.W., Solvent-free routes to clean technology. *Chem. Eur. J.* **7** (2001) pp. 2973-2978.
23. Shephard, D.S., Maschmeyer, T., Sankar, G., Thomas, J.M., Ozkaya, D., Johnson, B.F.G., Raja, R., Oldroyd, R.D. and Bell, R.G., Preparation, characterization and performance of encapsulated copper-ruthenium bimetallic catalysts derived from molecular cluster carbonyl precursors. *Chem. Eur. J.* **4** (1998) pp. 1214-1224.
24. Raja, R., Sankar, G., Hermans, S., Shephard, D.S., Bromley, S.T., Thomas, J.M., Maschmeyer, T. and Johnson, B.F.G., Preparation and characterization of a highly active bimetallic (Pd-Ru) nanoparticle heterogeneous catalyst. *Chem. Commun.* (1999) pp. 1571-1572.
25. Raja, R., Khimyak, T., Thomas, J.M., Hermans, S. and Johnson, B.F.G., Single-step, highly active and highly selective nanoparticle catalysts for the hydrogenation of key organic compounds. *Angew. Chem. Int. Ed. Engl.* **40** (2001) pp. 4638-4642.
26. Hermans, S., Raja, R., Thomas, J.M., Johnson, B.F.G., Sankar, G. and Gleeson, D., Solvent-free, low-temperature, selective hydrogenation of polyenes using a bimetallic nanoparticle Ru-Sn catalyst. *Angew. Chem. Int. Ed. Engl.* **40** (2001) pp. 1211-1215.

27. Sinfelt, J.H., Bimetallic Catalysts: Discoveries, concepts and applications (Exxon monograph, Wiley, New York, 1983).
28. Sinfelt, J. H., Ruthenium Copper – A model bimetallic system for studies of surface chemistry and catalysis. *Intl. Rev. Phys. Chem.* **7** (1988) pp. 281-315.
29. Hermans, S., Khimyak, T. and Johnson, B.F.G., High yield synthesis of Ru-Pt mixed-metal cluster compounds. *J. Chem. Soc., Dalton Trans.* (2001) pp. 3295-3302.
30. Ozkaya, D., Zhou, W.Z., Thomas, J.M., Midgley, P.A., Keast, V.J. and Hermans, S., High-resolution imaging of nanoparticle bimetallic catalysts supported on mesoporous silica. *Catal. Lett.* **1999,** 60, 113-120.
31. Midgley, P.A., Weyland, M., Thomas, J.M. and Johnson, B.F.G., Z-contrast tomography: A technique in 3-dimensional nanostructural analysis based on Rutherford scattering. *Chem. Commun.* **2001**, 907-908.
32. Weyland, M., Midgley, P.A. and Thomas, J.M., Electron tomography of nanoparticle catalysts on porous supports: A new technique based on Rutherford scattering. *J. Phys. Chem. B.* **2001,** 105, 7882-7886.
33. Ishida, H., Fukuoka, Y., Mitsui, O. and Kono, M., Liquid-phase hydration of cyclohexene with highly siliceous zeolites. *Stud. Surf. Sci. Catal.* **83** (1994) pp. 473-480.
34. Yang, H., Gao, H. R. and Angelici, R. J., *Organometallics* **19** (2000) pp. 622-629.
35. Haag, W.O., Lago, R.M. and Weisz, P.B., The active-site of acidic aluminosilicate catalysts. *Nature* **309** (1984) 589-591.
36. Thomas, J.M., Solid acid catalysts. *Sci. American.* **266** (1992) pp. 112-118.
37. Chen, J. and Thomas, J.M., MAPO-18 – a new family of catalysts for the conversion methanol to light olefins. *J. Chem. Soc. Chem. Commun.* (1994) pp. 603-604.
38. Wright, P.A., Natarajan, S., Thomas, J.M., Bell, R.G., Gai-Boyes, P.L., Jones, R.H. and Chen, J., Solving the structure of a metal-substituted aluminophosphate catalyst by electron microscopy, computer simulation and x-ray powder diffraction. *Angew. Chem. Int. Ed. Engl.* **31** (1992) pp. 1472-1475.
39. Lee, S.-O., Sankar, G., Kitchen, S.J., Dugal, M., Thomas, J.M. and Harris, K.D.M., Towards molecular sieve inorganic catalysts that are akin to enzymes: Studies of selective cyclodimerization over ferrierite at ambient temperature. *Catal. Lett.* **73** (2001) pp. 91-94.
40. Thomas, J.M., Raja, R., Sankar, G. and Bell, R.G., Molecular sieve catalysts for the regioselective and shape-selective oxyfunctionalization of alkanes in air. *Acc. Chem. Res.* **34** (2001) pp. 191-200.
41. Thomas, J.M., Raja, R., Sankar, G. and Bell, R.G., Molecular sieve catalysts for the selective oxidation of linear alkanes by molecular oxygen. *Nature* **398** (1999) pp. 227-230.

42. Raja, R. and Thomas, J.M., A manganese-containing molecular sieve catalyst designed for the terminal oxidation of dodecane in air. *J. Chem. Soc. Chem. Commun.* (1998) pp. 1841-1842.
43. Raja, R., Sankar, G. and Thomas, J.M., Designing a molecular sieve catalyst for the aerial oxidation of n-hexane to adipic acid. *Angew. Chem. Int. Ed. Engl.* **39** (2000) pp. 2313-2316.
44. Hill, C.L., ed.; *Activation and Functionalization of Alkanes*; (Wiley: Chichester, 1989), chaps 6-8 and references therein.
45. Raja, R., Sankar, G. and Thomas, J.M., Powerful redox molecular sieve catalysts for the selective oxidation of cyclohexane in air. *J. Am. Chem. Soc.* **121** (1999) 11926-11927.
46. Dugal, M., Sankar, G., Raja, R. and Thomas, J.M., Designing a heterogeneous catalyst for the production of adipic acid by aerial oxidation of cyclohexane. *Angew. Chem. Int. Ed. Engl.* **39** (2000) pp. 2310-2313.
47. Thomas, J.M., Maschmeyer, T., Johnson, B.F.G. and Shephard, D.S., Constrained chiral catalysts. J. Mol. Catal. A. **141** (1999) pp. 139-144.
48. Johnson, B.F.G., Raynor, S.A., Shephard, D.S., Maschmeyer, T., Thomas, J.M., Sankar, G., Bromley, S.T., Oldroyd, R.D., Gladden, L.G. and Mantle, M.D., Superior performance of a chiral catalyst confined within mesoporous silica. *Chem. Commun.* (1999) pp. 1167-1168.
49. Raynor, S.A., Thomas, J.M., Raja, R., Johnson, B.F.G., Bell, R.G. and Mantle, M.D., A one-step, enantioselective reduction of ethyl nicotinate to ethyl nipecotinate using a constrained, chiral, heterogeneous catalyst. *Chem. Commun.* (2000) pp. 1925-1926.
50. Thomas, J.M., Johnson, B.F.G., Raja, R., Sankar, G. and Midgley, P.A., High-performance nanocatalysts for single-step hydrogenations. *Acc. Chem. Res.* (2002) (in press).
51. Thomas, J.M., Raja, R., Johnson, B.F.G., Hermans, S., Jones, M.D. and Khimyak, T., Bimetallic catalysts and their relevance to the hydrogen economy. *Ind. Eng. Chem. Res.* (2002) (submitted).
52. Joo, S.H., Choi, S.J., Oh, J., Kwak, J., Liu, Z., Terasaki, O. and Ryoo, R., Ordered nanoporous arrays of carbon supporting high dispersions of platinum nanoparticle. *Nature,* **412** (2001) pp. 169-172.

NANOCOMPOSITES

WALTER CASERI

Institute of Polymers, Department of Materials, ETH Zentrum, Universitätsstr. 41, CH-8092 Zürich, Switzerland

Nanocomposites of polymers and isotropic inorganic particles can be prepared with a variety of methods. The inorganic particles are synthesized *in situ* in most procedures but the use of surface-modified colloids for nanocomposite fabrication is also established. Characteristic attributes of nanocomposites are an extremely high internal interface area and, if the particle diameters are below ca. 50 nm, a markedly reduced scattering of visible light. Such features may render nanocomposites attractive e.g. as materials with unusual catalytic or optical properties (e.g. photoconductivity, extreme refractive indices, or visually transparent UV filters). In most cases, a random dispersion of the nanoparticles is attempted in the polymer matrix but materials with ordered colloids have also been in the focus. For example, nanoparticles can assemble to vesicle-like structures or uniaxially oriented arrays, the latter inducing a pronounced dichroism in nanocomposites. Polymers containing a regular lattice of nanoparticles with diameters on the order of 100 nm can exhibit an exciting light scattering behavior. Generally, it has to be considered that physical properties of nanocomposites can depend on the size of the incorporated colloids. For example, the electrical conductivity, the color, the extent of light scattering, or the refractive index of metallic or semiconducting nanoparticles can vary with the particle diameter and markedly differ from the bulk values.

1 Introduction

This article deals with composites comprising polymers and isotropic inorganic particles with diameters up to ca. 100 nm. Particles with such dimensions are often designated as nanoparticles, nanosized particles, colloids, or ultrafine particles. Polymers containing particles of the addressed size are termed nanocomposites. As evident from the depictions below, materials properties of nanocomposites can substantially differ from those of respective composites with larger particles, for example if a large internal surface area plays an important role, if light scattering is an essential factor, or if physical constants of sufficiently small particles differ from the values of the bulk substances.

Nanocomposites can be far from equilibrium [1]. Due to the high surface free energy of the inorganic particles which amounts typically to 500-2000 mJ/m^2 [2], the inorganic particles should, in many systems, be fused in the thermodynamically most favorable state into a single sphere surrounded by the polymer. Hence, the formation of nanocomposites with randomly dispersed inorganic particles is often enabled by kinetically controlled processes. Although particle agglomeration can be a problem in the synthesis of nanocomposites, the mobility of the particles is usually negligible once the nanocomposites are formed, and therefore phase separation is commonly suppressed in the final products even over extended periods. For example, in nanocomposites consisting of an ethylene-methacrylic acid copolymer and PbS (ca. 3 % v/v), the particle sizes remained constant after extended exposure of samples to air, water, dichloromethane or hexane, and the period required for the agglomeration of the PbS particles was estimated to 10^5 years at 25 °C [3].

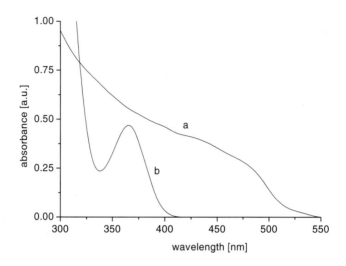

Figure 1. UV/vis spectra of bulk CdS (a) and CdS particles of a diameter of ca. 4 nm (b), sketch according to data of ref. [7,8].

It is important to be aware that physical properties of particles with diameters between 1 and 100 nm can markedly differ from those of the corresponding bulk materials, although in this size range the lattice parameters of crystalline particles are similar or identical to those of the bulk materials [4]. Properties such as the high electric conductivities and melting temperatures of bulk metals are believed to reflect the correlated interactions of a large number of atoms [5]. Accordingly, the melting temperature of bulk gold, 1063 °C, falls to ca. 700 °C for particles with 4 nm diameter and to ca. 200 °C at a diameter of 2.5 nm [5]. Optical attributes of semiconductors begin to change significantly below a certain particle diameter which markedly depends on the substance and amounts to ca. 2 nm for CdSe, 2.8 nm for GaAs, 20 nm for PbS, and 46 nm for PbSe [6]. In UV/vis spectra, shoulders or absorption maxima can be characteristic for nanoparticles of a certain size. As an example, the absorption spectrum of CdS particles of ca. 4 nm diameter with a narrow size distribution showed a well pronounced maximum at 368 nm [7] and clearly differed from the spectrum of bulk CdS [8] (Figure 1).

It should be pointed out that the inorganic nanoparticles embedded in a polymer matrix are usually not exactly of equal size, unless in rare cases of very small clusters consisting only of few formula units. Hence the term "monodisperse particles", which is occasionally found in the literature, is usually somewhat misleading since the related particles typically also show some polydispersity when the size distribution is carefully analyzed. Hence the expression "particles with narrow size distribution" better represents the status in such cases. The variation in sizes can sometimes be reduced by size-selective precipitation, as described for CdS particles prepared by reaction of cadmium acetate and thiourea in presence of thioglycerol in dimethylformamide at 120 °C [7], and it was assumed that the dispersion was stabilized by thioglycerol adsorbing on the particle surfaces. Upon careful addition of acetone, larger CdS particles precipitated first, and hence fractions with narrow particle size distribution could be gained. The particle diameters in the literature as well as in this article typically refer to number average particle diameters d_{na}, defined as

$$d_{na} = \frac{\sum_i n_i d_i}{\sum_i n_i}$$

with n_i the number of particles of diameter d_i. However, some physical properties, such as the refractive index of nanocomposites, depend on the volume fractions of the particles, i.e. the volume-weighted average diameters d_{va} might rather be appropriate, where

$$d_{va} = \sqrt[3]{\frac{\sum_i n_i d_i^3}{\sum_i n_i}}$$

The number-weighted and the volume-weighted average diameters of the particles differ often by 20-30 %, which is, fortunately, usually of little importance for the description of the studied phenomena.

2 Historic framework

Gold is frequently considered to be the first material which has been reported in the form of nanoparticles. The *aurum potabile* (potable gold) accounted by Paracelsus around 1570 [9,10] but also the *luna potabile* (potable silver) mentioned in 1677 [11] and other substances described centuries ago [12] obviously contained metal nanoparticles. Metal colloids have also been used to dye glass at least since the middle of the 16th century [13] and as paints for enamel, as described e.g. in great detail in 1765 [14]. The existence of tiny particles in such products was supposed long ago; for example, Macquer already assumed in 1774 that "all these gold tinctures (*aurum potabile*) are nothing but gold which is made extremely finely divided and floating in an oily fluid. They are therefore, properly speaking, no tinctures and they may be called drinkable gold only so far as we connect with this name no other idea than the one that the gold is swimming in a fluid and is made into particles so fine that it may itself be regarded as potable in the form of a fluid" [15]. Ostwald, a pioneer in colloid research, regarded Richter to be the first who found evidence for the elemental nature of gold in nanoparticles [16]. Among many other experiments, Richter observed in 1802 that gold powders with a golden gloss could be converted into violet powders by crumbling, that on the other hand violet gold powders acquired a golden color upon rubbing with a smooth stone or steel, that violet gold powders were insoluble in hydrochloric acid but soluble in *aqua regia*, that violet gold powders formed an amalgam with mercury, and that violet powders could be prepared from bulk gold by chemical reactions without a significant difference in mass. These experiments were in agreement with the assumption that gold nanoparticles indeed consisted of elemental gold regardless of the different colors of the nanoparticles and bulk elemental gold. Richter perceived that the color of gold powders could vary with the observation angle, and he deduced that the color of very small gold particles was determined by refraction of light. Nonetheless, there was still a lack of consensus in 1866 on the nature of gold nanoparticles, and besides the elemental state, gold oxide and a mixture of elemental gold and gold oxide were also discussed [17]. In this age, experiments and observations by Fischer [18] and Faraday [19] lead them to the conclusion that nanoparticles of gold indeed existed in spite of their color discrepancy to bulk gold (see also Chapter 4). Zsigmondy demonstrated in 1898 that gold colloids which appeared to be dissolved in water consisted in fact of particles since the gold was not able to penetrate dialysis membranes [20].

Diameters of gold colloids were measured first in 1903 in gold-glass nanocomposites by Siedentopf and Zsigmondy with the help of an ultramicroscope [21], which allowed to observe particles below the resolution limit of optical microscopes in dilute systems by scattering. A number of different glasses were examined. In those samples, average particle diameters between ca. 5 nm and 200 nm were frequently found. Values of nanoparticle diameters obtained by ultramicroscopy were confirmed by Scherrer in 1918 by evaluation of line widths in X-ray diffraction patterns [22]. Scherrer also demonstrated with X-ray diffraction that gold lattices in colloidal and bulk gold were identical even for particles as small as 4–5 unit cell lengths [22].

The preparation of a polymer nanocomposite is obvious from an abstract in 1835 [23]. First, bulk gold was dissolved in *aqua regia*. The water in the resulting solution was partially evaporated and the residual phase subjected to an additional work-up procedure. Upon addition of an aqueous gum arabic-tin(II) chloride solution, the gold ions were reduced to colloidal gold. Upon addition of ethanol, coprecipitation of a nanocomposite was observed in form of a purple solid. Another early report published in 1899 includes the formation of nanocomposites of gum and silver by coprecipitation with ethanol which was added to an aqueous phase containing gum and silver colloids [24]. A dichroic behavior of nanocomposites containing oriented aggregates of metal nanoparticles was recognized 100 years ago by Ambronn and Zsigmondy, as described in Chapter 5.2.

3 Preparation of nanocomposites

3.1 General remarks

As mentioned above, the specific specific surface energies of inorganic materials are high, thus rendering in isolated powders the primary particles to interact strongly with each other. The resulting agglomerates usually do not break into the individual primary particles during nanocomposite fabrication. Therefore, only rarely bare nanoparticles have been successfully used for the preparation of nanocomposites with primary particles that are well dispersed in the polymer matrix. As an example, silicon particles of 20-40 nm diameter were prepared by high-energy milling of silicon powders which were subsequently subjected in ethanolic suspension to ultrasound [25]. After sedimentation of the coarse fractions, the supernatant solutions were mixed with gelatin solutions and nanocomposites were prepared by spin coating.

A favorable method for the employment of nanoparticles is the use of colloids which are coated with a layer of organic molecules strongly bound to the particle surfaces. Such particles can often be isolated as viscous substances or solids which readily disperse as individual primary particles in water or organic solvents. In these cases, agglomeration is suppressed by the surface layer diminishing markedly the specific surface energy and thus decreasing the attraction tendency between the particles. In the last decade, a number of surface-modified colloids have been isolated, for example, gold or silver nanoparticles with phosphines or thiols attached to their surfaces [26-29] or CdS nanoparticles covered with aromatic thiols [8,30,31]. Corresponding colloids can be mixed with polymers in solution or in the melt (see below) and in this state further processed to nanocomposites.

Up to now, however, nanocomposites have been prepared most commonly by synthesis of the inorganic particles *in situ*, for instance in solution (see below) including micellar systems [30]. In the following, a selection of principles of nanocomposite manufacture is presented.

3.2 Formation of particles in situ, use of polymer in its final state

3.2.1 Formation of the particles in a liquid

Up to now, *in situ* formation of particles in a liquid medium is probably the most widely used method with a view to the preparation of polymer nanocomposites containing isotropic inorganic particles. Commonly, soluble inorganic or organometallic compounds are converted by chemical reactions to colloids in water or organic solvents. The polymer can be present already during colloid synthesis or otherwise added afterwards. The particle dispersion can be destabilized or stabilized by the polymer, depending on the system. In the former case, the nanocomposite forms by coprecipitation spontaneously after colloid formation, while in the latter case nanocomposites can be obtained by addition of a solvent acting as a coprecipitation agent, by casting followed by solvent evaporation, or by spin coating. Using casting and spin coating processes, it has to be considered that solid reaction side products from the colloid synthesis stay usually in the nanocomposites unless they crystallize e.g. at the surface of the resulting films from which they can be removed by washing, as described for gold-gelatin nanocomposites [32].

Example for colloid synthesis in organic solvents, polymer stabilizing the colloidal dispersion, and nanocomposite formation by coprecipitation: Addition of Li_2S to a solution of poly(aniline) and $Cd(CF_3SO_3)_2$ or $[Cu(CH_3CN)_4]CF_3SO_3$ in N-methylpyrrolidone caused the formation of CdS or Cu_2S nanoparticles, respectively, of 1-2 nm diameter [33]. Addition of ethanol caused coprecipitation yielding nanocomposites with poly(aniline) as the matrix polymer. Materials with 16-64% w/w (3.8-27% v/v) CdS or 17-53% w/w (3.5-17% v/v) Cu_2S, respectively, were prepared.

Example for colloid synthesis in water, polymer destabilizing the colloidal dispersion: Upon addition of a H_2S solution to a solution containing both $Pb(CH_3COO)_2$ and poly(ethylene oxide), nanocomposites with ca. 90% w/w (ca. 50% v/v) PbS particles of a diameter around 15 nm precipitated immediately after formation of PbS particles [34].

Example for colloid synthesis in water, polymer stabilizing the colloid dispersion, and nanocomposite formation by casting or spin coating: In presence of gelatin, PbS particles of 2-5 nm diameter formed in an aqueous $Pb(CH_3COO)_2$ solution upon addition of H_2S [35]. With casting, homogeneous films up to ca. 65% w/w (25% v/v) PbS were obtained after evaporation of the water. With spin coating, films up to 86.4% w/w (53% v/v) PbS were produced. Note that in both methods the reaction side product of the particle synthesis (acetic acid) is volatile and, therefore, can be removed from the nanocomposite by drying procedures.

3.2.2 Formation of the particles in a solid polymer matrix

Inorganic particles can be prepared *in situ* in solid polymer matrices, e.g. by thermal decomposition of incorporated precursors, reaction of incorporated compounds with gaseous species, or when polymer films containing an incorporated precursor are immersed in liquids containing the reactive species required for the formation of the desired colloids. If solid reaction byproducts arise from the particle synthesis, they can be embedded in the nanocomposites, hence the formation of volatile reaction side products which are able to leave polymer matrices should be preferred if possible.

Example for decomposition of an incorporated compound in a solid polymer matrix: Copper(II) formate and poly(2-vinylpyridine) were dissolved in methanol and the solvent was subsequently removed [36]. The films thus obtained were heated to temperatures above 125 °C, whereupon the copper ions were reduced in the polymer matrix to metallic copper particles (average diameter 3.5 nm) under release of the gaseous reaction side products H_2 and CO_2.

Example for a reaction of an incorporated compound with a gaseous reagent (ambient humidity): Upon immersion of sheets of a statistical copolymer of ethylene and vinyl acetate (EVA) in a diethyl zinc solution in hexane, diethyl zinc diffused into the polymer [37]. After removal of the films from the solution, the incorporated diethyl zinc hydrolyzed rapidly under the action of ambient humidity, and zinc oxide particles of diameters around 10 nm formed *in situ* in the polymer matrix. The reaction side product for the formation of the zinc oxide particles (ethane) is volatile and does not contaminate the nanocomposite.

Example for a reaction of an incorporated compound with a dissolved reagent: Nanocomposites of poly(aniline) (PANI) and PbS were prepared by deposition of PANI-Pb(NO$_3$)$_2$ films from aqueous solution followed by immersion of the films in 0.5 M sodium sulfide for 5 min, whereupon PbS particles of 2.6 nm diameter formed *in situ* in the films [38].

3.3 Use of particles in their final state, formation of polymer in situ

As an example, CdS nanoparticles with surface-bound 4-hydroxyphenol or 2-mercaptoethanol were prepared from $CdCl_2$ and Na_2S in reverse micellar systems [39]. The particles were collected by centrifugation and dispersed in dimethyl sulfoxide. Upon addition of ethylene glycol and toluenediyl-2,4-diisocyanate, a poly(urethane) was formed *in situ*. The resulting nanocomposites contained 0.07-0.14 mol CdS/kg.

3.4 Use of particles and polymer in their final states

3.4.1 Mixing of the components in dispersion/solution

Several powders of surface-modified colloids have been found to disperse well in liquids. Such dispersions can be mixed with dissolved polymers, and subsequently nanocomposites can readily be obtained by casting followed by solvent evaporation or by spin coating. The preparation of nanocomposites by diffusion of dispersed colloids in polymer films is also possible. Since suited particles were typically isolated without byproducts, the incorporation of reaction byproducts as in the cases described above for *in-situ*-prepared colloids does not afford problems.

Example for nanocomposite preparation with solvent casting: Dodecanethiol-coated gold particles of 2.2 nm diameter were dispersed in a solution of poly(ethylene) in *p*-xylene at 130 °C [40]. After casting and drying, gold-poly(ethylene) films with 2% w/w gold were obtained.

Example for nanocomposite preparation with spin coating: After dispersion of dodecanethiol-coated gold particles in a poly(phenylmethylsilane) solution in toluene, nanocomposites were obtained by spin coating [41].

3.4.2 Mixing of the components in a polymer melt

Some nanoparticles which can be isolated as powders disperse in polymer melts without pronounced agglomeration of the primary particles, especially colloids that are coated with a layer of organic molecules. This affords a simple technique for the preparation of nanocomposites by direct mixing of particles and polymers.

Example: Dodecanethiol-coated silver nanoparticles of 4.5 nm diameter were mixed with poly(ethylene) in an extruder at 180 °C for 10 min [29]. After compression molding of the extruded samples at 180 °C, homogeneous nanocomposites with 2-4% w/w silver particles were obtained.

3.4.3 Diffusion of nanoparticles in a polymer matrix

Nanocomposites have been rarely prepared by diffusion of colloids in a polymer matrix. This method is suited only for rather insoluble polymers with good swelling behavior or for very thin films.

Example: CdTe-poly(aniline) (PANI) nanocomposites were prepared by exposure of a PANI film of ca. 100 nm thickness to a 1 mM aqueous dispersion of thioglycerol-capped CdTe particles for 1 h [42]. The diameter of the employed CdTe particles was 2.5-4 nm.

3.5 *Formation of particles as well as polymer* in situ

The formation of both polymer and particles *in situ* is suited for a relatively limited number of systems. Again, it has to be considered that the reaction side products from colloid syntheses could be incorporated in the nanocomposites.

Example: Heating of a solution of tris(styrene)platinum(0) in styrene at 60 °C in presence of azobis(isobutyronitrile) (AIBN) as a radical initiator resulted in decomposition of the metal complex to colloidal platinum under concomitant formation of poly(styrene) [43]. The resulting platinum-poly(styrene) solids contained 1-8% w/w platinum particles of 1-2 nm diameter. The reaction side product, styrene, also serves as monomer for the formation of the desired polymer thus avoiding the problem of incorporation of reaction side products in the nanocomposite.

4 Optical properties of nanocomposites

The energy levels of the basic formula units can split upon formation of larger entities into more and more components with increasing number of atoms until they reach the quasi-continuous band structure of the bulk solid [44]. Therefore, optical properties of colloids, such as the color or the refractive index (see Chapter 5.1.3), can depend on the particle size and differ from those of the corresponding bulk materials. Such effects were recognized already by Richter in 1802 (see Chapter 2) and by Faraday in 1857 [19]. Faraday deduced that blue gold colloids were larger in size than red ones because blue gold dispersions could be prepared from red samples but not reverse and blue solids precipitated faster than red ones. The color of colloids can vary not only as the size of the primary particles but also as the distance to neighboring colloids changes. This was evident already in 1904 by experiments from Kirchner and Zsigmondy who reported that nanocomposites of colloidal silver [45] or gold [46] and gelatin reversibly changed their colors from blue to red upon swelling with water. Concomitantly, they observed upon drying of the gels a bathochromic shift of the absorption maximum in UV/vis spectra, which they attributed on theoretical considerations to a decrease in the distances between the embedded particles upon drying and not to a change in the primary size of the colloids [45,46]. This assumption was in agreement with the observation that swelling of the nanocomposites with water or acetic acid resulted in the same color changes which consequently were rather connected with the expanding volume of the materials than with chemical processes [45]. Around the same time, Maxwell Garnett explained the dependency of the color of colloidal systems on the particle diameters and volume contents of the nanoparticles on a theoretical basis [47,48]. Besides absorption of light by nanoparticles, scattering can become pronounced at larger particle diameters. For example, ca. 100 years ago Mie and his Ph.D. student Steubing measured quantitatively the scattered and the absorbed light intensity of illuminated dispersions of gold nanoparticles [49,50]. They found that the intensity loss of light transmitted through dispersed gold particles with diameters below 50 nm originated mainly from absorption (Figure 2) [49,50], but scattering became more prominent for larger particles [51].

This intensity loss of light caused by scattering depends in nanocomposites with spheric particles on the radius r and refractive index n_p of the incorporated particles, the refractive index n_m of the polymer matrix, the volume fraction ϕ_p of the particles, the wavelength λ of the incident light, and the thickness x of the nanocomposite. The intensity loss can be estimated with the equation [52]

$$\frac{I}{I_0} = e^{-\left[\frac{3\Phi_p x r^3}{4\lambda^4}\left(\frac{n_p}{n_m}-1\right)\right]}$$

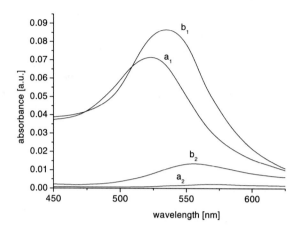

Figure 2. Intensity loss of light passing gold dispersions. Absorption (a_1) and scattering (a_2) by particles with ca. 20 nm diameter, and absorption (b_1) and scattering (b_2) by particles with 51 nm diameter; approximate representation according to Ref. [49].

where I is the intensity of the transmitted and I_0 of the incident light. As an example, the transmittance (I/I_0) is shown in Figure 3 as a function of the particle radius for x = 5 mm, ϕ_p = 0.1, λ = 500 nm, n_m=1.5, and various refractive indices of the particles (1.7, 2, 2.5, and 4). With those parameters, the intensity loss by scattering becomes pronounced for particle diameters above ca. 20-50 nm (r = 10-25 nm). It is, however, obvious that even relatively thick samples with large refractive index

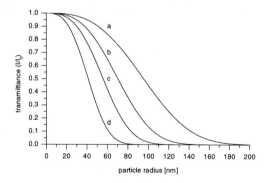

Figure 3. Calculated transmittance of nanocomposites as a function of the radius of the incorporated particles. Wavelength of the incident light: 500 nm, thickness of the nanocomposite: 5 mm, volume fraction of the particles: 0.1, refractive index of the matrix: 1.5, refractive index of the particles: 1.7 (a), 2.0 (b), 2.5 (c), and 4.0 (d).

differences between polymer and particles can be virtually transparent if the diameters of the colloids are below ca. 10 nm ($r = 5$ nm). Therefore, composites with very small particles can be particularly suited for optical applications. However, it should be noted that nanocomposites containing particles with diameters on the order of 100 nm can also exhibit unusual optical properties based on light scattering. If such particles are arranged in a regular lattice in polymer matrices, pronounced color effects can arise, as described in more detail in Chapter 5.4.

5 Structures and properties of selected nanocomposites

5.1 *Nanocomposites with randomly dispersed particles*

5.1.1 Nanocomposites which are active in photocatalysis

The high specific surface area of the colloids embedded in polymers can favor heterogeneous catalysis duo to the high number of active surface sites in the system. Further, in photocatalytic processes the diffusion of electrons and holes is probably more efficient in small particles [30]. This is illustrated by the photocatalytic hydrogen production from a 2-propanol/water mixture under the action of CdS [30]. While composites of a poly(thiourethane) and bulk CdS did not exhibit a significant photocatalytic activity, related materials containing CdS colloids with surface-bound benzylthiol (CdS-BT) or phenylthiol (CdS-PT) (Cd content 1-2 mol/kg) caused generation of hydrogen under illumination (Figure 4). After a delay period of a few hours, hydrogen evolved linearly with a rate of ca. 0.04 μmol/(μmol Cd·h) in presence of CdS-BT and ca. 0.025 μmol/(μmol Cd·h) of CdS-PT. Two effects could be responsible for the difference in effectivity of CdS-BT and CdS-PT. In contrast to CdS-BT, CdS-PT agglomerated during nanocomposite fabrication thus reducing the contact area for the catalytic reaction in the final nanocomposite [30]. Further, thiophenol quenched the fluorescence of the CdS particles, in opposition to benzylthiol, which might indicate that the transfer of the photogenerated electrons in CdS to the reactive species is suppressed in the catalytic process [30].

5.1.2 Nanocomposites exhibiting photoconductivity

The large internal interface area in nanocomposites enables an efficient separation of photo-induced charges, and the number of carrier traps caused by the so-called grain boundaries between particles and matrix is reduced [53,54]. Therefore, photoconducting nanocomposites can be created by combination of a p-type semiconducting polymer matrix, such as poly(aniline) (PANI), and n-type semiconducting nanoparticles, such as CdS or PbS. The charge carriers generated in the nanoparticles (electrons or holes) by irradiation of light move to the surface and are transported by the polymer matrix [54]. The amount of nanoparticles required for photoconductivity is typically a few weight percent [54]. Such particle contents are well below the percolation threshold of ca. 15% v/v [54], in particular when it is considered that the densities of inorganic nanoparticles are usually above those of polymers.

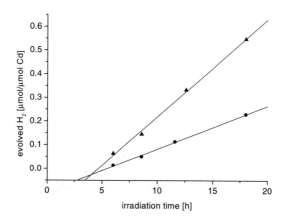

Figure 4. Amount of hydrogen produced with RSH-coated CdS particles embedded in a poly(thiourethane) under illumination, for R = phenyl (circles) and R = benzyl (triangles); draft according to data from Ref. [30].

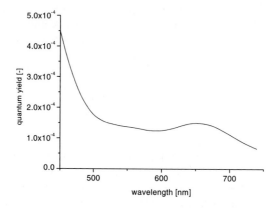

Figure 5. Quantum yield of a poly(aniline)-PbS nanocomposite upon irradiation; approximate representation according to Ref. [38].

As an example, in nanocomposites comprising poly(aniline) (PANI) and PbS of a diameter of 2.6 nm photoconductivity was observed in the entire visible spectrum with quantum yields on the order of 10^{-4} (Figure 5) [38]. Nanocomposites of PANI containing 11-61% w/w TiO_2 (30% rutile and 70% anatase) of a diameter of 21 nm displayed a photocurrent in a broad wavelength range [55]. The highest photoconductivity arose at irradiation at 337 nm; this wavelength corresponded to the absorption maximum of the TiO_2 particles. Upon irradiation, electrons appear to be transferred from PANI to TiO_2. Subsequently, the charges move to the respective electrodes. At a TiO_2 content of 33% w/w, irradiation with monochromatic light of a wavelength of 344 nm and an intensity of 19.5 $\mu W/cm^2$, a short-circuit current density of 3 $\mu A/cm^2$ and an open-circuit voltage of 790 mV was detected.

The photoconductivity σ of nanocomposite films of 3-5 μm thickness comprising poly(N-vinylcarbazole) (PVK) and CdS with diameters of 1-2 nm increased linearly with increasing illumination intensity I at an applied field of 54 V/μm and an irradiation wavelength of 514.5 nm [56]. This indicates an absence of a bimolecular recombination mechanism as in this case a dependency of σ on I^2 was expected. At $I = 3$ W/cm^2, a σ around $8 \cdot 10^{-11}$ S/cm was measured. The content of CdS of ca. 1% w/w lied well below the percolation threshold which implies that the photoconductivity mechanism was initiated with the absorption of a photon by a CdS particle and PVK was responsible for the charge transport (PVK is a hole-transporting polymer). Other nanocomposite films of PVK containing ca. 1% v/v CdS of average diameter 1.6 nm showed a low dark conductivity while upon irradiation at 340 nm photoconductivity with a charge generation efficiency of 0.16 was observed [31].

5.1.3 Nanocomposites with extreme refractive indices

Compared to inorganic materials, the refractive indices of most organic polymers lie in a relatively narrow range of 1.3 and 1.7 [57]. Inorganic materials, however, can possess refractive indices far below 1 or above 3 over a broad wavelength range [58]. Gold, silver, or copper are examples for low- and PbS, silicon, or germanium for high-refractive-index materials. It may, therefore, be anticipated that the incorporation of inorganic nanoparticles with extreme refractive indices in organic polymers results in composite materials with refractive indices outside of the typical range of organic polymers. Indeed, this has been demonstrated with nanocomposites containing 88% w/w (50% v/v) PbS particles of 19 nm diameter randomly dispersed in poly(ethylene oxide) [34]. Refractive indices determined by ellipsometry revealed values between 2.9 and 3.4 at wavelengths of 632.8 nm and 1295 nm; the indicated range also reflects the variations among the samples. Refractive index determinations from the evaluation of the reflectivity yielded typical values of 3.02-3.08 in the wavelength range of 1000 and 2500 nm (Figure 6).

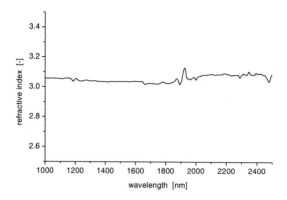

Figure 6. Refractive indices of a nanocomposite composed of poly(ethylene oxide) and 50% v/v PbS at different wavelengths [34].

The dependence of the refractive index on the particle content was investigated with films of 40 nm - 1 μm thickness with gelatin as the matrix polymer and PbS particles of diameters below 15-20 nm diameter [35]. Within the experimental precision, the refractive indices n of those films increased linearly with increasing PbS volume fractions from 1.54 for pure gelatin to ca. 2.5 for 55% v/v PbS, according to the equation

$$n = n_1\phi_1 + n_2\phi_2$$

with n_i and ϕ_i the refractive indices and volume fractions of the two components of the composites. Linear extrapolation to 100 % v/v PbS resulted in a refractive index around 3.4, i.e. clearly below that of bulk PbS (ca. 4.3 at a wavelength of 633 nm [58]). This difference appears to arise from a refractive index dependence on the particle size, as shown below. Linear relations of refractive indices following the above equation were also found in other systems [25,32,59]. According to theory, this behavior might be expected for many polymer composites and blends, but theoretical considerations suggest that non-linear relations should also be considered [25,56,60,61]. Because of the limited experimental precision of the determination of the refractive indices and the volume fractions of the particles, some deviation from the above linear relationship might also be consistent with experiments suggesting a linear behavior.

The size of PbS particles in poly(ethylene oxide) matrices was varied between 4 and 80 nm by addition of various amounts of acetic acid or sodium dodecylsulfate during the *in-situ*-synthesis of PbS from $Pb(CH_3COO)_2$ and H_2S [62]. Figure 7 shows the resulting refractive indices measured with ellipsometry at a wavelength of 632.8 nm; the values at 1295 nm were similar [62]. The nanocomposites containing particles above ca. 20 nm diameter possessed refractive indices of 3.5-3.8 at 632.8 nm and 3.3-3.7 at 1295 nm. In presence of smaller particle sizes, the refractive indices decreased continuously at 632.8 as well as at 1295 nm to 1.7-1.8 in the case of PbS particles with 4 nm diameter. Figure 7 also contains the refractive indices estimated for 100% PbS by linear extrapolation of the PbS volume fraction (see above). At 632.8 and 1300 nm, values of 4.0-4.3 were frequently obtained for PbS particles with sizes above ca. 20 nm. These values are in the range of those reported for bulk PbS (4.3 at 619.9 nm and 1300 nm [58]). However, the extrapolated refractive indices decreased at 632.8 and 1300 nm when the PbS diameters fell below 15-20 nm, as supposed by theoretical analyses [63,64]. For diameters of 4 nm, the extrapolated refractive indices for bulk PbS were as low as 2.3 at 632.8 nm and 2.0 at 1300 nm. It is, therefore, obvious that the refractive index of nanoparticles is an optical property which can depend markedly on the particle size.

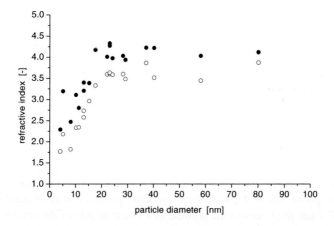

Figure 7. Refractive indices of poly(ethylene oxide)-PbS nanocomposites with particles of different diameters (open circles) and corresponding refractive indices extrapolated for pure PbS (filled circles) [62].

Very high refractive indices were also measured for nanocomposites of poly(ethylene oxide) and iron sulfides (2.6-2.8 at 632.8 nm and 1300 nm) [65] and nanocomposites of gelatin and silicon [25]. The refractive index of silicon, 5.57 at 400 nm and 3.91 at 620 nm [25], renders this material a very attractive component for nanocomposites of the type discussed in this chapter. Since colloidal silicon is difficult to achieve by chemical processes, related colloids were prepared by high-energy milling and processed to nanocomposites as described in Chapter 3.1. The refractive index of the resulting nanocomposites decreased from 3.3 at a wavelength of 400 nm to 2.5 at 1000 nm, in agreement with a linear dependence of the refractive indices of the nanocomposite and the volume fraction of silicon as mentioned above for the PbS-gelatin system.

The examples mentioned above suggest that the incorporation of low-refractive-index colloids should substantially diminish the refractive index of polymeric composites. Accordingly, films of gelatin and nanoparticles of gold, whose refractive index in the bulk is below 1.0 in the entire wavelength range of 500-2000 nm [58], were prepared with gold contents of 9.5-92.9 % w/w (0.7-48 % v/v). Indeed, exceptionally low refractive indices around 1.0 at 632.8 and 1295 nm were observed for a film containing 92.9 % w/w gold [32].

5.1.4 Transparent UV-absorbing nanocomposites

Ultraviolet light induces frequently undesired reactions in polymers which can affect mechanical properties or lead to discoloration. In order to suppress such processes, stabilizers are commonly added to polymers. However, over long periods of time the additives themselves can be consumed, in particular when the materials are exposed to intense light. As an alternative to organic compounds, stable inorganic substances such as ZnO or rutile might be used. These substances absorb UV light close to the visible wavelength range but are transparent in the visible wavelength range itself. Hence, ZnO or rutile could be useful in the protection of polymers from UV light or to produce nanocomposites as UV filters, provided the particle sizes are small enough to prevent significant light scattering.

Nanocomposites of ZnO and a statistical copolymer of ethylene and vinyl acetate (EVA) were prepared by immersion of EVA in a diethyl zinc solution [37]. Diethyl zinc diffused into the polymer, and after removal of the films from the solution, the incorporated diethyl zinc hydrolyzed rapidly under the action of ambient humidity. As a consequence, ZnO particles of ca. 10 nm diameter formed. Under appropriate reaction parameters, the ZnO particles were concentrated in a region of 1-2 µm thickness to the surface as a consequence of the limited diffusion rate of the diethyl zinc, i.e. the bulk mechanical properties of the polymer were not affected. The ZnO particles effectively absorbed UV light up to 370-380 nm while light scattering was virtually absent in the visible wavelength range.

Nanocomposite films containing dodecylbenzenesulfonic acid-coated nanoparticles of rutile, a crystal modification of TiO_2, were produced with poly(styrene) or a poly(carbonate) based on bisphenol A as the matrix [66]. The films of ca. 100 µm thickness contained 4% w/w surface-modified rutile colloids of 2 nm diameter. These nanocomposites appeared transparent to the eye and acted as visually transparent UV filters.

5.2 Nanocomposites with uniaxially oriented particle assemblies

More than 100 years ago, polymer nanocomposites with uniaxially oriented inorganic particles have been prepared [67,68]. Plant and animal fibrils were exposed to solutions of silver or gold salts and the metal ions were subsequently reduced *in situ* under the action of light [67]. These samples exhibited a pronounced dichroism. Also, dichroic films were obtained starting from gold chloride-treated gelatin which was subsequently drawn, dried and finally exposed to light [67]. Similar results were reported when gelatin was mixed with gold nanoparticles before drying and drawing [68]. It was suggested already in early reports that the dichroism was caused by linear assemblies of small particles [69-71] or by polycrystalline rod-like particles [72] and that the respective anisotropic distribution of the particles arose from their formation in uniaxially oriented spaces of the fibers. The average diameters of the silver and gold primary particles in such materials were between 5 and 14 nm as determined by evaluation of the line width in X-ray diffraction patterns [71].

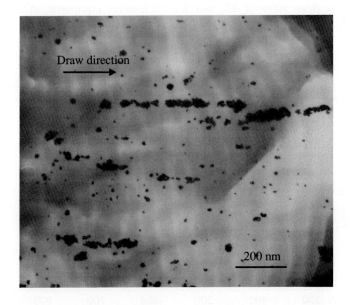

Figure 8. Transmission electron micrograph of a thin section of a drawn poly(ethylene)-gold nanocomposite containing arrays of gold particles oriented in the drawing direction [40].

More recently, anisotropic nanocomposites with gold or silver colloids of 2-10 nm average diameter and poly(ethylene) (PE) were prepared by drawing of isotropic films containing usually 2 or 4 % w/w particles [29,40,73]. The drawing procedure was performed on a hot stage at 120 °C, and typical draw ratios around 15 were employed. After drawing, transmission electron microscopy (TEM) images revealed arrays of nanoparticles which were aligned in the drawing direction (an example is presented in Figure 9). Obviously, the oriented aggregates of the colloids developed during solid-state drawing of the nanocomposites. The UV-vis spectra of the drawn nanocomposites strongly depended on the angle φ between the polarization direction of the incident light and the drawing direction of the films, as evident from the example in Figure 9. The absorption maximum of light vibrating parallel to the drawing axis (φ = 0°) was observed at higher wavelengths than that of light vibrating perpendicular to the drawing axis (φ = 90°). Isosbestic points as evident from Figure 9 appeared in all dichroic nanocomposites. When the gold content in PE was varied between 0.9 and 7.4 % w/w, most pronounced differences in the absorption maxima at φ = 0° and φ = 90° were observed at gold contents of 2-4 % w/w. The gold- and silver-containing films became dichroic already at low draw ratios. For example, a PE-gold specimen showed a $\Delta\lambda_{max}$ (difference of the absorption maxima at φ of 0° and 90°) of 65 nm at a draw ratio of 6 and a $\Delta\lambda_{max}$ of 85 nm at a draw ratio of 18, and for a PE-silver composite, $\Delta\lambda_{max}$ remained essentially constant (90 nm) between draw ratios of 6 and 22.

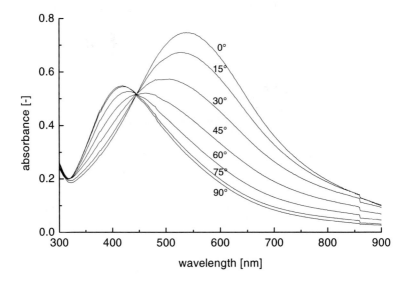

Figure 9. Absorption spectra in polarized light of a drawn poly(ethylene)-silver nanocomposite at different angles between polarization direction and draw direction [29].

As a consequence of the dependence of λ_{max} on φ, the color of the drawn nanocomposites strongly varied with φ. Depending on the size of the primary particles (4.5 or 10 nm), composites containing silver showed color changes from red to yellow or purple to amber at φ of 0° and 90°, respectively, and PE-gold specimen from blue to red. Such materials are potentially useful in Liquid Crystal Display (LCD) applications, and indeed bicolored displays were manufactured with the dichroic nanocomposites [40,73].

5.3   Nanocomposites with vesicle-like particle assemblies

Superstructures of nanoparticles in a polymer matrix were obtained upon radical polymerization of styrene containing dispersed platinum particles of 1-2 nm diameter and dissolved ammonium O,O'-dialkyldithiophosphate of various chain lengths ranging from ethyl to octadecyl [43]. The resulting poly(styrene)-platinum nanocomposites contained 1-8% w/w platinum. In particular in the case of phosphates with long alkyl groups and a dialkyldithiophosphate/platinum ratio of 1:2, superstructures of platinum colloids were frequently observed in transmission electron microscope (TEM) images (Figure 10). The TEM pictures of the superstructures resembled those of bilayer vesicles in aqueous solution. Since superstructures were never observed in absence of an ammonium O,O'-dialkyldithiophosphate, it appears that the formation of the superstructures was initiated by the dialkyldithiophosphates which are supposed to adsorb on the surfaces of the platinum particles. It was surmised that co-crystallization of alkyl chains adsorbed on adjacent platinum particles was the driving force for the arrangement of the superstructures.

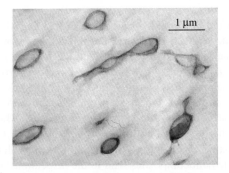

Figure 10. Transmission electron micrograph of superstructures of colloidal platinum particles embedded in a poly(styrene) matrix in presence of ammonium O,O'-dioctadecyldithiophosphate [43].

5.4 Nanocomposites with regular lattices of particles

Nanocomposites containing a regular lattice of particles were prepared with very uniform silica particles of diameters around 150 nm if not otherwise indicated [74,75]. The hydroxyl groups at the silica surfaces were converted with 3-(trimethoxysilyl)propyl methacrylate under formation of a surface-bound siloxane group and release of methanol [74,75]. The resulting organic surface layer decreased the specific surface energy and hence markedly reduced the agglomeration tendency of the particles. Dispersions of 35-40% w/w (19.5-22.7% v/v) of the silica colloids in methyl methacrylate (MMA) or 35-45% w/w in methyl acrylate (MA) were produced in liquid cells of 1 mm or 264 µm thickness. After some time, the dispersions became iridescent, and sharp peaks in UV/vis spectra recorded with the incident beam perpendicular to the face of the liquid cell indicated diffraction of light according to Bragg's equation [75]:

$$\lambda = 2nd \sin\theta$$

with λ the wavelength of diffracted light, d the interplanar spacing in the lattice formed by the particles, n the refractive index of the dispersion, and θ the Bragg angle. Such bands must originate in diffraction of light at the lattice formed by the particles because neither silica nor MMA absorb in the visible wavelength range. For instance, at a silica volume fraction ϕ of 0.195 in MMA, a band appeared at 554 nm, corresponding to an interplanar spacing d of 176 nm (n = 1.4210) [74]. As another example, at a silica weight fraction of 0.40 in MA, the maximum absorption wavelength λ_{max} decreased in UV/vis spectra from 600 nm for a particle diameter of 330 nm to λ_{max} of 490 nm for a particle diameter of 142 nm, accordingly the distance between the lattice planes decreased from 211 to 173 nm (calculated with Bragg's equation) [75].

Nanocomposites were obtained from silica-MA and silica-MMA dispersions by photopolymerization. It has been suggested, but not proven, that at least a fraction of surface-bound methacrylate groups reacted with the radicals at the endgroups of the growing polymer chains. Hence, it was expected that the particles were strongly connected with the final polymer matrix. The nanocomposites also exhibited the characteristic features in UV/vis spectra mentioned above for the dispersions. Absorption bands (Figure 11) appeared in the nanocomposites which were still quite sharp although somewhat broader than in the dispersions. Due to the shrinking of the volume of the organic phase upon polymerization, the volume fractions ϕ of silica were higher in the nanocomposites than in the dispersions they were prepared from. For instance, ϕ increased from 0.195 to 0.235 in a sample based on MMA [74]. As a consequence, λ_{max} shifted in UV/vis spectra upon polymerization, e.g. from 554 nm to 490 nm in poly(methyl methacrylate) (PMMA) with ϕ = 0.235, corresponding to a d of 190 nm calculated with Bragg's equation (n = 1.4819) [74]. The absorption maximum of the nanocomposites prepared from MA dispersions decreased with increasing silica content [75]. For particles of 153 nm diameter, λ_{max} decreased from 502 nm at $\phi = 0.35$ to 466 nm at $\phi = 0.45$. The corresponding distances of the lattice planes of 171 nm and 159 nm, respectively, were ca. 17% below the respective values before polymerization. Upon stretching of a film containing 40% w/w SiO_2, λ_{max} decreased from 486 to 440 nm. After the stress was released, the films retracted to their original lengths within 2-4 h and exhibited again the initial λ_{max} within the experimental precision. The maximum absorption wavelength was further influenced by the angle between the incident light and the sample surface. A film with λ_{max} of 490 nm at perpendicular incidence of light showed a λ_{max} of 472 nm at an incidence angle of 68°, in agreement with Bragg's law [75].

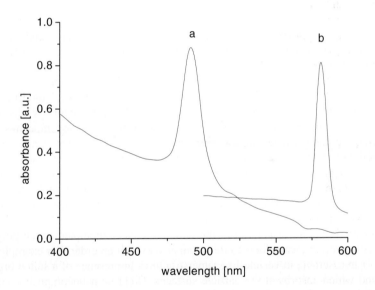

Figure 11. UV/vis spectra of nanocomposites containing silica (35% w/w) and poly(methyl methacrylate) (a) and poly(methyl acrylate) (b), approximate representations according to Refs. [74,75].

Scanning electron micrograph (SEM) images revealed a hexagonal pattern of the silica particles embedded in poly(methyl methacrylate) of poly(methylacrylate) [74,75]. The average distance between the centers or two adjacent particles amounted in the poly(methyl methacrylate) specimen to 234 nm and the surfaces of adjacent particles were separated by 82 nm [74]. Hence, the organization of the particles is rather driven by long-range interactions than to close packing of the silica spheres.

6 Conclusions

Nanoparticles have been prepared as early as in the 17th and polymer nanocomposites already in the 19th century. In the meantime, a variety of synthesis principles of nanocomposites comprising polymers and inorganic particles have been reported. Essentially, the nanoparticles are prepared *in situ* in most procedures but the use of surface-modified colloids for nanocomposite fabrication has also been established.

Fundamental analyses such as the determination of particle sizes, proof of the nature of the colloids and theoretical and experimental studies on optical properties had been published already around 1920. Compared to composites with larger particles, nanocomposites possess a very high internal surface area and intensity loss of transmitted light by scattering can often be neglected, in particular for colloid diameters below ca. 50 nm. Such attributes render nanocomposites attractive, e.g., for catalytic processes or optical studies and applications. For instance, the refractive index of polymers can be markedly modified by incorporation of inorganic colloids. As another example, transparent UV absorbing particles can introduce their particular optical properties to nanocomposites. The embedding of nanoparticles in polymers can also result in photoconductive or photocatalytic active materials. Since physical properties of metallic or semiconducting nanoparticles such as electrical conductivity, color, or refractive index can markedly differ from the bulk values and depend on the diameter of the colloids, properties of nanocomposites can also depend on the sizes of the incorporated particles.

Usually, a random dispersion of the nanoparticles is attempted in the polymer matrix. However, the embedded colloids can also exhibit an order. For example, the particles can assemble in vesicle-like superstructures in presence of a suited organic compound which adsorbs at the particle surfaces. Dichroic nanocomposites can be prepared by drawing of isotropic films containing metal nanoparticles. Upon drawing, the particles can arrange in arrays which are oriented in the drawing direction. In such composites, under the action of linearly polarized light the color and absorption maxima in UV/vis spectra vary with the angle between the polarization direction of the light and the orientation axis of the particle arrays. Of high interest are also particles with diameters on the order of 100 nm which are arranged as a regular lattice in a polymer. Such materials can selectively scatter visible light. The wavelengths which are preferentially scattered depend on the angle between the incident light and the surface of the nanocomposite.

Acknowledgements

I thank my colleagues which have invaluably contributed to the results presented from our laboratory, namely H.-J. Althaus, M. Büchler, C. Bastiaansen, C. Darribère, Y. Dirix, M. Gianini, W. Heffels, T. Kyprianidou-Leodidou, M. Müller, R. Nussbaumer, P. Smith, U. W. Suter, D. Vetter, P. Walther, E. Wehrli, M. Weibel, Y. Wyser, and L. Zimmermann.

References

1. Roy R., Purpose design of nanocomposites: Entire class of new materials. In *Ceramic Microstructures '86*, ed. By Pask J. A. and Evans, E. G. (Plenum Press, New York, 1987) pp. 25-32.
2. Kinloch A. J., *Adhesion and Adhesives* (Chapman and Hall, London, 1987).
3. Mahler W., Polymer-trapped semiconductor particles. *Inorg. Chem.* **27** (1988) pp. 435-436.
4. Godovski D. Yu, Electron behavior and magnetic properties of polymer nanocomposites. *Adv. Polym. Sci.* **119** (1995) pp. 79-122.
5. Israelachvili J., *Intermolecular and Surface Forces* (Academic Press, London, 1992).
6. Beecroft L. L. and Ober C. K., Nanocomposite materials for optical application. *Chem. Mater.* **9** (1997) pp. 1302-1317.
7. Chevreau A., Phillips B., Higgins B. G. and Risbud S. H., Processing and optical properties of spin-coated polystyrene films containing CdS nanoparticles. *J. Mater. Chem.* **6** (1996) pp. 1643-1647.
8. Herron N., Wang Y. and Eckert H., Synthesis and characterization of surface-capped, size-quantized CdS clusters. Chemical control of cluster size. *J. Am. Chem. Soc.* **112** (1990) pp. 1322-1326.
9. Paracelsus A. T., *Item zwen tractat von läme sampt gründtlicher gewisser irer cur* (Adam von Bodenstein, Basel, 1563); reprinted by Sudhoff K. (Ed.), *Theophrast von Hohenheim, gen. Paracelsus, Sämtliche Werke, 1. Abteilung, Medizinische, naturwissenschaftliche und philosophische Schriften, Vol. 2* (R. Oldenbourg, München, 1930).
10. von Hohenheim T., *Drey herrliche Schrifften Herrn Doctors/ Theophrasti/ von Hohenheim: Das erst/ vom geist des lebens und seiner krafft/ Das ander von krafft innerlicher/ geistlicher und leiblicher glider/ Das dritt/ von krafft eusserlicher glider/ unnd sterckung der inneren* (Adam von Bodenstein, Basel, 1572); reprinted by Sudhoff K. (Ed.), *Theophrast von Hohenheim, gen. Paracelsus, Sämtliche Werke, 1. Abteilung, Medizinische, naturwissenschaftliche und philosophische Schriften, Vol. 3* (R. Oldenbourg, München, 1930).
11. Valentinus B., *Fr. Basilii Valentini Benedictiner Ordens Chymische Schriften alle/ so viel derer verhanden/ anitzo Zum Ersten mahl zusammen*

gedruckt/auss vielen so wol geschriebenen als gedruckten Exemplaren vermehret und verbessert und in Zwey Theile verfasset, part 2 (Ander Theil, Das Vierdte Buch) (Johann Naumann und Georg Wolff, Hamburg, 1677; facsimile editon by Verlag Dr. H. A. Gerstenberg, Hildesheim, 1976)

12. Lösner, H., Zur Geschichte des kolloiden Goldes. *Z. Chem. Ind. Kolloide* **6** (1910) 1-3.
13. Cornejo A., Beiträge zur Geschichte des kolloiden Goldes. *Z. Chem. Ind. Kolloide* **12** (1913) pp. 1-6.
14. D'Arclais de Montamy, Traité des Couleurs pour la Peinture en Émail et sur la Porcelaine; Précédé de l'Art de Peindre sur l'Email, Et suivi de plusieurs Mémoirs sur différents sujets intéressants, tels que le travail de la Porcelaine, l'Art du Stuccateur, la maniere d'exécuter les Camées & les autres Pierres figurées, le moyen de perfectionner la composition du verre blanc & le travail des Glaces, &c. (G. Cavelier, Paris, 1765; facsimile edition by Georg Olms Verlag, Hildesheim, 1981)
15. Macquer, Dictionnaire de Chymie (Paris, 1774); cited by Svedberg T., Colloid Chemistry (The Chemical Catalog Company, New York, 1924, cf. also Neville R. G. and Smeaton W. A., Macquer dictionnaire de chymie - a bibliographical study. *Ann. Sci.* **38** (1981) 613-662).
16. Ostwald W., Zur Geschichte des kolloiden Goldes. *Z. Chem. Ind. Kolloide* **4** (1909) pp. 5-14.
17. Fischer J. C., Historische und praktische Untersuchungen über die Natur des Goldpurpurs (part 1). *Dingl. Polytech. J.* **182** (1866) pp. 31-40.
18. Fischer J. C., Historische und praktische Untersuchungen über die Natur des Goldpurpurs (part 2). *Dingl. Polytech. J.* **182** (1866) pp. 129-139.
19. Faraday M., Experimental researches in chemistry and physics (Taylor and Francis, Red Lion Court, 1859) pp. 391-443.
20. Zsigmondy, R., Ueber wässrige Lösungen metallischen Goldes. *Ann. Chem.* **301** (1898) pp. 29-54.
21. Siedentopf H. and Zsigmondy R., Ueber Sichtbarmachung und Grössenbestimmung ultramikroskopischer Teilchen, mit besonderer Anwendung auf Goldrubingläser. *Ann. Phys., vierte Folge (Drude's Ann.)* **10** (1903) pp. 1-39.
22. Scherrer P., Bestimmung der Grösse und der inneren Struktur von Kolloidteilchen mittels Röntgenstrahlen. *Nachr. Königl. Ges. Wiss. Göttingen, Math.-phys. Klasse* (1918) pp. 98-100.
23. Shepard C. U. (reviewer), Preparation of the Purple of Cassius for staining glass and enamels. *Sill. J. (Am. J. Sci. Arts)* **28(1)**, (1835) pp. 145-146.
24. Stoeckl K. and Vanino L., Ueber die Natur der sogenannten kolloidalen Metallösungen. *Z. Phys. Chem. Stöchiom. Verwandtschaftslehre* **30** (1899) pp. 98-112.

25. Papadimitrakopoulos F., Wisniecki P. and Bhagwagar, D. E., Mechanically attrited silicon for high refractive index nanocomposites. *Chem. Mater.* **9** (1997) pp. 2928-2933.
26. Schmid G., Large clusters and colloids - metals in the embryonic state. *Chem. Rev.* **92** (1992) pp. 1709-1727.
27. Schmid G., Peschel S. and Sawitowski T., Two-dimensional arrangements of gold clusters and gold colloids on various surfaces. *Z. anorg. allg. Chem.* **623** (1997) pp. 719-723.
28. Brust, M., Walker, M., Bethell, D., Schiffrin, D. J. and Whyman, R., synthesis of thiol-derivatized gold nanoparticles in a 2-phase liquid-liquid system. *J. Chem. Soc., Chem. Commun.* (1994) pp. 801-802.
29. Dirix Y., Bastiaansen C., Caseri W. and Smith P., Preparation, structure and properties of uniaxially oriented polyethylene-silver nanocomposites. *J. Mater. Sci.* **34** (1999) pp. 3859-3866.
30. Hirai T., Miyamoto M., Watanabe T., Shiojiri S. and Komasawa I., Effects of thiol on photocatalytic properties of nano-CdS-polythiourethane composite particles. *J. Chem. Eng. Japan* **31** (1998) pp. 1003-1006.
31. Wang Y. and Herron N., Photoconductivity of CdS nanocluster-doped polymers. *Chem. Phys. Lett.* **200** (1992) pp. 71-75.
32. Zimmermann L., Weibel M., Caseri W., Suter U. W. and Walther P., Polymer nanocomposies with „ultralow" refractive index. *Polym. Adv. Technol.* **4** (1993) pp. 1-7.
33. Godovsky D. Yu., Varfolomeev A. E., Zaretsky D. F., Chandrakanthi R. L. N., Kündig A., Weder C. and Caseri W., Preparation of nanocomposites of polyaniline and inorganic semiconductors. *J. Mater. Chem.* **11** (2001) pp. 2465-2469.
34. Weibel M., Caseri W., Suter U. W., Kiess H. and Wehrli E., Preparation of polymer nanocomposites with "ultrahigh" refractive index. *Polym. Adv. Technol.* **2** (1991) pp. 75-80.
35. Zimmermann L., Weibel M., Caseri W. and Suter U. W., High refractive index films of polymer nanocomposites. *J. Mater. Res.* 8 (1993) pp. 1742-1748.
36. Lyons A. M., Nakahara S., Marcus M. A., Pearce E. M. and Waszczak J. V., Preparation of copper-poly(2-vinylpyridine) nanocomposites. *J. Phys. Chem.* **95** (1991) 1098-1105.
37. Kyprianidou-Leodidou T., Margraf P., Caseri W., Suter U. W. and Walther P., Polymer sheets with a thin nanocomposite layer acting as a UV filter, *Polym. Adv. Technol.* **8** (1997) pp. 505-512.
38. Gaponik N. P. and Sviridov D. V., Synthesis and characterization of PbS quantum dots embedded in the polyaniline film. *Ber. Bunsenges. Phys. Chem.* **101** (1997) pp. 1657-1659.
39. Hirai T., Miyamoto M., Komosawa I., Composite nano-CdS-polyurethane transparent films. *J. Mater. Chem.* **9** (1999) pp. 1217-1219.

40. Dirix Y., Darribère C., Heffels W., Bastiaansen C., Caseri W. and Smith P., Optically anisotropic polyethylene-gold nanocomposites. *Appl. Optics* **38** (1999) pp. 6581-6586.
41. Gonsalves K. E., Carlson G., Kumar J., Aranda F. and Jose-Yacaman M., Polymer composites of nanostructured gold and their third-order nonlinear optical properties. *ACS Symp. Ser.* **622** (1996) pp. 151-161.
42. Gaponik N. P., Talapin D. V. and Rogach A. L., A light-emitting device based on CdTe nanocrystal/polyaniline composite. *Phys. Chem. Chem. Phys* **1** (1999) pp. 1787-1789.
43. Gianini M., Caseri W. R. and Suter U. W., Polymer nanocomposites containing superstructures of self-organized platinum colloids. *J. Phys. Chem. B* **105** (**2001**) 7399-7404.
44. Kreibig U. and Vollmer M., Optical properties of metal clusters (Springer, Berlin, 1995).
45. Kirchner F., Über beobachtete Absorptions- und Farbenänderungen infolge von Abstandsänderungen der absorbierenden Teilchen. *Ber. Königl. Sächs. Ges. Wiss., Math.-Phys. Klasse* **54** (1902) pp. 261-266.
46. Kirchner F. and Zsigmondy R., Über die Ursachen der Farbenänderungen von Gold-Gelatinepräparaten. *Ann. Phys., vierte Folge (Drude's Ann.)* **15** (1904) pp. 573-595.
47. Maxwell Garnett J. C., Colours in metal glasses, in metallic films, and in metallic solutions. *Phil. Trans. Royal Soc. London A* **205** (1906) pp. 237-288.
48. Maxwell Garnett J. C., Colours in metal glasses and in metallic films. *Phil. Trans. Royal Soc. London A* **203** (1904) pp. 385-420.
49. Mie G., Die optischen Eigenschaften kolloider Goldlösungen. *Z. Chem. Ind. Kolloide* **2** (1907) pp. 129-133.
50. Steubing W., Über die optischen Eigenschaften kolloidaler Goldlösungen. *Ann. Phys., vierte Folge (Drude's Ann.)* **26** (1908) pp. 329-371.
51. Mie G., Beiträge zur Optik trüber Medien, speziell kolloidaler Metallösungen. *Ann. Phys., vierte Folge (Drude's Ann.)* 25 (1908) pp. 377-445.
52. Novak B. M., Hybrid nanocomposite materials - between inorganic glasses and organic polymers. *Adv. Mater.* **5** (1993) pp. 422-433.
53. Wang Y. and Herron N., Semiconductor nanocrystals in carrier-transporting polymers. Charge generation and charge transport. *J. Lumin.* **70** (1996) pp. 48-59.
54. Wang Y., Semiconductor nanocrystals in photoconductive polymers: Charge generation and charge transport. *Studies Surf. Sci. Catal.* **103** (1996) pp. 277-295.
55. Feng W., Sun E., Fujii A., Wu H., Niihara K. and Yoshino K., Synthesis and characterization of photoconducting polyaniline-TiO_2 nanocomposite. *Bull. Chem. Soc. Jpn.* **73** (2000) pp. 2627-2633.

56. Winiarz J. G., Zhang L., Lal M., Friend C. S. and Prasad P. N., Observation of the photorefractive effect in a hybrid organic-inorganic nanocomposite. *J. Am. Chem. Soc.* **121** (1999) pp. 5287-5295.
57. Brandrup J. and Immergut E. H. (eds.), Polymer Handbook (John Wiley and Sons, New York, 1989).
58. Palik E. D. (ed.), Handbook of Optical Constants of Solids (Academic Press, Orlando, 1985).
59. Pope E. J. A., Asami M. and Mackenzie J. D., Transparent silica gel-PMMA composites. *J. Mater. Res.* **4** (1989) pp. 1016-1026.
60. Seferis J. C. and Samuels R. J., Coupling of optical and mechanical properties in crystalline polymers. *Polym. Eng. Sci.* **19** (1979) pp. 975-994.
61. Wedgewood A. R. and Seferis J. C., A quantitative description for the optical properties of crystalline polymers applied to polyethylene. *Polym. Eng. Sci.* **24** (1984) pp. 328-344.
62. Kyprianidou-Leodidou T., Caseri W. and Suter U. W., Size variation of PbS particles in high-refractive-index nanocomposites. *J. Phys. Chem.* **98** (1994) pp. 8992-8997.
63. Wang Y. and Herron N., Nanometer-sized semiconductor clusters - materials synthesis, quantum size effects, and photophysical properties. *J. Phys. Chem.* **95** (1991) pp. 525-532.
64. Schmitt-Rink S., Miller D. A. B. and Chemla D. S., Theory of the linear and nonlinear optical properties of semiconductor microcrystallites. *Phys. Rev. B* **35** (1987) pp. 8113-8125.
65. Kyprianidou-Leodidou T., Althaus H.-J., Wyser Y., Vetter D., Büchler M., Caseri W. and Suter U. W., High refractive index materials of iron sulfides and poly(ethylene oxide). *J. Mater. Res.* **12** (1997) pp. 2198-2206.
66. Nussbaumer R., Caseri W. and Smith P., Synthesis and characterization of surface-modified rutile particles and transparent polymer composites thereof. *J. Nanoparticle Res.*, in print.
67. Ambronn H., Ueber Pleochroismus pflanzlicher und thierischer Fasern, die mit Silber- und Goldsalzen gefärbt sind. *Königl. Sächs. Ges. Wiss.* **8** (1896) pp. 613-628.
68. Ambronn H. and Zsigmondy R., Ueber Pleochroismus doppelbrechender Gelatine nach Färbung mit Gold- und Silberlösungen. *Ber. Verhandlungen Sächs. Ges. Wiss. Leipzig* **51** (1899) pp. 13-15.
69. Frey A., Das Wesen der Chlorzinkjodreaktion und das Problem des Faserdichroismus. *Jahrbuch wiss. Botanik* **67** (1927) pp. 597-635.
70. Ambronn H., Ueber pleochroitische Silberkristalle und die Färbung mit Metallen. *Z. wiss. Mikrosk.* **22** (1905) pp. 349-355.
71. Frey-Wyssling A., Röntgenometrische Vermessung der submikroskopischen Räume in Gerüstsubstanzen. *Protoplasma* **27** (1937) pp. 372-411.
72. Frey-Wyssling A., Ultramikroskopische Untersuchung der submikroskopischen Räume in Gerüstsubstanzen. *Protoplasma* **27** (1937) pp. 563-571.

73. Dirix Y., Bastiaansen C., Caseri W. and Smith P., Oriented pearl-necklace arrays of metallic nanoparticles in polymers: A new route toward polarization-dependent color filters. *Adv. Mater.* **11** (1999) pp. 223-227.
74. Sunkara H. B., Jethmalani J. M. and Ford W. T., Composite of colloidal crystals of silica in poly(methyl methacrylate). *Chem. Mater.* **6** (1994) pp. 362-364.
75. Jethmalani J. M. and Ford W. T., Diffraction of visible light by ordered monodisperse silica-poly(methyl acrylate) composite films. *Chem. Mater.* **8**, (1996) pp. 2138-2146.